Communications
in Computer and Information Science 2436

Series Editors

Gang Li ⓘ, *School of Information Technology, Deakin University, Burwood, VIC, Australia*

Joaquim Filipe ⓘ, *Polytechnic Institute of Setúbal, Setúbal, Portugal*

Zhiwei Xu, *Chinese Academy of Sciences, Beijing, China*

Rationale

The CCIS series is devoted to the publication of proceedings of computer science conferences. Its aim is to efficiently disseminate original research results in informatics in printed and electronic form. While the focus is on publication of peer-reviewed full papers presenting mature work, inclusion of reviewed short papers reporting on work in progress is welcome, too. Besides globally relevant meetings with internationally representative program committees guaranteeing a strict peer-reviewing and paper selection process, conferences run by societies or of high regional or national relevance are also considered for publication.

Topics

The topical scope of CCIS spans the entire spectrum of informatics ranging from foundational topics in the theory of computing to information and communications science and technology and a broad variety of interdisciplinary application fields.

Information for Volume Editors and Authors

Publication in CCIS is free of charge. No royalties are paid, however, we offer registered conference participants temporary free access to the online version of the conference proceedings on SpringerLink (http://link.springer.com) by means of an http referrer from the conference website and/or a number of complimentary printed copies, as specified in the official acceptance email of the event.

CCIS proceedings can be published in time for distribution at conferences or as post-proceedings, and delivered in the form of printed books and/or electronically as USBs and/or e-content licenses for accessing proceedings at SpringerLink. Furthermore, CCIS proceedings are included in the CCIS electronic book series hosted in the SpringerLink digital library at http://link.springer.com/bookseries/7899. Conferences publishing in CCIS are allowed to use Online Conference Service (OCS) for managing the whole proceedings lifecycle (from submission and reviewing to preparing for publication) free of charge.

Publication process

The language of publication is exclusively English. Authors publishing in CCIS have to sign the Springer CCIS copyright transfer form, however, they are free to use their material published in CCIS for substantially changed, more elaborate subsequent publications elsewhere. For the preparation of the camera-ready papers/files, authors have to strictly adhere to the Springer CCIS Authors' Instructions and are strongly encouraged to use the CCIS LaTeX style files or templates.

Abstracting/Indexing

CCIS is abstracted/indexed in DBLP, Google Scholar, EI-Compendex, Mathematical Reviews, SCImago, Scopus. CCIS volumes are also submitted for the inclusion in ISI Proceedings.

How to start

To start the evaluation of your proposal for inclusion in the CCIS series, please send an e-mail to ccis@springer.com

Biljana Risteska Stojkoska ·
Smilka Janeska Sarkanjac
Editors

ICT Innovations 2024

TechConvergence: AI, Business, and Startup Synergy

16th International Conference, ICT Innovations 2024
Ohrid, North Macedonia, September 28–30, 2024
Proceedings

Editors
Biljana Risteska Stojkoska ⓘ
Ss. Cyril and Methodius University in Skopje
Skopje, North Macedonia

Smilka Janeska Sarkanjac ⓘ
Ss. Cyril and Methodius University in Skopje
Skopje, North Macedonia

ISSN 1865-0929 ISSN 1865-0937 (electronic)
Communications in Computer and Information Science
ISBN 978-3-031-86161-1 ISBN 978-3-031-86162-8 (eBook)
https://doi.org/10.1007/978-3-031-86162-8

© The Editor(s) (if applicable) and The Author(s), under exclusive license
to Springer Nature Switzerland AG 2025

This work is subject to copyright. All rights are solely and exclusively licensed by the Publisher, whether the whole or part of the material is concerned, specifically the rights of translation, reprinting, reuse of illustrations, recitation, broadcasting, reproduction on microfilms or in any other physical way, and transmission or information storage and retrieval, electronic adaptation, computer software, or by similar or dissimilar methodology now known or hereafter developed.
The use of general descriptive names, registered names, trademarks, service marks, etc. in this publication does not imply, even in the absence of a specific statement, that such names are exempt from the relevant protective laws and regulations and therefore free for general use.
The publisher, the authors and the editors are safe to assume that the advice and information in this book are believed to be true and accurate at the date of publication. Neither the publisher nor the authors or the editors give a warranty, expressed or implied, with respect to the material contained herein or for any errors or omissions that may have been made. The publisher remains neutral with regard to jurisdictional claims in published maps and institutional affiliations.

This Springer imprint is published by the registered company Springer Nature Switzerland AG
The registered company address is: Gewerbestrasse 11, 6330 Cham, Switzerland

If disposing of this product, please recycle the paper.

Preface

We are proud to present the proceedings of the 16th International Conference ICT Innovations 2024, held in Ohrid, North Macedonia, from September 28-30, 2024. This year's conference marked another significant milestone in our ongoing commitment to fostering groundbreaking research and innovation in information and communication technologies (ICT). Under the theme, "TechConvergence: AI, Business, and Start-Up Synergy," the conference brought together over 120 researchers, practitioners, and industry leaders from across the globe to explore the synergies between AI, business intelligence, and entrepreneurship.

The ICT Innovations conference series, organized by the Macedonian Society of Information and Communication Technologies (ICT-ACT) and supported by the Faculty of Computer Science and Engineering (FCSE) in Skopje, has built a distinguished reputation as a platform for presenting both fundamental and applied research. Over the years, this conference has become a critical venue for sharing innovative solutions and scientific discoveries, continually addressing the most pressing challenges and opportunities within the field of ICT.

In this 2024 edition, we turned our focus to how data science intersects with the business and start-up world, encouraging interdisciplinary collaboration and knowledge-sharing. As technology evolves rapidly, these collaborations become essential for harnessing the transformative potential of innovations. The central focus of this event revolved around the convergence of data science and business strategies, with particular emphasis on how these technologies shape the future of entrepreneurship.

This volume contains 21 full papers (plus 21 short papers in the Web Proceedings edition), which were carefully reviewed and selected from 80 high-quality submissions. These papers cover a wide range of topics, including machine learning, network science, digital transformation, natural language processing, and more. The review process was rigorous, with about 100 reviewers from 35 countries providing detailed feedback. Each submission was evaluated by at least three experts in the field, ensuring that the selected papers meet the high standards of academic excellence and originality that this conference is known for.

The program featured two distinguished keynote speakers who demonstrated the convergence of technology, innovation, and business growth. Their presentations showcased how advancements in ICT transform industries and drive entrepreneurial success. Paul Kayne discussed his work at Palatin Technologies, exploring how ICT and big data are used to understand genomes and the melanocortin system, presenting new approaches to treating inflammatory diseases. He also proposed ways to share genomics data while safeguarding intellectual property. Dejan Zvekic, a key figure in the regional IT start-up scene, shared his entrepreneurial journey, detailing how he transformed the fashion industry at Material Exchange, expanded Plugin76, and successfully integrated it into PTC Inc. His presentation highlighted the importance of strategic insight, market awareness, and leveraging opportunities for growth.

Alongside the main conference, participants had the opportunity to engage in seven specialized workshops focused on entrepreneurship, human-machine collaboration, and data analytics for business intelligence: Innovations in anti-drone technologies; Black-box and explainable artificial intelligence methods; Blended research on air pollution; Interactive data science; CyberMACS; and CHATMED.

At the workshop titled "Women in STEM", a diverse group of speakers, including Aneta Antova Pesheva, Eliot Bytyci, and Afrodita Shalja, addressed the critical issue of underrepresentation of women in Informatics across all educational and professional levels. The speakers highlighted the slow progress in increasing female participation in STEM fields, despite ongoing efforts in Europe. They encouraged researchers to submit papers on initiatives aimed at engaging and retaining female students and professionals, fostering a supportive environment for women pursuing careers in these disciplines.

All these workshops provided a hands-on experience, allowing researchers and practitioners to collaborate and explore new ideas in an interactive setting. The conference also offered a variety of social events aimed at fostering connections among participants, a tradition that has been highly valued since the conference's inception.

We extend our heartfelt thanks to all the authors who contributed their work to this year's proceedings, to the reviewers who ensured a fair and thorough evaluation process, and to all the participants who enriched the conference with their knowledge and expertise. Special thanks go to our generous sponsors, companies Netcetera and Ultra Computing; also to the organizing committee and the technical support team at FCSE, whose dedication and hard work were instrumental in making this conference a success. We are also deeply grateful to Ilinka Ivanoska for her invaluable assistance throughout the organization of the event.

As we conclude this edition of ICT Innovations, we look ahead with excitement to future conferences, where we will continue to explore the frontiers of ICT and foster innovation across industries and disciplines. We invite you to join us at the 17th ICT Innovations conference in 2025, where we will continue this journey of scientific discovery and collaboration.

Sincerely,

November 2024

Biljana Risteska Stojkoska
Smilka Janeska Sarkanjac

Organization

General Chairs

Biljana Risteska Stojkoska   Ss. Cyril and Methodius University in Skopje, North Macedonia

Smilka Janeska Sarkanjac   Ss. Cyril and Methodius University in Skopje, North Macedonia

Program Commitee Chairs

Biljana Risteska Stojkoska   Ss. Cyril and Methodius University in Skopje, North Macedonia

Smilka Janeska Sarkanjac   Ss. Cyril and Methodius University in Skopje, North Macedonia

Ilinka Ivanoska   Ss. Cyril and Methodius University in Skopje, North Macedonia

Program Committee

Aleksandar Bojchevski	University of Cologne, Germany
Aleksandar Stojmenski	Ss. Cyril and Methodius University in Skopje, North Macedonia
Aleksandra Mileva	Goce Delčev University of Štip, North Macedonia
Alessandro Cantelli-Forti	Lab RaSS National Laboratory - CNIT, Italy
Amelia Badica	University of Craiova, Romania
Ana Madevska Bogdanova	Ss. Cyril and Methodius University in Skopje, North Macedonia
Andrea Kulakov	Ss. Cyril and Methodius University in Skopje, North Macedonia
Andrej Brodnik	University of Ljubljana, Slovenia
Andreja Naumoski	Ss. Cyril and Methodius University in Skopje, North Macedonia
Antonio De Nicola	ENEA, Italy
Antun Balaz	Institute of Physics Belgrade, Serbia
Arianit Kurti	Linnaeus University, Sweden
Betim Cico	EPOKA University, Albania

Biljana Risteska Stojkoska	Ss. Cyril and Methodius University in Skopje, North Macedonia
Biljana Mileva Boshkoska	Jožef Stefan Institute, Slovenia
Blagoj Ristevski	St. Kliment Ohridski University Bitola, North Macedonia
Bojan Ilijoski	Ss. Cyril and Methodius University in Skopje, North Macedonia
Bojana Koteska	Ss. Cyril and Methodius University in Skopje, North Macedonia
Boris Delibašić	University of Belgrade, Serbia
Dejan Gjorgjevikj	Ss. Cyril and Methodius University in Skopje, North Macedonia
Dejan Spasov	Ss. Cyril and Methodius University in Skopje, North Macedonia
Dilip Patel	London South Bank University, UK
Dimitar Trajanov	Ss. Cyril and Methodius University in Skopje, North Macedonia
Edmond Jajaga	University for Business and Technology, Kosovo
Eftim Zdravevski	Ss. Cyril and Methodius University in Skopje, North Macedonia
Elena Vlahu-Gjorgievska	University of Wollongong, Australia
Elinda Kajo Mece	Polytechnic University of Tirana, Albania
Eliot Bytyçi	University of Prishtina, Kosovo
Francesco Mancuso	University of Pisa and CNIT, Italy
Fu-Shiung Hsieh	Chaoyang University of Technology, Taiwan
Georgina Mirceva	Ss. Cyril and Methodius University in Skopje, North Macedonia
Giacomo Longo	University of Genoa, Italy
Giulio Meucci	Consorzio Nazionale Iteruniversitario per le Telecomunicazioni - Laboratorio RaSS, Italy
Gjorgji Madjarov	Ss. Cyril and Methodius University in Skopje, North Macedonia
Goce Armenski	Ss. Cyril and Methodius University in Skopje, North Macedonia
Hrachya Astsatryan	Institute for Informatics and Automation Problems, National Academy of Sciences of Armenia, Armenia
Hristina Mihajloska	Ss. Cyril and Methodius University in Skopje, North Macedonia
Igor Mishkovski	Ss. Cyril and Methodius University in Skopje, North Macedonia
Igor Ljubi	University of Zagreb, Croatia
Ilche Georgievski	University of Stuttgart, Germany

Ilinka Ivanoska	Ss. Cyril and Methodius University in Skopje, North Macedonia
Ivan Kitanovski	Ss. Cyril and Methodius University in Skopje, North Macedonia
Ivan Chorbev	Ss. Cyril and Methodius University in Skopje, North Macedonia
Jatinderkumar Saini	Symbiosis Institute of Computer Studies and Research, India
Josep Silva	Universitat Politècnica de València, Spain
Jugoslav Achkoski	Military Academy General Mihailo Apostolski, North Macedonia
Katarina Trojachanec Dineva	Ss. Cyril and Methodius University in Skopje, North Macedonia
Katerina Zdravkova	Ss. Cyril and Methodius University in Skopje, North Macedonia
Kire Trivodaliev	Ss. Cyril and Methodius University in Skopje, North Macedonia
Kostadin Mishev	Ss. Cyril and Methodius University in Skopje, North Macedonia
Ladislav Huraj	University of SS. Cyril and Methodius in Trnava, Slovakia
Lasko Basnarkov	Ss. Cyril and Methodius University in Skopje, North Macedonia
Ljiljana Trajkovic	Simon Fraser University, Canada
Ljupcho Antovski	Ss. Cyril and Methodius University in Skopje, North Macedonia
Loren Schwiebert	Wayne State University, USA
Luis Alvarez Sabucedo	Universidade de Vigo, Spain
Marcin Michalak	Silesian University of Technology, Poland
Marco Porta	University of Pavia, Italy
Marjan Gusev	Ss. Cyril and Methodius University in Skopje, North Macedonia
Martin Drlik	Constantine the Philosopher University in Nitra, Slovakia
Massimiliano Zanin	IFISC (CSIC-UIB), Spain
Matus Pleva	Technical University of Košice, Slovakia
Melanija Mitrović	University of Niš, Serbia
Mile Jovanov	Ss. Cyril and Methodius University in Skopje, North Macedonia
Milos Jovanovik	Ss. Cyril and Methodius University in Skopje, North Macedonia
Milos Stojanovic	Visoka tehnicka skola Nis, Serbia

Miroslav Mirchev	Ss. Cyril and Methodius University in Skopje, North Macedonia
Monika Simjanoska	Ss. Cyril and Methodius University in Skopje, North Macedonia
Natasha Ilievska	Ss. Cyril and Methodius University in Skopje, North Macedonia
Natasha Stojkovikj	Goce Delčev University of Štip, North Macedonia
Nevena Ackovska	Ss. Cyril and Methodius University in Skopje, North Macedonia
Novica Nosović	University of Sarajevo, Bosnia and Herzegovina
Özge Büyükdağlı	International University of Sarajevo, Bosnia and Herzegovina
Pance Ribarski	Ss. Cyril and Methodius University in Skopje, North Macedonia
Pece Mitrevski	University St. Kliment Ohridski, North Macedonia
Periklis Chatzimisios	International Hellenic University, Greece
Petar Sokoloski	Ss. Cyril and Methodius University in Skopje, North Macedonia
Petre Lameski	Ss. Cyril and Methodius University in Skopje, North Macedonia
Riste Stojanov	Ss. Cyril and Methodius University in Skopje, North Macedonia
Rossitza Goleva	New Bulgarian University, Bulgaria
Sashko Ristov	University of Innsbruck, Austria
Sasho Gramatikov	Ss. Cyril and Methodius University in Skopje, North Macedonia
Sergio Ilarri	University of Zaragoza, Spain
Shuxiang Xu	University of Tasmania, Australia
Simona Samardjiska	Radboud University, The Netherlands
Slobodan Kalajdziski	Ss. Cyril and Methodius University in Skopje, North Macedonia
Smilka Janeska Sarkanjac	Ss. Cyril and Methodius University in Skopje, North Macedonia
Snezana Savoska	University St. Kliment Ohridski Bitola, North Macedonia
Stanimir Stoyanov	University of Plovdiv "Paisii Hilendarski", Bulgaria
Suzana Loshkovska	Ss. Cyril and Methodius University in Skopje, North Macedonia
Tarik Namas	International University of Sarajevo, Bosnia and Herzegovina
Ustijana Rechkoska-Shikoska	UIST - Ohrid, North Macedonia

Vacius Jusas	Kaunas University of Technology, Lithuania
Verica Bakeva	Ss. Cyril and Methodius University in Skopje, North Macedonia
Vesna Dimitrievska Ristovska	Ss. Cyril and Methodius University in Skopje, North Macedonia
Vesna Dimitrova	Ss. Cyril and Methodius University in Skopje, North Macedonia
Vesna Dimitrievska	Silicon Austria Labs, Austria
Vladimir Trajkovik	Ss. Cyril and Methodius University in Skopje, North Macedonia
Vladimír Siládi	Matej Bel University, Slovakia
Zlatko Varbanov	Veliko Tarnovo University, Bulgaria

Technical Committee

Ilinka Ivanoska	Ss. Cyril and Methodius University in Skopje, North Macedonia
Ana Todorovska	Ss. Cyril and Methodius University in Skopje, North Macedonia
Mila Dodevska	Ss. Cyril and Methodius University in Skopje, North Macedonia
Marija Taneska	Ss. Cyril and Methodius University in Skopje, North Macedonia
Zorica Karapancheva	Ss. Cyril and Methodius University in Skopje, North Macedonia
Stefan Andonov	Ss. Cyril and Methodius University in Skopje, North Macedonia

Contents

Session 1

Evaluation of Vector Databases and LLMs in RAG-Based Multi-document Question Answering 3
   Elena Filipovska, Ana Mladenovska, Jovana Dobreva, Dimitar Kitanovski, Goran Mitrov, Petre Lameski, and Eftim Zdravevski

Aligning Food Ingredients with Multiple Semantic Resources 19
   Darko Sasanski, Andrej Todorovski, Bojan Trpeski, Dimitar Trajanov, Tome Eftimov, and Riste Stojanov

Crossword Generation as a Constraint Satisfaction Problem Using Parallel Processing and Lemmatization 34
   David Arsov, Teo Kitanovski, and Mile Jovanov

Session 2

Comprehensive Examination of Network Access, Logging, and Auditing Strategies in Public and Private Institutions: Safeguarding Information Security, Resilience, and Compliance in the Digital Era 51
   Elissa Mollakuqe, Vesna Dimitrova, Hasan Dag, and Simon Atanasovski

Benefits of Parallelization in CPU Rendering: Quantitative Analysis Using a Custom 3D Rendering Engine 63
   Admir Huseini, Art Saiti, and Kiril Avramovski

Simulation of the Quasigroup Redundancy Check Code's Ability to Detect Errors 81
   Natasha Ilievska

Session 3

YOLOv8 Oriented Bounding Box (OBB) Model for Waymo Open Dataset 95
   Atanasko Boris Mitrev and Georgina Mirceva

Deep Multimodal Fusion for Semantic Segmentation of Remote Sensing Earth Observation Data 106
   Ivica Dimitrovski, Vlatko Spasev, and Ivan Kitanovski

Transfer Learning with Yolo for Object Detection in Remote Sensing 121
   Ema Pandilova, Marko Petrov, Vlatko Spasev, Ivica Dimitrovski,
   and Ivan Kitanovski

Comparison of On-Board and Off-Board Processing Power Consumption
for Drone Camera Images 136
   Atanasko Boris Mitrev and Biljana Risteska Stojkoska

Classification of Some Cosmological Images Using Deep Learning
and Persistent Homology 147
   Petar Sekuloski and Vesna Dimitrievska Ristovska

Session 4

Mushroom Classification Using Machine Learning 159
   Gulce Berfin Ercan, Melis Baran, Ecem Konca, Ilhan Mert Cetin,
   and Ilker Korkmaz

Towards a Framework for Promoting Student Engagement to Maximize
Learning in Higher Education: A Case Study 174
   Livinus Obiora Nweke and Rania El-Gazzar

Session 5

Blood Oxygen Saturation Estimation Using PPG Signals
from the MIMIC-III Database 195
   Nenad Petrovikj, Bojana Mishkovska, Bojana Koteska,
   and Ana Madevska Bogdanova

Novel Methodology for Gaining New Insights Into the Pharmacological
Mechanisms of *Cannabis sativa* and *Alzheimer's Disease* Through
Signaling Pathway Analysis Using Bioinformatics Tools 206
   Filip Donev, Nevena Ackovska, and Ana Madevska Bogdanova

Session 6, 7

AI Cardiologist: Arrhythmia Detection by Transformer-Based Language
Model 223
   Marjan Gusev

Ambient Assisted Living Sensor-Based Solution for Elderly
Self-monitoring 238
   Ivan Kuzmanov and Nevena Ackovska

Classification of Autism and Typical Development Children Based
on EEG Signals 253
   Aleksandar Tenev, Silvana Markovska-Simoska, and Igor Mishkovski

NATO Workshop

Detecting the Unseen: Exploiting Radar-Sonar Sensor Fusion for Visual
Detection of Low-Profile Naval Drones 263
   Giacomo Longo, Alessandro Cantelli-Forti, and Enrico Russo

Evaluating Killer Drone Defense: NATO SPS Project "Anti-Drones" Field
Trials 277
   Alberto Lupidi, Francesco Mancuso, Giulio Meucci, Edmond Jajaga,
   Veton Rushiti, and Alessandro Cantelli-Forti

STEM Workshop

Academic Career and Gender Balance Perceptions Among Bachelor's
Students in Computer Science: A Case Study 295
   Ozge Buyukdagli, Amal Mersni, and Merjem Talic

Author Index 311

Session 1

Evaluation of Vector Databases and LLMs in RAG-Based Multi-document Question Answering

Elena Filipovska[2], Ana Mladenovska[1,2], Jovana Dobreva[1,2], Dimitar Kitanovski[1,2], Goran Mitrov[1,2], Petre Lameski[1,2], and Eftim Zdravevski[1,2(✉)]

[1] Faculty of Computer Science and Engineering, Ss. Cyril and Methodius University in Skopje, Skopje, Macedonia
eftim.zdravevski@finki.ukim.mk

[2] Magix.AI, Skopje, Macedonia

Abstract. With the explosion of use of Generative AI, deploying large language models (LLMs) in private environments and making them understand proprietary data has become a necessity. Retrieval augmented generation (RAG) in combination with vector databases have become the leading technique for enhancing LLMs with enterprise data. In this paper, we evaluate different vector databases (Chroma, Qdrant, FAISS, and Pinecone) combined with LLM models, such as OpenAI's GPT-4o mini, in answering multiple questions based on multiple documents. Specifically, we process the data privacy policies of over 100 EdTech providers and use an LLM to answer 45 questions related to GDPR with a yes/no. By evaluating the speed of the vector databases, their technical requirements, and the consistency of their responses, this study provides practical guidelines in selecting the vector database for several use-cases. The results show that Chroma and Qdrant are significantly faster than Pinecone and FAISS, while being consistent in the selection of text chunks based on which questions are answered.

Keywords: OpenAI · RAG · GPT · QA · LLM · Large Language Model · Multi-document Q&A · Qdrant · Pinecone · Chroma · FAISS

1 Introduction

With the continuous growth of information, the real challenge is no longer accessing data but finding the correct answers quickly and accurately. Traditional databases, designed for structured data, struggle to handle unstructured data, such as text, and cannot meet the demands of modern scale. The development of new methods and algorithms that capture the meaning of words and phrases through vector embeddings has created a need for databases that can process this type of information. This shift has led to the rise of vector databases, which provide a more efficient way to store and retrieve meaningful information from vast amounts of unstructured data [1].

Understanding and processing human language is a complex task, influenced by grammatical rules, context, and subtle nuances. With the rise of large language models (LLMs), we are witnessing a breakthrough in handling these complexities, as these models demonstrate astonishing performance in understanding and generating human-like text. This explosion of LLM capabilities has opened up a wide range of new possibilities. Complementing this progress, vector databases enhance LLM functionality by providing efficient storage, retrieval, and management of high-dimensional vector representations, further empowering advanced tasks like question answering and knowledge retrieval [2].

While LLMs demonstrate remarkable abilities, these models still face challenges, including hallucinations and the reliance on static, pre-trained knowledge. To mitigate these issues, Retrieval-augmented generation (RAG) methods combine LLMs with external data sources, enhancing both the accuracy and relevance of their responses [3]. The combination of RAG with LLMs and vector databases is now a leading technique for effectively leveraging enterprise data, offering both precision and scalability in business applications.

Motivated by the growing role of vector databases in enhancing the capabilities of LLMs and the potential of vector databases and LLMs in information retrieval, this study evaluates their application in question-answering tasks. We focus on four popular vector databases (ChromaDB, Qdrant, FAISS, and Pinecone) and assess their performance in terms of speed, retrieval efficiency, answer accuracy, and consistency. Our experiments involve a series of use cases where these databases are integrated with a large language model to answer questions drawn from over 100 data privacy policies of EdTech providers. Specifically, we test 45 GDPR-related questions, each requiring a binary Yes or No response, to assess how effectively the databases can provide accurate and consistent answers in a practical regulatory compliance scenario.

In our investigation, we integrate these vector databases with OpenAI's GPT-4o-mini, a smaller and resource-efficient variant of GPT-4o that maintains high performance. Using a Retrieval-Augmented Generation (RAG) system, we retrieve relevant information from the document repository and feed it to GPT-4o-mini to generate answers. Our evaluation focuses on three key aspects: the execution time of each query, the quality and consistency of the retrieved information, and the associated computational costs of these executions.

The remainder of this paper is organized as follows: Sect. 2 reviews the foundational concepts of vector databases, LLMs, and RAG, along with relevant studies by other researchers in this field. Section 3 describes our research methodology, including a detailed explanation of the dataset, use cases, evaluation metrics, and how we integrate vector databases with the RAG system. In Sect. 4, we present the outcomes of our experiments and analyze the performance of each vector database. Finally, Sect. 5, summarizes our findings and suggests potential directions for future research in this rapidly evolving field.

2 Related Work

Question Answering (QA) is a key subfield of Natural Language Processing (NLP) that focuses on building systems capable of automatically answering human questions in natural language. Traditionally, QA systems relied on structured techniques like Named Entity Recognition (NER), post tagging, tokenization, and relation finding, but these methods struggle with scalability and complexity [4]. A study by Alanazi et al. highlights that Question Answering systems remain a highly active area of research, with scholars employing diverse approaches, particularly deep learning, word embeddings, and knowledge-based systems, which have proven to be the most effective methods [5].

With the emergence of large language models (LLMs), which are pretrained on vast text corpora and fine-tuned on QA-labeled datasets, the field of QA has seen a significant shift. As highlighted by Hadi et al., the linguistic knowledge gained during pretraining enables LLMs to understand question semantics while fine-tuning on QA data equips them to identify relevant context, making QA one of the most promising areas where LLMs can have a transformative impact [6]. In a survey by Chang et al., it was found that numerous researchers have evaluated the performance of LLMs on QA tasks, with the results showing that LLMs generally exhibit outstanding performance in this area [7]. One of the most extensively researched areas is the application of LLMs in answering medical questions. Studies have shown that LLMs can tackle complex medical questions and demonstrate rapid progress toward achieving physician-level performance in this domain [8,9]. These promising results pave the way for the use of LLMs in QA across various other fields.

As LLMs demonstrate capabilities in various domains, including medical question answering, there is growing interest in applying them to question answering in the context of policy documents. Krafft et al. [10] discuss the differences in how researchers and policymakers define AI, with researchers focusing on technology and policymakers on human behavior. This misalignment can lead to regulatory gaps, particularly when addressing current AI systems. In response to these challenges, Ravichander et al. introduce PrivacyQA, a dataset of questions about privacy policies aimed at improving QA systems for better understanding complex policy documents [11].

Retrieval-augmented generation (RAG) combines retrieval techniques with deep learning advancements to overcome the static nature of large language models (LLMs) by dynamically incorporating up-to-date external information. Primarily used in the text domain, this approach offers a cost-effective way to mitigate the generation of plausible yet potentially incorrect responses by LLMs, improving the accuracy and reliability of outputs through real-world data integration [12]. In a recent article, Muludi et al. applied the RAG approach to document question answering using LLMs, concluding that the results demonstrated the superiority of RAG in improving the performance and accuracy of QA systems [13]. Wiratunga et al. demonstrated that combining Case-Based Reasoning (CBR) with RAG enhances the accuracy and reliability of LLMs in tasks such as legal question-answering. By leveraging relevant cases as context and utilizing indexing vocabularies and similarity measures, their approach improves LLM inputs and enables validation against a knowledge base, resulting in more precise and contextually accurate answers [14].

Building on these ideas, Devjeet Roy et al. [15] explored the use of LLMs for automating root cause analysis (RCA) in cloud incident management, demonstrating that fine-tuned LLMs could identify root causes using limited incident information. Their system, RCACopilot, echanced LLM-based RCA by integrating retrieval-augmented generation and diagnostic data collection tools. Their approach allows the predictions to be more context-aware and as such they allow the system to predict specific types of incidents.

Privacy and security are critical concerns in AI development, especially when handling sensitive data. Hasal et al. [16] examine the risks involved in chatbots processing and storing personal data, emphasizing the need for strict data protection measures and secure communication channels to safeguard user privacy and protect systems from cyber threats. Chen et al. [17] examine the reliability of models such as ChatGPT and Bard in specialized domains. They focus on security and privacy (S&P), and their study reveals that both models demonstrate high error rates when attempting to deny misconceptions. These rates suggest that LLMs, while powerful, can struggle to maintain accuracy and reliability when handling domain-specific content.

Pan et al. [18] provide an in-depth analysis of state-of-the-art VDBMS technologies, highlighting the importance of optimizing data indexing, storage, and retrieval for managing vector embeddings. They show that these improvements are crucial for enhancing retrieval speed and accuracy, supporting LLMs in generating contextually relevant responses. Jing et al. [2] propose an adaptive learning algorithm that leverages vector databases to optimize exploration in multi-armed bandit problems. By efficiently storing and retrieving vector representations of past actions and rewards, their method maintains high performance in non-stationary environments, demonstrating the potential of vector databases in applications like online advertising, personalized recommendations, and dynamic pricing.

In [19], authors explore the application of large language models (LLMs) in document analysis. They highlight their potential to improve accuracy and efficiency in document processing in educational institutions and online platforms. The study applies various approaches, including direct API calls and AI frameworks like LangChain and RAG systems, to determine optimal practices for document analysis, providing insights into their effectiveness based on execution time, cost, and response quality.

3 Methods

3.1 Dataset

The data used for this research combines a number of 100 publicly available data privacy policies of EdTech providers, and the dataset used is available at https://github.com/admin-magix/edtech-policies/tree/main/dataset, [19]. The selected policies have varying text lengths. Therefore, we investigate the impact of the text lenth on the retrieval performance and the consistency of the answers related to the 45 GDPR-questions. These questions were manually crafted to investigate how compliant is a certain data privacy policy with different GDPR acpects. Finally, the data used for evaluating the results attained from the LLMs and their respective vector databases consists of the questions and the answers gained for them for each policy used.

3.2 Vector Databases

Vector databases are specialized data storage systems designed to handle and manage vector embeddings, which are numerical representations of data that capture semantic meanings or patterns in high-dimensional space. Unlike traditional databases that rely on structured data formats such as tables, vector databases are optimized for storing and retrieving data points in the form of dense vectors, which are typically generated by machine learning models or natural language processing (NLP) techniques. This unique capability makes vector databases suitable for applications involving similarity search, recommendation systems, and real-time data retrieval in high-dimensional spaces.

The primary function of a vector database is to efficiently store vector embeddings and support fast, approximate nearest neighbor (ANN) search. This is helpful when finding similar data points, such as images, documents, or user profiles, is necessary. In practice, a vector database takes in a query in the form of a vector and retrieves the most similar vectors from the database. Indexing techniques such as locality-sensitive hashing (LSH) or tree-based methods like KD-Trees and Ball Trees are used to achieve high efficiency during the retrieval process [20].

Vector databases offer significant advantages for AI applications, but they also present challenges. Scalability is a key issue, as maintaining fast search speeds in growing databases requires efficient indexing and search techniques, often involving trade-offs between speed and accuracy. These databases also demand high computational resources, especially when managing dynamic data with frequent updates. Additionally, integrating vector databases with existing systems can be complex, requiring changes to data pipelines and architectures. Despite these challenges, vector databases are essential for advanced search and retrieval tasks, and their role will likely grow as demand for intelligent, context-aware applications increases [2].

In this research, each privacy policy undergoes a chunking process, dividing the text into smaller segments with a slight overlap. For each chunk, the text is embedded using an OpenAI embedding model, and the resulting embeddings, along with the original text and a link as an ID, are stored in the vector database. During the retrieval phase, the question is also embedded using the same model, and a similarity search is performed. The closest matching chunk is retrieved and provided as context to the prompt to generate the answer. The same approach is consistently applied across all use cases described later on to evaluate the performance of the vector database.

Chroma. Chroma is a vector database optimized for managing and retrieving high-dimensional data through semantic similarity searches. It transforms data into vector embeddings using advanced algorithms, enabling context-aware queries across large datasets. Chroma offers fast, scalable, and efficient data retrieval by comparing vector representations rather than relying on keyword searches. Its high-performance architecture and integration capabilities via APIs make it suitable for various applications, providing developers with an easy way to incorporate its search and indexing features into different use cases without significant changes [20].

In Fig. 1, we have a visual representation of the data retrieval process with Chroma. During the retrieval phase, the policy's id is used to filter the chunks against which the similarity search is evaluated.

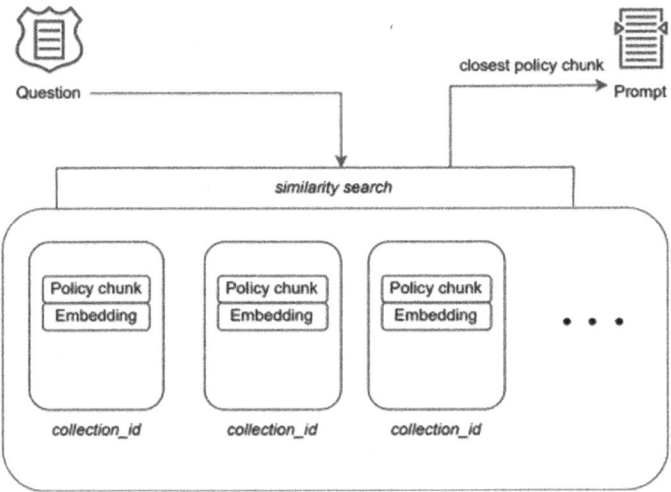

Fig. 1. Data retrieval process with Chroma and Qdrant

The performance of Chroma is measured against several metrics, including the quality of the search results and the efficiency of data retrieval. These metrics provide insights into the effectiveness of Chroma in tasks that involve processing and managing large-scale data. By utilizing these performance indicators, Chroma can be further fine-tuned so that it meets specific application needs while maintaining efficient data retrieval.

Qdrant. Qdrant is a high-performance vector search engine optimized for storing, indexing, and retrieving high-dimensional data using vector embeddings. It performs semantic similarity searches, making it ideal for applications in machine learning, natural language processing, and recommendation systems. By leveraging vector-based methods, Qdrant retrieves data based on contextual meaning rather than traditional keyword matching. Its indexing methods ensure fast queries and scalability, making it efficient for large datasets. Additionally, Qdrant's flexible architecture, supported by APIs, allows for easy integration into various applications, enhancing its versatility across multiple domains and use cases. [21] The architecture of Qdrant's data retrieval process is identical to Chroma's, and it is depicted in Fig. 1.

Qdrant's effectiveness is evaluated through several performance metrics, including search accuracy, retrieval speed, and system scalability, which help assess how well Qdrant performs in real-world scenarios, guiding ongoing improvements and optimizations. By focusing on these key performance indicators, Qdrant ensures that it consistently delivers high-quality results and meets the demands of modern data-driven applications.

FAISS. FAISS, developed by Facebook AI Research, is a vector search index library optimized for managing similarity searches in large-scale datasets [22]. It supports various nearest neighbor search and clustering algorithms, ranging from exact to approximate methods, allowing users to balance accuracy and efficiency. FAISS's indexing structures, including flat, inverted file, and quantization-based indexes, enable fast processing of high-dimensional data, making it suitable for real-time and large-scale applications. Additionally, FAISS offers GPU acceleration for enhanced performance and can be easily integrated into existing systems through its flexible APIs, supporting a range of applications and environments. A visual representation of the architecture of the data retrieval process using FAISS is represented in Fig. 2.

Performance evaluation of FAISS involves metrics such as search accuracy, query response time, and system scalability. These metrics are crucial for assessing how effectively FAISS handles large-scale vector data and ensures efficient retrieval operations. Continuous optimization and refinement based on these performance indicators help maintain FAISS's robustness and adaptability in various real-world scenarios, delivering high-quality results consistently.

Pinecone. Pinecone is a vector database designed for indexing and searching vector data, making it ideal for machine learning applications. It uses advanced indexing systems, such as HNSW (Hierarchical Navigable Small World) and IVF (Inverted File Index), to efficiently perform similarity searches across large vector embeddings, focusing on context rather than keywords. Pinecone's architecture is built for scalability and flexibility, offering both exact and approximate search methods depending on the user's needs. Its APIs facilitate easy integration into various systems, enabling seamless deployment of its indexing and search capabilities. Pinecone's indexing system aligns closely with that used by FAISS, and a visual representation of the data retrieval process in showcased in Fig. 2.

Pinecone's performance is evaluated based on key metrics such as the accuracy of its indexing algorithms, search latency, and scalability. By concentrating on these metrics, Pinecone consistently delivers high-quality results, meeting the demands of modern data-driven applications. In this paper, the evaluation is con-

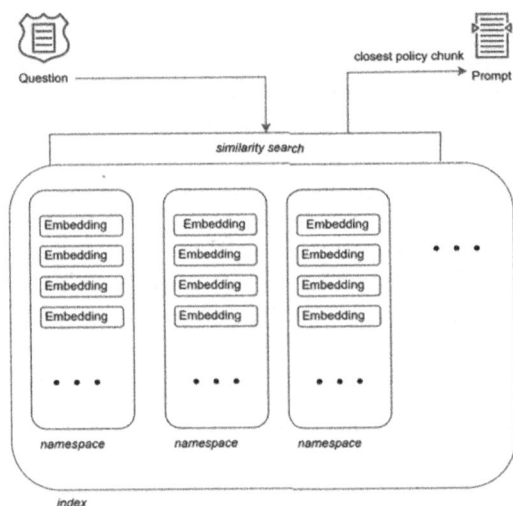

Fig. 2. Data retrieval process with Pinecone and FAISS

ducted using Pinecone's free tier plan, which introduces certain limitations that affect its overall effectiveness [20].

3.3 Retrieval-Augmented Generation (RAG) Systems

A Retrieval-Augmented Generation (RAG) system is an advanced framework that integrates retrieval-based and generative models to enhance information processing and question-answering capabilities. [12] The retrieval component of a RAG system is responsible for searching a large-scale database or knowledge repository to fetch relevant information based on a given query. This component uses vector search techniques or other indexing methods to locate and retrieve relevant data efficiently. The retrieval process addresses the limitations of standalone models, which may lack current or specific information. Additionally, RAG systems generate responses customized to the needs of the query, resulting in more relevant outputs [23].

These systems can also be integrated into existing applications via APIs or cloud-based services, letting developers use retrieval and generative features without needing to build custom solutions.

The effectiveness of a RAG system is usually measured against different metrics, which are based on the quality of the retrieved information as well as the generated response.

3.4 RAG with an OpenAI

There are various services that utilize an LLM, and in fact, these models serve as the foundation for such systems, and they are usually accessed through a specialized API.

This API serves as a bridge, allowing seamless incorporation of the model's capabilities into external projects without the need to develop models from scratch. GPT-4o-mini represents an LLM, which has been fine-tuned to be able to handle various tasks. It can be used with the help of an API or it can be furhter fine-tuned so that it to meets some additional needs, and thus offers great flexibility. For implementing GPT-4o-mini within a Retrieval-Augmented Generation (RAG) system, the process involves a few key steps:

1. **Retrieval Process:** Initially, the RAG system retrieves relevant information from a knowledge base or document repository based on the input query. This retrieval step ensures that the generative model has access to pertinent context and data.
2. **Integration with GPT-4o-mini:** Once the relevant information is retrieved, it is fed into GPT-4o-mini, allowing the model to then generates a response that incorporates both the retrieved data and its own generative capabilities, with the ultimat comprehensive and contextually accurate output.
3. **Evaluation:** The performance of GPT-4o-mini in the RAG system is evaluated by examining the accuracy of the retrieved information, the relevance of the generated responses, and the overall efficiency of the process. This evaluation helps in refining the system to ensure it meets specific application needs and delivers high-quality results.

Overall, the use of GPT-4o-mini within a RAG system enhances the model's ability to provide precise and contextually relevant answers by combining advanced retrieval and generation techniques.

Prompt Used for All Use Cases Tested:

"""""For the following documents:""""" + f"{closest_chunk}" + """""answer the following question with a "Yes"/"No" answer and if the answer is "Yes" then provide an extract from the document that is LONGER than 200 characters and best fits the answer, otherwise return an empty string, nothing else that differs from this. Make sure that your response is only in a JSON format like this and DO NOT PROVIDE ANY ADDITIONAL TEXT: "'Answer": "Your answer", "Extract": "Extract from the document"', where " Your Answer" represents your answer to the question, "Extract from the document" is the best fit extract (make sure it is composed of whole sentences and also don't include any quotation marks). Here is the question: """"" f"{question_content['Text']}"; where {question_content['Text']} denotes the text of the question to be answered

3.5 Evaluation

Use Cases. Three distinct use cases were defined to test the effectiveness of the methods. These use cases range from simple to more complex, and each has different computational power and resource needs. Since the testing dataset consists of 100 policies, with the median being 30,000 characters, it gets split into two groups of 50 policies. As such, the use cases are defined based on the policy character length. Longer texts result in more chunks, which affect the retrieval executed by the specific vector database, and this serves as the basis for the use case definition. Below is a detailed overview of the identified use cases:

1. **Use Case 1:** Using 50 policies that do not exceed the text length of 30,000 characters and all 45 questions while retrieving only the first closest chunk
2. **Use Case 2:** Using 50 policies that exceed the text length of 30,000 characters and all 45 questions while retrieving only the first closest chunk
3. **Use Case 3:** Using 50 policies that exceed the text length of 30,000 characters and all 45 questions while retrieving the first three closest chunks

Metrics. The defined use cases are evaluated against several metrics to note the effectiveness of the vector database under investigation. Namely, the retrieval time (the time the vector database takes to retrieve relevant results) is taken as a key measure to estimate the efficiency of vector database. Another evaluation considers the chunk overlap. This scenario only compares the consistency of retrieved chunks by fastest vector databases. All of these metrics are assessed against a specific use case in detail in the Results section. The policies and questions follow the same format throughout all use cases in a given approach.

4 Results and Discussion

The analysis yielded results that were further evaluated to assess the performance of each vector database. Specifically, the total execution time was examined for each database, focusing on the differences in retrieved chunks across various questions in the context of the use case. The following section discusses the total execution times for specific use cases, highlighting the observed similarities and differences in retrieval performance across the vector databases using the GPT-4o-mini model.

Statistical analyses were conducted based on the results obtained during the evaluation process and are presented in several figures. The analysis began by extracting key data points, including policies, policy lengths, questions, question lengths, vector databases, and the execution times associated with each vector database. For all 100 policies and 45 questions, every possible combination was analyzed, resulting in 1,800 data points for further evaluation.

The primary objective is to identify which database has the fastest and slowest execution times overall. In (Fig. 3), we present the average execution time across all policy documents and questions for each database. As shown in the figure, the Chroma database emerges as the fastest, closely followed by the

Qdrant database. In contrast, the Pinecone database and the Faiss search index library lag significantly, with Pinecone being approximately 15 times slower and Faiss around 30 times slower on average.

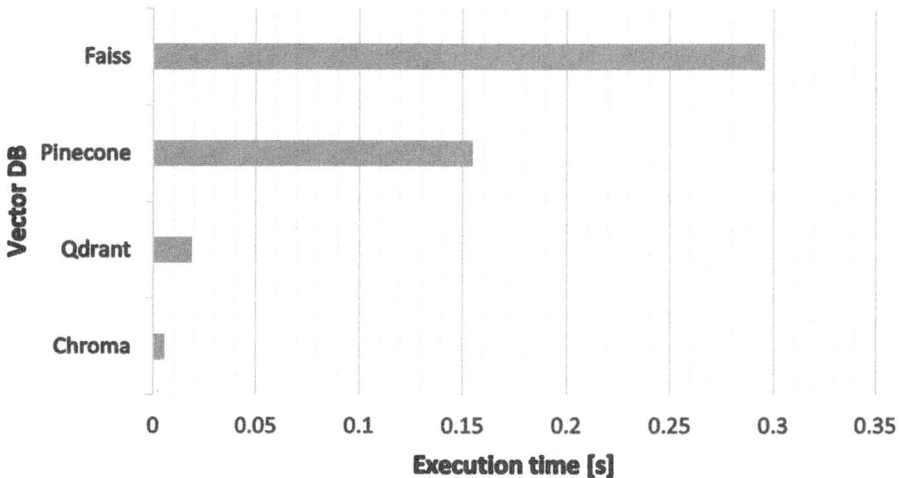

Fig. 3. Comparison of execution times across vector databases

Next, we analyze the vector database execution times for each policy, represented by the length of the policy. In (Fig. 4), the policies are ordered by length on the x-axis, and the results shown are based solely on use case 1. From the figure, we observe that Chroma and Qdrant maintain relatively stable execution times across all policy lengths, consistently staying below 0.05 s. In contrast, the execution times for Pinecone and Faiss exhibit greater variability depending on the policy length.

In (Fig. 5), the x-axis represents the question length, ordered in ascending order, while the y-axis shows the cumulative sum of execution times. This highlights the significant difference in performance when scaling up the number of queries. From the figure, we observe that the cumulative execution time for Faiss across all questions reaches approximately 1,250 s, while for the fastest databases, Chroma and Qdrant, it is around 50 s-about 25 times faster. This difference is crucial to note, as in practice, more processing is required than just a single query, making execution time a key factor.

After evaluating the execution times, we proceeded to compare the two fastest vector databases, Chroma and Qdrant, across additional metrics. In the first and second use cases, where only the first relevant chunk was extracted, we compared whether both databases retrieved the same chunk. For the third use case, where the three closest chunks were extracted, we compared the results as lists to assess the similarities between the databases.

The databases extracted identical results when they retrieved the first closest chunk across all questions and policies. In the case of retrieving the three

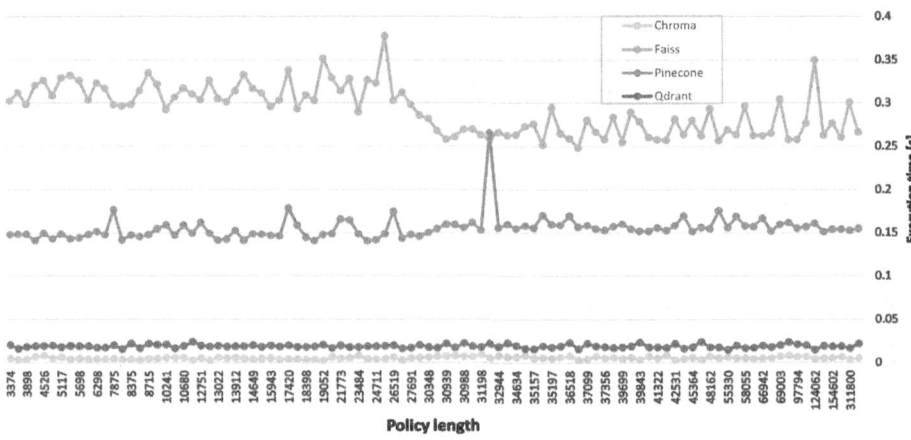

Fig. 4. Vector database execution time based on policies ordered by length

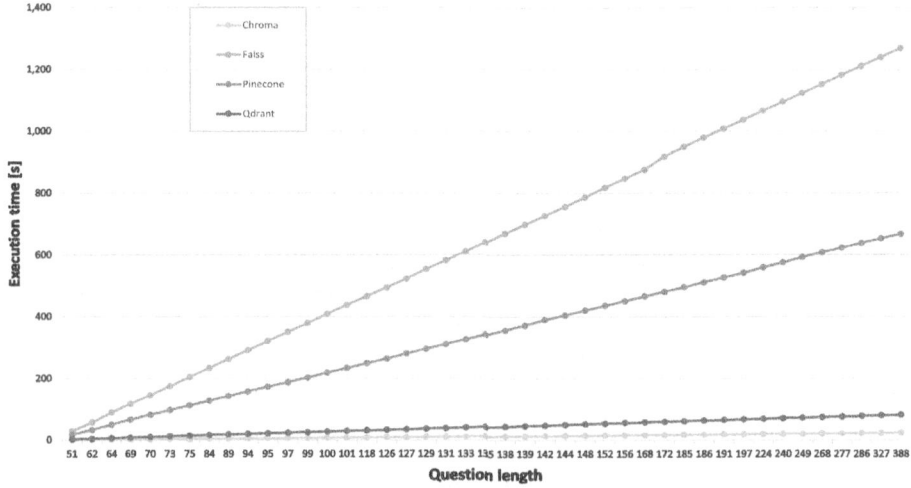

Fig. 5. Vector database cumulative execution time based on questions ordered by length

closest chunks, the comparison included various match options: exact match, two matches, one match, or no matches. In (Fig. 6), the x-axis represents the policies, while the y-axis shows the number of questions corresponding to each match option. The figure shows that the results remain consistent, with both vector databases typically returning the same chunks for most policies, except for one case where there was a notable discrepancy.

After analyzing the two best databases based on their response times, we now proceed to compare the complete execution process of the cases we are evaluating. So far, we have focused primarily on the vector databases and the

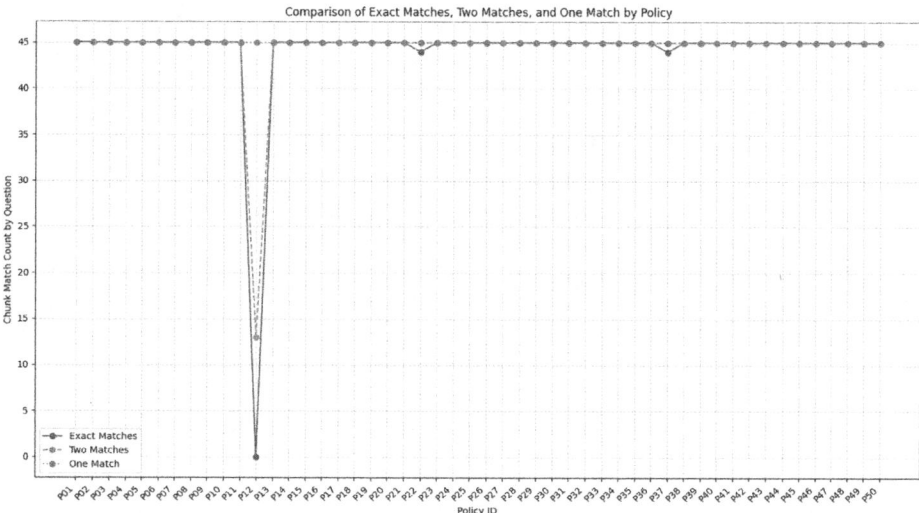

Fig. 6. Count of matches between Chroma and Qdrant when retrieving 3 chunks

results they generated. Now, we will present the total execution time, which includes the time taken to run each policy for each question and to obtain the final answer. As shown in Table 1, vector databases Chroma and Qdrant once again lead in terms of execution speed and providing answers across all fields and questions, while the other two databases consistently yield poorer results. Regarding the cost of the entire process, it can be observed that the highest-ranked Chroma database also comes with the highest price, whereas the lowest-ranked Faiss, despite its lower performance, offers the cheapest option when considering the full analysis and resources consumed.

Handling the volume and complexity of data poses considerable challenges, particularly in the areas of storage, retrieval, and processing. As LLMs grow in size and capability, the computational costs associated with managing and querying large-scale vector data can become considerable, and these costs can further increase when dealing with vast datasets, making the efficient use of resources a concern. To mitigate these issues, our research focuses on smaller models like GPT-4o mini and limits test case scope to reduce runtime and computational overhead while still providing valuable insights. This approach allowed us to maintain a balance between performance and resource utilization, emphasizing the importance of optimizing model scale and use case complexity practical applications.

Table 1. Comparison of retrieval between different vector databases

Vector Database	Use case	Total Execution Time	Total Cost
Chroma	UC1	51 min 48 s	$1.10
	UC2	56 min 48.7 s	$0.86
	UC3	64 min 21.1 s	$2.34
FAISS	UC1	58 min 40.5 s	$0.46
	UC2	61 min 58 s	$0.54
	UC3	98 min 10.6 s	$0.78
Pinecone	UC1	50 min 35.2 s	$0.62
	UC2	71 min 2.2 s	$0.84
	UC3	92 min 2.1 s	$2.21
Qdrant	UC1	42 min 31.1 s	$0.62
	UC2	42 min 57 s	$0.81
	UC3	68 min 28.8 s	$2.08

5 Conclusion

This research addresses a few challenges in AI, in particular the need for efficient and scalable data management solutions that can keep up with the increasing complexity of LLM, enabling faster and more accurate searching of high-dimensional data, and the vector databases that play key role in improving LLM performance in various natural language processing tasks. The findings of this paper highlight the importance of continuing to develop and refine technologies that support LLMs, including vector databases. Although there are challenges to overcome, such as managing data volume and ensuring compatibility with evolving models, the potential benefits are significant. With the continuous advancement of AI, the integration of vector databases and LLM will likely become even more integral to the development of intelligent systems capable of understanding and generating human language with high accuracy. This paper provides a complete overview of the current state of four-vector database integration with LLMs, offering insights that are valuable to both researchers and practitioners in the field. Future research should focus on further optimizing this integration, exploring new techniques and technologies that can push the boundaries of what is possible with artificial intelligence.

References

1. Han, Y., Liu, C., Wang, P.: A comprehensive survey on vector database: storage and retrieval technique, challenge (2023)
2. Jing, Z., et al.: When large language models meet vector databases: a survey (2024)
3. Gao, Y., et al.: Retrieval-augmented generation for large language models: a survey (2024)
4. Bakır, D., Aktas, M.S.: A Systematic Literature Review of Question Answering: Research Trends, Datasets, Methods, pp. 47–62. Springer, Cham (2022)
5. Alanazi, S.S., Elfadil, N., Jarajreh, M., Algarni, S.: Question answering systems: a systematic literature review. Int. J. Adv. Comput. Sci. Appl. **12**(3) (2021)
6. Hadi, M.U., et al.: A survey on large language models: applications, challenges, limitations, and practical usage. Authorea Preprints (2023)
7. Chang, Y., et al.: A survey on evaluation of large language models. ACM Trans. Intell. Syst. Technol. **15**(3), 1–45 (2024)
8. Liévin, V., Hother, C.E., Motzfeldt, A.G., Winther, O.: Can large language models reason about medical questions? Patterns **5**(3), 100943 (2024)
9. Singhal, K., et al.: Towards expert-level medical question answering with large language models. arXiv preprint arXiv:2305.09617 (2023)
10. Krafft, P.M., Young, M., Katell, M., Huang, K., Bugingo, G.: Defining AI in policy versus practice. In: Proceedings of the AAAI/ACM Conference on AI, Ethics, and Society, pp. 72–78 (2020)
11. Ravichander, A., Black, A.W., Wilson, S., Norton, T., Sadeh, N.: Question answering for privacy policies: combining computational and legal perspectives. In: Inui, K., Jiang, J., Ng, V., Wan, X. (eds.) Proceedings of the 2019 Conference on Empirical Methods in Natural Language Processing and the 9th International Joint Conference on Natural Language Processing (EMNLP-IJCNLP), Hong Kong, China, pp. 4947–4958. Association for Computational Linguistics (2019)
12. Huang, Y., Huang, J.: A survey on retrieval-augmented text generation for large language models (2024)
13. Muludi, K., Fitria, K.M., Triloka, J., Sutedi: Retrieval-augmented generation approach: document question answering using large language model. Int. J. Adv. Comput. Sci. Appl. **15**(3) (2024)
14. Wiratunga, N., et al.: CBR-RAG: case-based reasoning for retrieval augmented generation in LLMs for legal question answering. In: International Conference on Case-Based Reasoning, pp. 445–460. Springer, Cham (2024)
15. Roy, D., et al.: Exploring LLM-based agents for root cause analysis. In: Companion Proceedings of the 32nd ACM International Conference on the Foundations of Software Engineering, pp. 208–219 (2024)
16. Hasal, M., Nowaková, J., Ahmed Saghair, K., Abdulla, H., Snášel, V., Ogiela, L.: Chatbots: security, privacy, data protection, and social aspects. Concurr. Comput. Pract. Exp. **33**(19), e6426 (2021)
17. Chen, Y., Arunasalam, A., Celik, Z.B.: Can large language models provide security & privacy advice? Measuring the ability of LLMs to refute misconceptions. In: Proceedings of the 39th Annual Computer Security Applications Conference, pp. 366–378 (2023)
18. Pan, J.J., Wang, J., Li, G.: Survey of vector database management systems. VLDB J. 1–25 (2024)
19. Filipovska, E., et al.: Benchmarking openai's APIs and other large language models for repeatable and efficient question answering across multiple documents. In: 2023

18th Conference on Computer Science and Intelligence Systems (FedCSIS). IEEE (2024)
20. Xie, X., Liu, H., Hou, W., Huang, H.: A brief survey of vector databases. In: 2023 9th International Conference on Big Data and Information Analytics (BigDIA), pp. 364–371. IEEE (2023)
21. Zovak, B.C.J.: Learned indexing in vector database management systems (2024)
22. Douze, M., et al.: The faiss library (2024)
23. Petroni, F., Siciliano, F., Silvestri, F., Trappolini, G.: IR-RAG@ SIGIR24: information retrieval's role in rag systems. In: Proceedings of the 47th International ACM SIGIR Conference on Research and Development in Information Retrieval, pp. 3036–3039 (2024)

Aligning Food Ingredients with Multiple Semantic Resources

Darko Sasanski[1](✉), Andrej Todorovski[1](✉), Bojan Trpeski[1](✉),
Dimitar Trajanov[1](✉), Tome Eftimov[2](✉), and Riste Stojanov[1](✉)

[1] Faculty of Computer Science and Engineering, Ss. Cyril and Methodius University, Skopje, North Macedonia
{darko.sasanski,andrej.todorovski,bojan.trpeski}@students.finki.ukim.mk,
{dimitar.trajanov,riste.stojanov}@finki.ukim.mk
[2] Computer Systems Department, Jozef Stefan Institute, Ljubljana, Slovenia
tome.eftimov@ijs.si

Abstract. To address the lack of integrated resources that combine nutritional, chemical, and semantic information, we have integrated the Recipe1M+ dataset with the USDA National Nutrient Database, FooDB, and the FoodOn ontology. This integration brings significant advantages by enriching the dataset with comprehensive information across multiple domains. By linking Recipe1M+ ingredients to the USDA database, we provide detailed nutritional profiles that enable a comprehensive analysis of recipes, supporting researchers and dietitians in developing personalized nutrition plans based on accurate nutrient data. Furthermore, incorporating FooDB enhances the dataset with in-depth chemical compositions and health effects of food constituents, facilitating research on functional foods and their role in disease prevention and health promotion. Mapping ingredients to FoodOn further expands the dataset's semantic context, encompassing food products, production, agriculture, and environmental impacts.

This integration promotes interdisciplinary research and the creation of comprehensive knowledge graphs, bridging gaps between nutrition, food science, agriculture, and health domains. Our methodology has been validated, demonstrating significantly better results with a precision of 76.25%. Adhering to the FAIR principles, we ensure the data is findable, accessible, interoperable, and reusable. The enriched Recipe1M+ dataset supports advanced applications, such as predictive modeling of dietary impacts, food recommendation systems, and sustainable food system studies, making it a valuable resource for the research community and beyond (The data and code are publicly available at the following link).

Keywords: Entity Linking · Name Normalization · Nutritional Profiles · Semantic Resources · Functional Foods · Knowledge Graphs

1 Introduction

Food is a crucial environmental factor affecting human health. Even healthy and eco-friendly foods can cause health issues when consumed with certain drugs or during specific diseases. Comprehensive dietary assessments are essential to understand food's impact on health. Automating food entity detection is vital for applications such as identifying food-drug interactions and health issues related to diet.

Computer science can significantly contribute to this research, especially through ML, NLP, and data analysis. Data collected in studies carry important information, which is not easily extracted due to different data formats (structured, semi-structured, and unstructured) and domains (food, medicine, pharmacy, ecology, and agriculture). Extracting this information enables the creation of knowledge graphs [1], which integrate and unify data from various sources.

One significant problem is the fragmentation within research communities, which often work in isolation and produce overlapping results. This is evident in the existence of multiple ontologies and datasets such as USDA, FoodOn and FooDB, which are not mapped to each other. Even though there are such initiatives in the computer science community [2-4], we are far from complete mapping of the resources. Recipe1M+ is a similar example of a dataset that is not integrated with these ontologies and datasets. Recipe1M+ is a large-scale dataset containing over one million recipes, but it lacks connections to other food and nutrition datasets, limiting its utility. The authors of FoodKG [5] try to bridge this gap, but the mappings they provide are unreliable.

Recipe1M+ [6] consists of recipes with associated ingredients, instructions, and images. The USDA National Nutrient Database [7] focuses on nutritional data for a wide range of food items, providing detailed information on nutrients and food composition. FoodOn [8] is an ontology designed to represent food-related entities, including food products, processes, and nutrition. FooDB [9] is a comprehensive database of food constituents, detailing the chemical composition of foods and their effects on human health.

Linking Recipe1M+ ingredients with the USDA, FoodOn, and FooDB ontologies and datasets can enable several beneficial use cases. For instance, it can facilitate personalized nutrition recommendations by combining recipe data with detailed nutritional information. This integration can also support research on the environmental impacts of food choices by linking recipes to data on food production processes. Ultimately, such efforts can promote healthier and more sustainable diets, contributing to global food system sustainability.

The objective of this paper is to extend the Recipe1M+ dataset by linking its ingredients with established food ontologies and datasets such as USDA, FoodOn, and FooDB. By integrating these resources, we aim to enhance the dataset's utility for research in food science, nutrition, and health. This linkage will facilitate comprehensive dietary assessments, personalized nutrition recommendations, and the study of food's environmental impacts. Ultimately, our goal is to promote healthier and more sustainable food systems by providing a more structured and interconnected dataset for the research community.

2 Related Work

2.1 Entity Linking

Entity linking identifies and links data records referring to the same real-world entity within a single or multiple data sources. This process typically involves three phases: recognition, resolution, and relationship detection [10].

In the recognition phase, data is cleansed, standardized, and quality-checked to maintain data integrity. A key component of this phase is Data Quality Management (DQM), which optimizes data quality through rules for repairing, cleaning, and standardizing identity data. These rules ensure correct formatting and address errors and inaccuracies, particularly in name normalization.

The resolution phase involves comparing cleansed data values to existing entities using search algorithms. Candidate lists are generated, and resolution rules with scoring methods determine if incoming identities match existing entities based on set thresholds.

The final phase detects relationships between resolved identities and entities, triggering alerts for relevant relationships. However, this phase is not utilized in our approach as the relationships are predefined.

The terms entity resolution and entity linking are often used interchangeably to describe the process of identifying and linking mentions of real-world entities across datasets. While there are subtle distinctions, both tasks aim to disambiguate entities by connecting them to a reference point. In entity linking, this reference point is typically a knowledge base (KB) with unique identifiers. Entity resolution, on the other hand, involves linking entities within a single dataset or across multiple datasets, where the matched entity may not necessarily be linked to a KB identifier.

2.2 Food Entity Linking And Recognition

In recent years, there has been substantial progress in addressing the challenges related to food NER and NEL, particularly the scarcity of annotated datasets and specialized NLP methods. Initially, much of the focus was on rule-based approaches. For instance, drNER [11] was developed to extract dietary recommendations grounded in evidence, while FoodIE [12] focused on extracting food-related data from recipe collections. StandFood [13] contributed to this by introducing a classification technique that utilized lexical similarities between food entities to link them to the FoodEx2 database maintained by EFSA.

The introduction of the FoodBase corpus [14] marked a turning point, leading to the development of machine learning-based methods. One of the early examples was BuTTER [15], a bidirectional LSTM model that utilized FoodBase to identify food entities. Building on this, FoodNER [16] was released, leveraging BERT [17] to accurately extract and categorize food entities across multiple tasks, including distinguishing between food groups. To support food specialists in navigating the complexity of food classification systems, FoodViz [18] introduced a web-based platform designed for the semantic labeling and annotation of food entities. More recent innovations have looked into leveraging deep learning architectures for extracting food-related entities from recipe datasets [19], alongside utilizing large language models (LLMs) to advance named entity recognition within agricultural domains [20].

2.3 FoodKG

FoodKG [5] addresses the challenge of disparate food ontologies and datasets by creating an integrated knowledge graph for food. It facilitates tasks like recommending healthier food alternatives and menu items by integrating data from sources like the USDA [7], Recipe1M+ [6], and FoodOn [8]. Using techniques like lexical similarity and string matching, FoodKG links datasets to form a cohesive knowledge graph.

FoodKG aims to serve as a resource for food recommendations and a benchmark for evaluating entity resolution and semantic linking methods. However, inconsistencies in the linking process prompted the development of our own food knowledge graph, enhancing ingredient-class mappings between Recipe1M+, USDA, and FoodOn, and incorporating additional datasets for richer ingredient information.

These inconsistencies became apparent upon closer examination, as discrepancies emerged in the links between the Recipe1M+ dataset and USDA classes, as demonstrated by the inconsistent mappings of different cheese types shown in Table 1.

Table 1. Inconsistent Mappings for Cheese Ingredients

Ingredient	USDA Class
1% fat cottage cheese	Blue Cheese
2% cheddar cheese	Blue Cheese
Anejo cheese	Blue Cheese
Better Than Cream Cheese cream cheese substitute	Blue Cheese
Cotija cheese	Blue Cheese
Greek feta cheese	Blue Cheese

FoodKG's methodology significantly aided our data collection, allowing us to start with parsed data and accelerate our project.

2.4 Nutritional Resources

FoodOn [8] is an ontology developed by a consortium for food-related terminology. It covers a wide range of food areas such as food products, food production, agriculture, environment, food safety, food quality, nutrition, and diet.

FoodOn is part of the open-source OBO Foundry [21], a collection of interoperable ontologies in life sciences. This gives FoodOn a broad scope, including consumer demographics, animal and plant anatomy, chemical composition, and disease phenotypes. It is widely used in research and datasets across academia and government.

The USDA National Nutrient Database [7] is a vital resource providing detailed nutritional information on thousands of foods. It includes data on the nutrient content of various food items, such as vitamins, minerals, macronutrients, and more. This database supports research and applications in nutrition and dietetics by offering a reliable source of nutrient data.

FooDB [9] is a comprehensive database containing detailed information about food constituents, including their chemical composition and health effects. It provides data on various foods, their nutrients, and bioactive compounds, facilitating research in food science and nutrition.

FoodOntoMap [2] is a dataset of unique food concepts extracted from recipes described in eight different ontologies, including the Hansard corpus, FoodOn, parts of SNOMED CT, and OntoFood. It also includes a single dataset with mappings of concepts between these ontologies.

Recipe1M+ [6] dataset is a comprehensive collection of recipes. Each recipe includes a list of ingredients, quantities, and nutritional information per ingredient and per 100 g. The dataset provides valuable insights into the nutritional content and health implications of different dishes. Ingredients are listed with detailed sub-characteristics, such as 'Greek,' 'plain,' and 'nonfat' for 'yogurt'. Nutritional information is provided in two lists: one detailing values per ingredient (including fats, energy, protein, saturates, sodium, and sugars) and the other per 100 g of the entire recipe (including fats, protein, salt, saturates, and sugars).

3 Alignment of Nutritional Resources Through Normalization Process

To resolve the inconsistencies in the FoodKG mappings, a custom entity linking process, which is initially used for matching ingredients with FooDB classes, is implemented. This process is then adapted to generate new mappings between ingredients and USDA classes, ensuring more accurate and comprehensive matches.

The normalization process involves several steps to standardize ingredient names while retaining critical nutritional information. These steps include converting ingredient names to lowercase, removing irrelevant characters, selectively retaining important stop words, and lemmatizing the ingredient names. Additionally, ingredient names across the datasets are standardized to address any remaining inconsistencies. Finally, branded foods are specifically handled, and extra words are removed from the ingredient names to ensure consistency and clarity.

Depending on the dataset or resource being handled, different normalization steps have been introduced. Datasets such as FooDB typically feature broader, less specific classes, whereas datasets like USDA offer finer, more detailed classifications. This variation influences the normalization approach, specifying which aspects of the ingredient names are considered relevant for effective entity linking.

Figure 1 illustrates the normalization process used to match ingredients from the Recipe1M+ dataset with classes from the USDA dataset. As an illustrative example, the ingredient '1% low-fat Chocolate Milk' is used to demonstrate this process.

The normalization process is initiated by applying lexical normalization steps using the NLTK library [22]. This involves converting all uppercase letters to lowercase and removing specific irrelevant characters.

The selection of irrelevant characters is outlined in Table 2. For instance, when aligning with the USDA dataset, numerical data is preserved to utilize information such as fat percentages or lean content, enhancing alignment accuracy. Conversely, when aligning with the FooDB dataset, both numerical and non-alphanumeric characters, including punctuation, are excluded.

Table 2. Selection of irrelevant characters

	USDA	FoodOn	FooDB
Punctuation	yes	yes	yes
Non-alphanumeric	yes	yes	yes
Numeric	no	yes	yes

After these initial steps, our example ingredient is represented as '1 low fat chocolate milk', eliminating uppercase letters, hyphens, and percentage signs.

Additionally, the normalization process is further refined by excluding stopwords (e.g., 'a', 'the', 'is', 'are', etc.) from ingredient names. This step aims to eliminate non-essential words that, due to their frequent occurrence, could introduce bias into the matching process. The strategy for retaining specific stopwords is detailed in Table 3. For example, when aligning with the USDA dataset, stopwords such as 'no' and 'with' are retained to preserve meaningful context in ingredient names, which helps differentiate between phrases like 'with no salt' and 'with salt'. Omitting these stopwords could result in misleading similarities between distinctly different ingredients.

Table 3. Retention of stop words

Resource	Stopwords retained
USDA	'no' and 'with'
FoodOn	None
FooDB	None

At the end of the lexical normalization process, lemmatization is employed to further reduce naming differences. Following these final steps, there are no changes to our example ingredient name.

After the completion of lexical normalization, the need for further refining ingredient names to ensure more reliable links was recognized. It was identified that standardizing how ingredients are named across the two datasets of interest is crucial, as discrepancies such as 'milkshake' versus 'milk shake', and 'lowfat' versus 'low fat' were found to create significant challenges during the linking process.

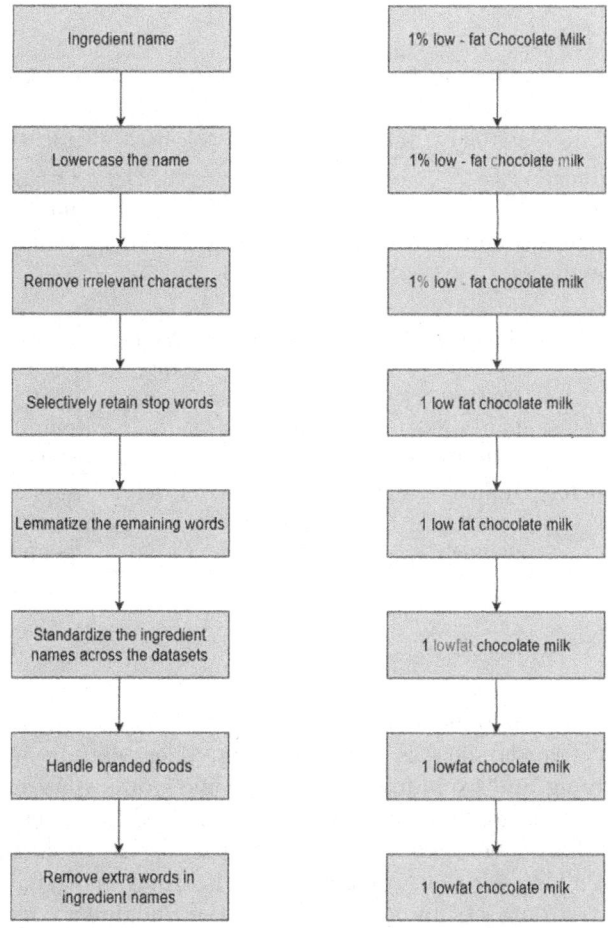

Fig. 1. Normalization steps for USDA matching

To address these inconsistencies, a method that utilizes bi-grams generated from each ingredient name in both datasets is employed. The frequency of each bi-gram appearing as a uni-gram in the opposite dataset is assessed. Whenever a bi-gram exhibits high frequency as a uni-gram, all occurrences of the bi-gram are replaced with its corresponding uni-gram. Following this process, the normalized name of our ingredient becomes '1 lowfat chocolate milk', as the bi-gram 'low fat' is replaced with 'lowfat'.

The next step in the normalization process addresses ingredients from the Recipe1M+ dataset whose names closely resemble those of companies or brands without additional descriptive information, such as '7-up', 'M&M's', and 'Coca-Cola'. Approximately 15–20 ingredients out of 11,463 fall into this category. These are handled by replacing their original names with a generic term of our choosing; for example, '7-up' is renamed to 'soda'.

Table 4. Entity Linking Results Comparison based on n-grams and Smaller Threshold

Name	g3t020	g4t020	g3t023	g4t023	g3t018	g4t018
unsweetened orange juice concentrate	/	/	/	/	Orange mint	Orange mint
fat-free sugar-free instant chocolate pudding mix	Pudding	Pudding	Pudding	/	Pudding	Pudding
delicata squash	Sunburst squash (patty-pan squash)	Winter squash	Sunburst squash (patty-pan squash)	Winter squash	Sunburst squash (patty-pan squash)	Winter squash
liquid red pepper seasoning	Italian sweet red pepper	Pepper	Italian sweet red pepper	Pepper	Italian sweet red pepper	Pepper
unsweetened tart red cherries	/	/	/	/	Red tea	Red tea

To further refine ingredient names, an additional step is introduced to filter out unnecessary words such as measurements, sizes, and brand names. This involves identifying and excluding dataset-specific words that are not common to both datasets. Our goal is to ensure that normalized ingredient names consist only of terms shared between the datasets, thereby minimizing noise in the linking process. This approach enhances the consistency and reliability of linking ingredients across datasets by standardizing the vocabulary used to describe them.

The chosen ingredient for demonstrating the normalization process is not a 'branded' food, and its name does not include any 'unwanted' words. Consequently, the final normalized name is '1 lowfat chocolate milk', reflecting the result of applying all the steps in our normalization procedure.

To perform entity linking, the CountVectorizer [23] technique with n-grams [24] is leveraged, drawing insights from the FoodKG paper, to generate vector representations for ingredient names.

Several string similarity metrics were evaluated, including cosine similarity [25], Jaccard similarity [26–28], and Levenshtein distance [29]. Each metric offers unique advantages and faces specific limitations depending on our data's nature and task requirements. Cosine similarity was selected as the most suitable choice since it effectively captures and leverages contextual and lexical similarities within our data, particularly when complemented by n-grams through CountVectorizer. In contrast, Jaccard similarity overlooks term frequency, and Levenshtein distance focuses solely on literal character changes, missing semantic similarities.

Table 5. Entity Linking Results Comparison based on n-grams and Bigger Threshold

Name	g3t020	g4t020	g3t023	g4t023	g3t030	g4t030
85% lean ground beef	Cattle (Beef, Veal)	Cattle (Beef, Veal)	Cattle (Beef, Veal)	Cattle (Beef, Veal)	/	/
Baker's Special Dry Milk	Milk and milk products	Milk and milk products	Milk and milk products	Milk and milk products	/	/
korean red pepper paste	Italian sweet red pepper	Pepper	Italian sweet red pepper	Pepper	Italian sweet red pepper	Pepper
cooked lean ground beef	/	/	/	/	/	/
fat - free mayonnaise	Fats and oils	Fats and oils	Fats and oils	Fats and oils	/	/

Analysis of the distribution of word counts in normalized ingredient names from Recipe1M+ revealed that the majority of normalized names consist of one to three words. Based on these findings, it was decided to employ an n-gram range of (1, 3). This range enables maximum lexical and contextual information to be captured from each entry, which is particularly beneficial for shorter texts. Additionally, focusing on tri-grams allows a more concise and informative feature set to be achieved, ensuring computational efficiency without unnecessary complexity.

In Tables 4 and 5, examples that influenced our parameter selection for the linking process are presented, specifically regarding the length of n-grams and the similarity threshold. These tables compare the outcomes of the linking process using maximum n-gram lengths of three and four, along with varying similarity thresholds for cosine similarity. Column labels, such as 'g3t020,' indicate the parameters used-'g3' denotes a maximum n-gram length of three, and 't020' signifies a cosine similarity threshold of 0.20.

After careful analysis and evaluation, it was determined that cosine similarity with a threshold of 0.23 and a maximum n-gram range of three produced optimal results for our entity linking task. Also, the importance of setting a threshold was recognised; however, lowering it resulted in increased occurrences of false positives (e.g., 'unsweetened tart red cherries' incorrectly matched with 'red tea'), as illustrated in Table 4. Conversely, higher thresholds led to false negatives (e.g., '85% lean ground beef' not matched with 'Cattle (Beef, Veal)'), as shown in the analysis of Table 5.

Table 6. New USDA Mappings for Cheese Ingredients

Ingredient	USDA Class
1% fat cottage cheese	Cheese, cottage, lowfat, 1% milkfat
2% cheddar cheese	Cheese spread, American or Cheddar cheese base, reduced fat
Anejo cheese	Cheese, mexican, queso anejo
Better Than Cream Cheese cream cheese substitute	Cheese, cream
Cotija cheese	Cheese, mexican, queso cotija
Greek feta cheese	Cheese, feta

In summary, our approach involves a normalization process and custom entity linking to reconstruct the knowledge graph with accurate and reliable ingredient mappings. This effort has significantly improved consistency and precision, particularly in nutritional analysis.

Our improved mappings are showcased in Table 6, demonstrating accurate alignment with USDA classes and ensuring accurate nutritional data.

Table 7. Revised Datasets Linkage Statistics

Statistics	Count
Linkages	
Number of USDA links made	10417
Number of FoodOn links made	10586
Nubmer of FooDB links made	9377
Additional linkage stats	
Percentage ingredients linked to USDA	90.87%
Percentage ingredients linked to FoodOn	92.35%
Percentage ingredients linked to FooDB	81.80%

4 Human Evaluation of the Normalization Process

The need to validate the new ingredient mappings beyond internal assessment was acknowledged. To achieve this, a group of students was enlisted to review and compare the new mappings with those of FoodKG, ensuring the robustness of the findings and their comprehensibility to a broader audience.

To evaluate the precision of our ingredient mappings against previously published ones, a straightforward annotation process was devised. Each student reviewer was assigned a subset of ingredient mappings, which included the ingredient name from the Recipe1M+ dataset, the corresponding USDA classification from FoodKG, our proposed USDA classification, and a column for their annotation results.

The students' task was to review each ingredient and its candidate mappings to determine the most accurate one. They recorded their assessments using a simple rating system:

0 - None of the mappings are correct.
1 - FoodKG's mapping is better.
2 - Our mapping is better.
3 - Both mappings are correct.

This method allows us to gather clear and structured feedback on the perceived precision of the mappings from an unbiased group.

The analysis of student annotations reveals a clear preference for our revised mappings, with an observed precision of 76.26% for our new entity linking process, compared to 41.41% for the FoodKG's process. These results are detailed in Table 8.

Table 8. Comparison of the mappings

	Our Mapping	FoodKG's Mapping
True Positives	1985	1078
False Positives	618	1525
Precision (%)	76.25	41.41

Table 8 summarizes the effectiveness of our approach compared to FoodKG's in accurately linking entities. Precision, which measures the proportion of correctly identified mappings out of all proposed mappings, highlights the superior performance of our method in this evaluation.

The inclusion of student reviewers is justified for several reasons. Firstly, students were available, motivated to help, and had sufficient knowledge to evaluate similar outcomes, offering valuable insights into the usability and clarity of the mappings. Secondly, the annotation process was intentionally designed to be straightforward and accessible, requiring no specialized expertise, making it suitable for non-experts. Furthermore, involving multiple students ensures that each reviewer evaluates a manageable number of ingredients, enabling them to maintain focus and thoroughness in their assessments. In contrast, relying solely on a small group of experts may result in fatigue and potentially less meticulous validation, as they might approach the task in a routine manner.

The validation process is ongoing, with 10 students having completed the initial validation. In the future, it is planned for multiple students to validate each ingredient to ensure consistency and reliability. Ultimately, the results will be confirmed by experts, combining the benefits of extensive student reviews with expert validation to achieve the highest accuracy and comprehensibility of the mappings.

5 Conclusion

Mapping the ingredients from Recipe1M+ to the USDA, FooDB, and FoodOn ontologies and datasets offers several significant advantages. Firstly, linking Recipe1M+ ingredients to the USDA National Nutrient Database provides detailed nutritional information for each ingredient. This enables comprehensive nutritional analysis of recipes, helping users understand nutrient content, such as vitamins, minerals, and macronutrients. Researchers and dietitians can use this data to develop personalized nutrition plans and dietary recommendations based on accurate nutrient profiles.

Secondly, integrating Recipe1M+ with FooDB enriches the dataset with detailed chemical compositions and health effects of food constituents. This integration allows for a more thorough understanding of how different ingredients impact health, supporting research on food's role in disease prevention and health promotion. It also facilitates the identification of bioactive compounds in recipes, contributing to studies on functional foods and nutraceuticals.

Additionally, mapping Recipe1M+ ingredients to FoodOn provides a broader semantic context, covering various food-related areas, including food products, production, agriculture, and the environment. This connection allows for the incorporation of additional data such as food safety standards, environmental impacts of food production, and quality attributes. Researchers can explore these dimensions to promote sustainable and safe food systems.

Moreover, mapping ingredients to these ontologies and datasets bridges the gap between different research domains such as nutrition, food science, agriculture, and health. It enables interdisciplinary studies and the development of comprehensive knowledge graphs that integrate diverse data sources. Such knowledge graphs can support advanced applications like predictive modeling of dietary impacts on health, food recommendation systems, and the study of food-environment interactions.

Finally, aligning Recipe1M+ ingredients with established ontologies and datasets like USDA, FooDB, and FoodOn promotes standardization and interoperability. This standardization ensures that data is consistently represented, facilitating data sharing and collaboration across different platforms and research communities. It also supports the development of universal tools and applications that can leverage the integrated data for various purposes, from academic research to commercial food products.

In addition to these benefits, our approach adheres to the FAIR (Findable, Accessible, Interoperable, and Reusable) principles [30]. By integrating Recipe1M+ with established ontologies and datasets, we ensure that the data is well-organized and easily searchable (Findable). Publishing the data as a semantic web resource guarantees that it can be accessed by various users and systems, promoting openness and transparency (Accessible). The use of standard ontologies and consistent data representation ensures that our dataset can be combined with other datasets and used in various applications (Interoperable). Finally, by providing detailed metadata and clear documentation, we make it easier for others to understand and reuse the data for their research and applications (Reusable).

Overall, our process enhances the nutritional, chemical, and semantic richness of Recipe1M+ while adhering to FAIR principles. This supports comprehensive dietary assessments, interdisciplinary research, and the development of innovative applications aimed at promoting healthier and more sustainable diets. The resulting dataset will be a valuable resource for the research community and beyond, fostering collaboration and driving advancements in food science and nutrition.

Acknowledgment. The author(s) declare financial support was received for the research, authorship, and/or publication of this article. This research work was financially supported by the Slovenian Research Agency under programme P2-0098, the European Union's Horizon 2020 research and innovation programme [grant agreement 101005259] (COMFOCUS), and FoodMarketMap project selected as an innovator within the FOODITY project that has received funding from the European Union's Horizon Europe research and innovation programme under grant agreement No. 101086105.

References

1. Fensel, D., et al.: Introduction: what is a knowledge graph? In: Knowledge Graphs: Methodology, Tools and Selected Use Cases, pp. 1–10 (2020)
2. Popovski, G., Korousic-Seljak, B., Eftimov, T.: Foodontomap: linking food concepts across different food ontologies. In: KEOD, pp. 195–202 (2019)
3. Mijalcheva, V., Davcheva, A., Gramatikov, S., Jovanovik, M., Trajanov, D., Stojanov, R.: Learning robust food ontology alignment. In: 2022 IEEE International Conference on Big Data (Big Data), pp. 4097–4104. IEEE (2022)
4. Stojanov, R., Kocev, I., Gramatikov, S., Popovski, G., Seljak, B.K., Eftimov, T.: Toward robust food ontology mapping. In: 2020 IEEE International Conference on Big Data (Big Data), pp. 3596–3601. IEEE (2020)
5. Haussmann, S., et al.: FoodKG: a semantics-driven knowledge graph for food recommendation. In: Ghidini, C., et al. (eds.) ISWC 2019. LNCS, vol. 11779, pp. 146–162. Springer, Cham (2019). https://doi.org/10.1007/978-3-030-30796-7_10

6. Marın, J., et al.: Recipe1m+: a dataset for learning cross-modal embeddings for cooking recipes and food images. IEEE Trans. Pattern Anal. Mach. Intell. **43**(1), 187–203 (2021)
7. Haytowitz, D., et al.: USDA national nutrient database for standard reference, release 24. US Department of Agriculture, Washington, DC, USA (2011)
8. Dooley, D.M., et al.: Foodon: a harmonized food ontology to increase global food traceability, quality control and data integration. NPJ Sci. Food **2**(1), 23 (2018)
9. Wishart, D.: FooDB: the food database (2018)
10. IBM Docs: Entity Resolution. Accessed 5 May 2024
11. Eftimov, T., Koroušić Seljak, B., Korošec, P.: A rule-based named-entity recognition method for knowledge extraction of evidence-based dietary recommendations. PLoS ONE **12**(6), 0179488 (2017)
12. Popovski, G., Kochev, S., Korousic-Seljak, B., Eftimov, T.: Foodie: a rule-based named-entity recognition method for food information extraction. ICPRAM **12**, 915 (2019)
13. Eftimov, T., Korošec, P., Koroušić Seljak, B.: Standfood: standardization of foods using a semi-automatic system for classifying and describing foods according to foodex2. Nutrients **9**(6), 542 (2017)
14. Popovski, G., Seljak, B.K., Eftimov, T.: Foodbase corpus: a new resource of annotated food entities. Database **2019**, 121 (2019)
15. Cenikj, G., Popovski, G., Stojanov, R., Seljak, B.K., Eftimov, T.: Butter: bidirectional LSTM for food named-entity recognition. In: 2020 IEEE International Conference on Big Data (Big Data), pp. 3550–3556. IEEE (2020)
16. Stojanov, R., Popovski, G., Cenikj, G., Koroušić Seljak, B., Eftimov, T.: A fine-tuned bidirectional encoder representations from transformers model for food named-entity recognition: algorithm development and validation. J. Med. Internet Res. **23**(8), 28229 (2021)
17. Devlin, J.: Bert: pre-training of deep bidirectional transformers for language understanding. arXiv preprint arXiv:1810.04805 (2018)
18. Stojanov, R., Popovski, G., Jofce, N., Trajanov, D., Seljak, B.K., Eftimov, T.: FoodViz: visualization of food entities linked across different standards. In: Nicosia, G., et al. (eds.) LOD 2020. LNCS, vol. 12566, pp. 28–38. Springer, Cham (2020). https://doi.org/10.1007/978-3-030-64580-9_4
19. Agarwal, A., et al.: Deep learning based named entity recognition models for recipes. In: Proceedings of the 2024 Joint International Conference on Computational Linguistics, Language Resources and Evaluation (LREC-COLING 2024), pp. 4542–4554 (2024)
20. Chebbi, A., Kniesel, G., Abdennadher, N., Dimarzo, G.: Enhancing named entity recognition for agricultural commodity monitoring with large language models. In: Proceedings of the 4th Workshop on Machine Learning and Systems, pp. 208–213 (2024)
21. Smith, B., et al.: The obo foundry: coordinated evolution of ontologies to support biomedical data integration. Nat. Biotechnol. **25**(11), 1251–1255 (2007)
22. Loper, E., Bird, S.: NLTK: the natural language toolkit. arXiv preprint cs/0205028 (2002)
23. Pedregosa, F., et al.: Scikit-learn: machine learning in python. J. Mach. Learn. Res. **12**, 2825–2830 (2011)
24. Manning, C., Schutze, H.: Foundations of Statistical Natural Language Processing. MIT Press (1999)

25. Salton, G., Buckley, C.: Term-weighting approaches in automatic text retrieval. Inf. Process. Manag. **24**(5), 513–523 (1988). https://doi.org/10.1016/0306-4573(88)90021-0
26. Jaccard, P.: Distribution de la flore alpine dans le bassin des dranses et dans quelques régions voisines. Bull. Soc. Vaudoise Sci. Nat. **37**, 241–272 (1901)
27. Wikipedia: Jaccard Index. Accessed 26 Feb 2024
28. Fr-Academic.com: Paul Jaccard. Accessed 26 Feb 2024
29. Levenshtein, V.I.: Binary codes capable of correcting deletions, insertions, and reversals. Soviet Phys. Doklady **10**, 707–710 (1965)
30. Wilkinson, M.D., et al.: The fair guiding principles for scientific data management and stewardship. Sci. Data **3**(1), 1–9 (2016)

Crossword Generation as a Constraint Satisfaction Problem Using Parallel Processing and Lemmatization

David Arsov[1], Teo Kitanovski[2](✉), and Mile Jovanov[1]

[1] Faculty of Computer Science and Engineering, Ss. Cyril and Methodius University, Skopje, Macedonia
david.arsov@students.finki.ukim.mk, mile.jovanov@finki.ukim.mk
[2] School of Engineering, Vanderbilt University, Nashville, USA
teo.kitanovski@vanderbilt.edu

Abstract. Crossword puzzles pose a challenging problem in vocabulary and logic, and automatic generation of these puzzles is complex due to their combinatorial nature. This paper will explore their compilation by representing it as a Constraint Satisfaction Problem (CSP), as well as novel methods of generating crossword dictionaries, layouts and controlling puzzle difficulty. We will present a model for representing crossword generation as a CSP, including preprocessing steps, dictionary organization, and constraint modeling. Various optimization techniques, including heuristics, constraint propagation, and parallel processing are discussed. Furthermore, this work is the first of its kind for the Macedonian language, utilizing NLP and existing databases to determine the frequency of various words and control difficulty. Finally, we will use experimental tests and comparisons to existing puzzles to demonstrate the effectiveness of these techniques in generating crossword puzzles of varying sizes.

Keywords: Artificial Intelligence · Constraint Satisfaction · Crossword · Parallel Processing · Lemmatization

1 Introduction

As a beloved word game, crossword puzzles challenge both vocabulary and logic skills, and have been popular for a long time as a pastime activity, or even a learning tool [1]. There are several variations in terms of the subject of the words and the layout of the grid, but all of them are united by the method of solving, that is, placing the corresponding words in the empty spaces of the diagram according to the definition, while paying attention to the equality of the intersecting characters [2]. In Macedonia, these puzzles are usually found in specialized printed magazines. However, in the absence of a tool for automatic and randomized generation of grid layouts with different words, we are left with familiar templates and repetitive words from a small dictionary. Nevertheless, the automatic generation of these puzzles is a complex combinatorial and linguistic problem [3]. This paper explores the ways of application of Constraint Satisfaction Problems (CSPs) to address the placement of appropriate words on the grid, as well as using lemmatization and cross-referenced dictionaries in order to construct a dictionary of sufficient quality. Although this implementation focuses on the Macedonian language, the ideas and conclusions can also be utilized for any other languages.

CSPs are a powerful framework for solving problems where variables, their possible values (domains), and constraints that must be satisfied should be defined for a solution to be valid [4]. In the context of crossword generation, variables can represent the actual words in the grid, with domains containing valid words, respectively. Constraints ensure the interlinking of the puzzle, guaranteeing letter intersections create valid words both across and down, and also that the words satisfy the length of the field.

This paper provides a look at modeling crossword generation as a Constraint Satisfaction Problem (CSP), with specifically modeled variables, domains, and constraints involved. It uses different approaches to solving the generated CSP, potentially utilizing backtracking algorithms with various heuristics to efficiently navigate the search space and identify a valid crossword puzzle.

Firstly, the construction of a dictionary with all words to be used will be discussed. There are various online databases of Macedonian words, but none of them are complete and all contain various errors. Therefore, we will look at ways to ensure only high-quality words are being used, as well as determine each word's frequency in order to be able to control the puzzle's difficulty.

We will also present a new method of constructing the layouts of crossword puzzles, by starting construction with a layout with a (unrealistically) low number of words that is to be expanded by partitioning words that seem to cause issues. Furthermore, we will explore using parallelization in order to make this process more efficient.

Finally, a short section describes testing methods, as well as the potential advantages and drawbacks of the different approaches. This approach offers the potential to generate diverse and challenging crosswords. Also, the exploration of optimization techniques aims to improve the quality of the generated crosswords and improve the time efficiency for the generation.

2 Related Work

This paper includes a novel algorithm for crossword generation with advanced word preprocessing (to include an efficient, but large and high-quality dictionary, as well as consider word frequency), and a semi-random way of constructing the puzzle grid (utilizing parallel processing) instead of generating random layouts that could often lead to no puzzle being generated. It is an extension of our previous work, presented at the Conference on Informatics and Information Technologies in April 2024 [5], which did not include the implementation of these advancements.

Seeking through the research community, we came across several implementations of automatic crossword generation algorithms, all of them based on the Constraint Satisfaction Problem framework. The beginnings were set by Mazlac in 1976, where backtracking is implemented as letter-by-letter solving, which is quite inefficient taking into consideration the size of the input and search space [3]. Furthermore, in 1990, Ginsberg et al. implemented a similar approach with word-by-word backtracking [6]. The more advanced and most widely acclaimed solution is the one offered by Rigutini et al. in 2008 [7], where a NLP model is included in the processing in order to automatically generate definitions. Our approach differs in terms of the dictionary representation and the constraint propagation (arc consistency) technique which will be presented later.

These acclaimed implementations also do not allow for the difficulty of the puzzle to be controlled, although there are some other publications that include that constraint too, such as the work of Benda et al. [8] and Arbiser [9]. These works are also based on imposing a limit on the ratio of occurrences of rare (less known) words. Nevertheless, this paper is the first implementation of this idea for words in the Macedonian language, for which there is significantly less and lower-quality data regarding word frequencies.

3 Preprocessing

3.1 Dictionary Construction

The adequate choice of a dictionary and the fast processing of words is key in the compilation of the puzzle in order for the CSP to converge as well as to produce a high-quality puzzle. Since crosswords with less words are considered to be of higher quality (due to them containing less one-, two-, or three-letter words) [10], the goal is to have the dictionary be as large as possible, but it is also important to make sure that all strings in it are standard, logical words. To this end, we used multiple databases, including the Official Digital Macedonian Language Dictionary [11], the Digital Dictionary of the Macedonian Language [12], and Leipzig University's Corpora datasets for the Macedonian Language [13]. There are multiple Macedonian Corpora datasets, of which we included "newscrawl" (words from web news articles), "community" (words from various online sources), and "wiki" (words from the Macedonian Wikpedia). Although only the Official Digital Macedonian Language Dictionary was used in the preliminary version of our work, we opted to use these additional sources in order to include more derivative words that are often longer.

However, the Corpora datasets also included a sizable proportion of non-standard words or typing errors. Since they include the frequency (number of mentions) of every word, we excluded every word with less than 5 mentions. Further, we cross-referenced the datasets to include only words that appeared in at least two of them (this did not apply to words in the official Macedonian dictionaries).

This process yielded a database of 100,780 high-quality words. In order to be able to control the difficulty of the puzzles, we were also interested in the rank (frequency) of each word. Some words in the official dictionary included this data, but they represented only 28.19% of all words. Therefore, we also used the number of mentions of each word in the Corpora datasets to determine its frequency. However, since all datasets were of varying size and quality, we only split words into two categories: common and rare, instead of assigning them a continuous value.

While manually analyzing the dataset, we noticed many of the words that are derivatives are flagged as rare even though their root word (lemma) is obvious and common. Thus we also applied lemmatization to consider as common all words with common lemmas. We used the SpaCy NLP library by Todosovska and Georgievski [14], as well as the CLASSLA library by Ljubešić et al. [15] [16] to find lemmas (initially, only SpaCy was used, but the CLASSLA library was added to reduce potential errors that resulted in non-existent words being returned as lemmas). Essentially, for each word, its dictionary rang and its Corpora frequency in each dataset were considered in order to determine its rarity; if one of these flagged it as common, it is considered common. Later, in case it is flagged as rare, its lemmas are also considered as a final check.

Overall, in this dictionary, there are 72,370 (71.81%) common and 28,409 (28.19%) rare words.

3.2 Dictionary and Variables Data Structure

In contrast to existing puzzles where the patterns on the matrix repeat, we implemented a semi-random generation of the empty positions, i.e. the word positioning. At the beginning, a substantial number of positions are randomly inserted, but more positions are added during the backtracking process to ensure a higher rate of constructed puzzles. Each position is represented by an object with x and y coordinates, length and a direction of the word's extension. The objects are stored in a custom implementation of a HashSet where the validity of a position is checked and the direction is calculated every time an object is added. The set implements limitations regarding the size and number of words to be assigned. The process begins by adding positions in the first row or column, and later assigning positions with randomly generated coordinates. It is also possible to allow for empty space to insert a photo or advertisement. A visualization of a generated matrix is given in Fig. 1.

After the positions are added to the set, a check is conducted that verifies the length and intersection of words, from which the general constraints for the CSP's solution later arise. The words that extend horizontally and vertically are considered different cases, and the length is calculated as the distance to the following (blocked) position in the specific row or column. Intersecting words are stored in a map with the key being the position in the current word, and the value being an object including the intersecting word and intersecting character position in that word. This calculation is conducted through the mapping of words extending in the opposite direction of the currently processed word to the applicable representation of an intersection stated above.

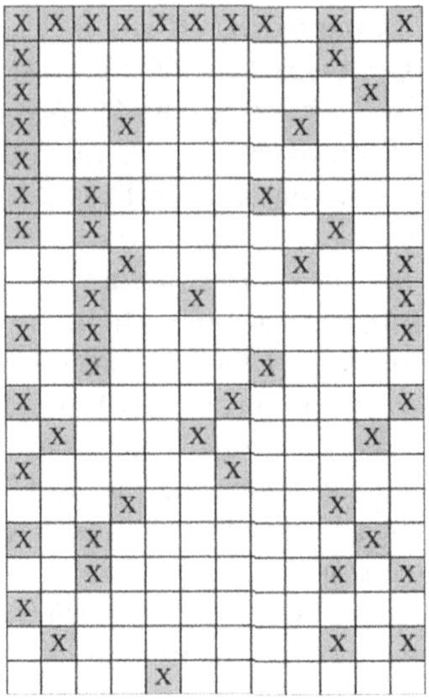

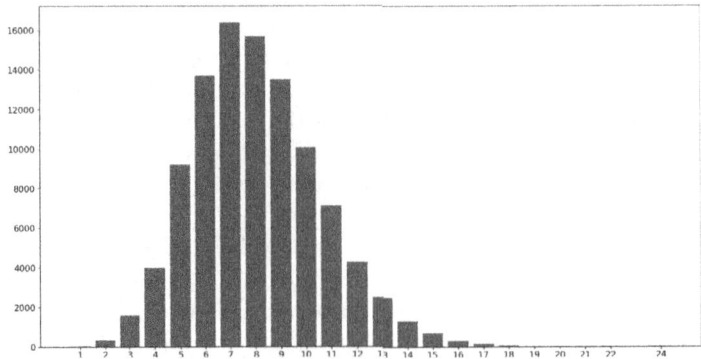

Fig. 1. Representation of generated crossword grid.

From the distribution in Fig. 2, it can be concluded that the existence of a small number of words with over 10 characters leads to problems in terms of the distribution of positions and the number of words in the matrix, since bad positioning of the words and/or the presence of a large number of words with over 10 characters could make the puzzle practically impossible to be generated due to the many constraints that the few dictionary words could not fulfill.

Fig. 2. Distribution of words with given length

For the efficient access of words according to their length and the presence of a certain character at a specific position, the dictionary is represented as a multi-level map (visualized in Fig. 3) where firstly, the key is the length of the word, which leads to the best pruning, and the value is a higher-level map where the keys are the letters of the alphabet and a final map as a value, that includes the position of a certain letter in the word as the key and an ordered collection (set) of the actual objects representing the words as its value. This structure allows a set of words satisfying specific restrictions to be retrieved in constant time. Additionally, the repeated existence of an object representing a word in multiple different sets does not significantly increase memory complexity since they are passed by reference in the Java programming language. The final memory complexity of this dictionary representation is O(n), where n is the total number of words.

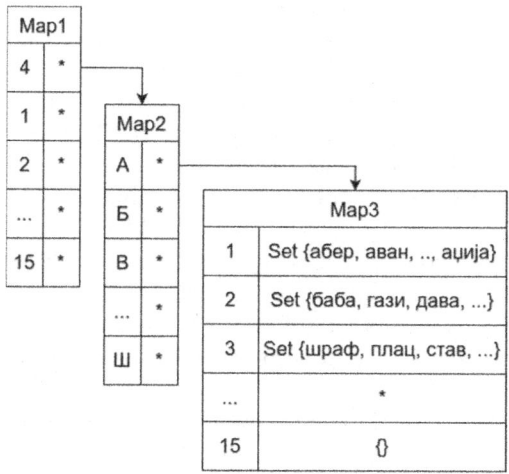

Fig. 3. Visual representation of the data structure for the dictionary

4 CSP Model

The manual solving of a crossword puzzle actually represents the placing of appropriate words in the matrix's empty spaces according to their length while also maintaining consistency in intersections with other words. Thus, the nature of this task implies that the problem can be defined as a CSP [7].

In our model, variables are defined as positions that store information regarding the coordinates of the starting letter in the matrix, the direction of extension, length, and positions of intersections with other words. Length constraints are applied to each variable and are not changed throughout the course of the pruning. Additionally, it is important to assure that duplicate words are not placed, which is implemented as an additional constraint. In order to control difficulty, a difficulty constraint does not allow the percentage of rare words in the final puzzle to rise above a fixed constant. Finally, a word intersection constraint is also used. The implementation of all constraints is rather trivial, and words with one, two and three letters are treated differently and exempt from the duplicates constraint. The definition of the problem continues with the definition of domains, where at first, all variables utilize a full set, filtered only by words' length.

Variables in the problem are stored in a priority queue, where the priority is defined by the heuristics used to solve the CSP, with Minimum Remaining Values being the primary one and Least Constraining Value used as a tie breaker [17], allowing the next variable to be processed to be accessed in logarithmic time.

4.1 Backtracking Search

In this subsection we will present two approaches, two backtracking algorithms that we used in our research. The backtracking search verifies all constraints in all phases of the solving, and partial solutions are kept in the execution stack due to the recursive nature. For each partial solution, a verification of constraints is conducted on all variables with assigned values. The idea is to verify if the partial solution could be part of the final one, i.e. conduct a forward check in order to maintain consistency and prevent assigning a value to a variable that violates one or more of the given constraints. The second approach includes filtering the variables' domains, a technique known as constraint propagation and usually utilizing the concept of arc consistency [18]. If the domain of one of the variables turns into an empty set, the partial solution is considered invalid and backtracking to the last valid step follows, unless the conditions for words partitioning as described in the following section are fulfilled. This is a notable optimization since the goal is to filter out invalid partial solutions as soon as possible; futile checks on a solution that could be known to be invalid use up a lot of resources (both time and computing power) due to the high number of iterations necessary to complete. In a problem with real-world data like this, where some words with higher length introduce many constraints on the other words, a solution with no constraint propagation would not converge.

Using the aforementioned representation of the dictionary, the propagation is executed in linear time. Namely, as soon as a partial solution is confirmed to be valid, each of the other variables that it constraints have their domains pruned (implemented in Java using the retainAll method on a set) to effectively conduct a cross section finding operation very efficiently.

4.2 Words Partitioning

Taking into account the functionality of the previous model, when a partial solution goes into a state where for at least one variable the domain is empty and it is rejected, we introduced an optimization regarding the arrangement of word positions. Namely, when the partial solution is found in the state described above, a calculation is made for the possible partitioning of the word that is currently being processed into two smaller ones, taking into account that the newly obtained words have a length greater than one and that the division position is not already occupied by another word that it extends in the opposite direction.

The positioning of the new word is done in such a way that a new instance of the problem is created which is completely independent from the previous one, and at the same time, it continues execution from the previous partial solution. To maintain consistency and to satisfy all constraints, the intersecting characters and word lengths are changed beforehand.

Partitioning is only conducted if the number of words is below a previously set constant limit. This restriction may seem to negatively impact performance, since longer words can be partitioned in many different ways, and are also more likely to be chosen first, but in fact leads to a solution sooner and greatly improves the quality of the puzzle since it results in a lower total number of words compared to a random generation of a static crossword grid.

4.3 Parallel Processing

The principle presented in the previous section has opened up possibilities for further optimization. The independence of each of the newly created instances of the solution after the partitioning allowed us to explore the possibility of parallel execution, which would potentially yield significant time savings and reduce the number of solutions that are marked as invalid at an early stage. In our implementation, we utilized 3 threads to coordinate the entire process. The first thread performs the preprocessing operations, creating the data structure for the dictionary, generating and asigning initial solution grids. The second thread acted as a scheduler, that is, it distributes the tasks that have arrived in the queue depending on the available resources. And the third thread, which practically monitors the progress, checks the solutions from each execution and in case a valid solution is found, stops the program. In our case, to parallelize the execution of the newly obtained instances after the division, we used an additional 8 threads. This approach (as fully visualized in Fig. 4) significantly reduced the proportion of invalid solutions, from 60% in the original version to 40% (for a 15×15 layout). Furthermore, we observed a reduction in the time needed to generate a solution, which is discussed in the results section below.

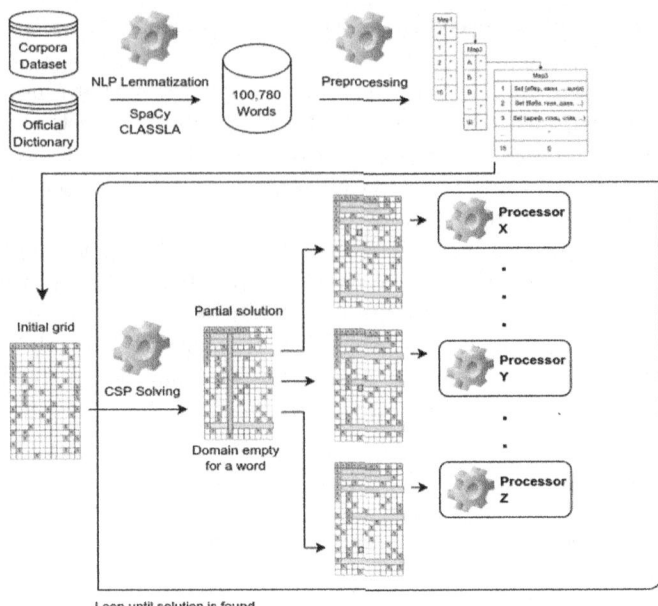

Fig. 4. Visualisation of the process.

5 Experimental Results and Analysis

The ideas presented in this paper were implemented in a Java program and various approaches were tested on the full dictionary of 100,780 words, and experiments were run on crossword matrices of size 5 × 5, 8 × 8, 10 × 10, 15 × 15 and 20 × 15, with 10 samples in each test series. All experiments were run on an AMD Ryzen 7 7735U CPU with 16 GB of memory and a time limit of 600 s for each instance. The experimental results presented below are average values of 10 independent executions of each instance.

5.1 Heuristic Choice and Constraint Propagation Testing

Initially, we performed testing using an implementation of a naive backtracking search, generating the puzzles without using any heuristics for the selection of the next variable to be processed and without constraint propagation. The results of testing a total of 50 instances (10 of each size) are shown in Table 1. Although the solution does converge for the matrix of size 5 × 5, with an execution time of 2573ms, it is certainly not efficient enough, due to the complexity of the problem and the high number of edges that arise from the size of the larger grids. Therefore, this approach is not sufficient, but additional optimizations can be applied to it in order to reduce the execution time.

Table 2 shows the results of testing a total of 50 instances, 10 of each size, this time using the Least Remaining Values heuristic as a first condition for the

Table 1. Effect of no heuristics and no arc consistency on time taken

Grid Size	Time taken (ms)
5 × 5	2573
8 × 8	N/A
10 × 10	N/A
15 × 15	N/A
20 × 15	N/A

selection of the variable whose value is to be assigned in the partial solution, and the Most Constraining Value heuristic as a tie-breaker. Although notable reductions in execution time are visible for the matrix of size 5 × 5, the results for larger matrices are the same, i.e. no solution was found within the time limit.

Table 2. Effect of LRV, MCV and no arc consistency on time taken

Grid Size	Time taken (ms)
5 × 5	1559
8 × 8	N/A
10 × 10	N/A
15 × 15	N/A
20 × 15	N/A

Applying the arc consistency constraint propagation technique, the results shown in Table 3, with the same number of tests conducted, show drastic improvement in execution time for the matrix of size 5 × 5 and convergence for the 8 × 8 matrix. The problems that are more complex are not convergent yet, which may be due to a suboptimal ordering of the variables chosen to have their values assigned. Since longer words are used in bigger matrices, their frequencies in the dictionary are lower while they also place constraints on many variables, so it is important to ensure that they are processed earlier in order to optimize execution time.

Table 3. Effect of LRV, MCV with arc consistency on time taken

Grid Size	Time taken (ms)
5 × 5	9
8 × 8	90
10 × 10	894
15 × 15	N/A
20 × 15	N/A

Using both the heuristics (LRV and MCV) and the constraint propagation technique, the problem converges for all matrix dimensions. The green curve in Fig. 5 shows the average execution time for the various matrix sizes. The matrices of 10×10 and 15×15 notably stand out with longer execution times, which is to be expected, since larger matrix sizes also utilize longer words, and, as shown in Fig. 2, words with 10 characters or more have considerably lower frequencies.

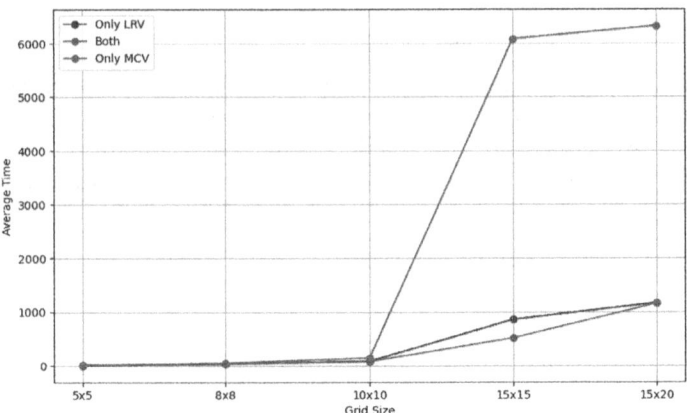

Fig. 5. Time needed to complete grids of various sizes with application of only LRV (blue), only MCV (red), and both (green) with Arc consistency (Color figure online)

Additionally, Fig. 5 also shows the results of the test with only one of the two heuristics utilized. It is evident the curve of the results while utilizing the Most Constraining Variable heuristic only, largely deviates from the green one, representing the results utilizing both heuristics, which supports using this heuristic as secondary (a tie-breaker heuristic) instead of a dominant one when choosing variables. The curve of the results while utilizing the Least Remaining Values heuristic only nearly matches the green curve in Fig. 5, which additionally confirms its efficiency and choice as a primary heuristic.

These results strongly suggest the appropriate choice of heuristics and appropriate constraint propagation are key in the convergence of a problem of this type.

5.2 Words Partitioning Testing

We performed 30 rounds of testing on different grid sizes (10×10, 15×15 and 20×20) and on average we observed the results shown in Fig. 6. Partitioning potentially invalid solutions as described in Sect. 4.2 allows for the reduction of the lowest number of words by an average of 10, which is a significant improvement compared to the total number of words in the grid.

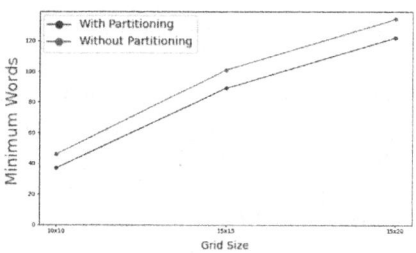
(a) Minimum number of words with and without partitioning

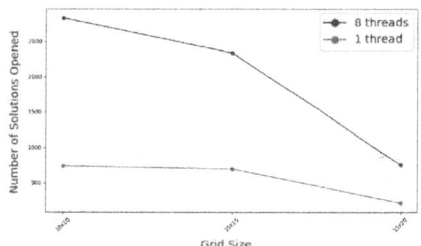
(b) Number of solutions opened with and without parallel processing

Fig. 6. Words partitioning results

Furthermore, the benefits of optimizing the process through parallelization (as described in Sect. 4.3) are evident from the graph in Fig. 6(b). *The parallel execution of sub-solutions with differently partitioned positions resulted in a three-fold increase in the number of solutions processed within a time span of 3 min.* This improvement opens the potential for further improvements by scaling up resources, not only during the partitioning phase but also throughout the entire solution process.

5.3 Dictionary Quality and Difficulty Control Testing

We manually extracted the words from 20 different crosswords from 3 Macedonian magazines in order to test the quality of our dictionary as a tool for generating puzzles similar to those. For each crossword, we counted the number of words and flagged each word as either common, rare, or not existent in our database (Table 4). To interpret these results, one needs to consider the assumption that a high-quality generated crossword puzzle has as few words as possible, as many familiar (found in other puzzles) words as possible, and as few rare words as possible (since it is easier to control difficulty through definitions).

Table 4. Proportions of various word types in existing puzzles

Measurement	Average Share
Common Words	59.74%
Rare Words	13.69%
Not found in database	26.57%

Overall, 73.43% of the words were found in our database. While this statistic is somewhat encouraging, there is certainly space for improvement. Through manual analysis we observed most of the missing words are names of people and places, so future work could improve this result by looking for data sources that include these words.

On average, common words represented 59.74% of all words, or 81.36% of words that were found in the database. Compared to the 71.81% of words that common words actually represent (in our dictionary), this further confirms the notion that it is ideal to minimize the amount of rare words. In the crosswords that we generated (of equal size), *common words could represent at most 91% of words, meaning we can not only manage to compile puzzles of equal difficulty, but even ones that are easier, without impacting the word count.*

Regarding the coverage (percentage of non-empty positions), *these 20 puzzles had an average 74.56% of the positions filled, while our algorithm produced puzzles with as much as 82.22%*, further showing the quality of the generated puzzles but also supporting the findings in the previous section, i.e. the efficacy of the word partitoning strategy.

6 Conclusion and Future Work

In this paper, we demonstrated the effectiveness of modelling crossword generation as a Constraint Satisfaction Problem, providing a concept of both the theoretical framework and the practical implementation. By utilizing efficient preprocessing and advanced optimization techniques such as constraint propagation through arc consistency, we achieved good efficiency and convergence of the crossword compilation problem.

We have also constructed a high-quality database of Macedonian words and their frequency ("common" or "rare") that is cross-referenced from multiple datasets and can be used to control the difficulty of the generated crossword puzzle according to the needs of the user. Since the Macedonian language has many words derived from others by adding prefixes and suffixes, we also used NLP lemmatization in order to consider the frequency of each word's lemma (root word) when determining its rarity. This approach can also be utilized in other languages.

The experimental results show the importance of adequate heuristic selection and constraint propagation when generating crossword puzzles of any size. Specifically, utilizing the Least Remaining Values heuristic first and the Most Constraining Variable as a tie-breaker, along with narrowing the domain for each variable (word to be placed) during backtracking can lead to efficient puzzle generation, even in large layouts (20×20 or more).

Furthermore, our novel method of generating the crossword layout by adding additional empty positions when no solution is found was shown to aid in generating puzzles with less words that are considered of higher quality and more interesting to the final user. Utilizing parallel processing allowed us to further advance this idea and make the processing significantly more efficient, increasing the number of processed solutions by three times.

While this approach has yielded promising results, there are many more concepts to be explored and improved; investigating other advanced heuristics and optimization strategies may also lead to even greater efficiency gains and a wider application of the solution. Furthermore, incorporating other user preferences and adjustments, such as themed puzzles and automatically generated custom word lists could improve the practical utility of the solution. Finally, there is also likely more space for improvement in the construction of the dictionary, especially in terms of identifying ways to include more words that appear in existing crossword puzzles, such as names.

Acknowledgement. The research presented in this paper is partly supported by the Faculty of Computer Science and Engineering, at the Ss. Cyril and Methodius University in Skopje.

References

1. Wise, A.: Web-based puzzle program to assist students' understanding of research methods. Act. Learn. High. Educ. **4**(2), 193–202 (2003). https://doi.org/10.1177/1469787403004002007
2. Crossword puzzle - Definition, History, & Facts - britannica.com. https://www.britannica.com/topic/crossword-puzzle. Accessed 03 May 2024
3. Mazlack, L.J.: Machine selection of elements in crossword puzzles: an application of computational linguistics. SIAM J. Comput. **5**(1), 51–72 (1976). https://doi.org/10.1137/0205004
4. Brailsford, S.C., Potts, C.N., Smith, B.M.: Constraint satisfaction problems: algorithms and applications. Eur. J. Oper. Res. **119**(3), 557–581 (1999). https://doi.org/10.1016/S0377-2217(98)00364-6
5. Arsov, D., Kitanovski, T., Jovanov, M.: Crossword generation as a constraint satisfaction problem. In: Proceedings of the International Conference on Informatics and Information Technologies (CIIT) (2024)
6. Ginsberg, M.L., Frank, M.C., Halpin, M.P., Torrance, M.C.: Search lessons learned from crossword puzzles. In: AAAI Conference on Artificial Intelligence (1990). https://api.semanticscholar.org/CorpusID:399327
7. Rigutini, L., Diligenti, M., Maggini, M., Gori, M.: Automatic generation of crossword puzzles. Int. J. Artif. Intell. Tools **21**(03), 1250014 (2012). https://doi.org/10.1142/S0218213012500145
8. Benda, A.: Crossword generator using web data. Master's thesis, Faculty of Information Technology CTU in Prague (2023)
9. Arbiser, A.: Practical crossword generation with checkpoint search. In: Proceedings of IADIS International Conference on Applied Computing 2005 (2005)
10. Niculescu, V., Ştefănică, R.: Tries-based parallel solutions for generating perfect crosswords grids. Algorithms **15**, 22 (2022). https://doi.org/10.3390/a15010022
11. Official Digital Dictionary of the Macedonian Language. https://makedonski.gov.mk/. Accessed 12 Mar 2024
12. Digital Dictionary of the Macedonian Language. http://drmj.eu/. Accessed 19 May 2024
13. Goldhahn, D., Eckart, T., Quasthoff, U.: Building large monolingual dictionaries at the leipzig corpora collection: from 100 to 200 languages. In: Proceedings of the Eighth International Conference on Language Resources and Evaluation (LREC'12) (2012)

14. Todosovska, M., Georgievski, B.: Macedonian spaCy. https://blog.netcetera.com/macedonian-spacy-f3c85484777f. Accessed 20 May 2024
15. Ljubešić, N., Dobrovoljc, K.: What does neural bring? Analysing improvements in morphosyntactic annotation and lemmatisation of Slovenian, Croatian and Serbian. In: Proceedings of the 7th Workshop on Balto-Slavic Natural Language Processing, Florence, Italy, pp. 29–34. Association for Computational Linguistics (2019). https://doi.org/10.18653/v1/W19-3704
16. Terčon, L., Ljubešić, N.: CLASSLA-Stanza: The Next Step for Linguistic Processing of South Slavic Languages (2023)
17. Russell, S.J., Norvig, P.: Artificial Intelligence: A Modern Approach, 4th edn, pp. 384–385 (2021)
18. Van Hentenryck, P., Deville, Y., Teng, C.-M.: A generic arc-consistency algorithm and its specializations. Artif. Intell. **57**(2), 291–321 (1992). https://doi.org/10.1016/0004-3702(92)90020-X

Session 2

Comprehensive Examination of Network Access, Logging, and Auditing Strategies in Public and Private Institutions: Safeguarding Information Security, Resilience, and Compliance in the Digital Era

Elissa Mollakuqe[1(✉)], Vesna Dimitrova[2], Hasan Dag[1], and Simon Atanasovski[2]

[1] Faculty of Management Information Systems, Kadir Has University, Istanbul, Turkey
elissamollakuqe@gmail.com, hasan.dag@khas.edu.tr
[2] Faculty of Information Sciences and Computer Engineering, Skopje, North Macedonia
{vesna.dimitrova,simon.atanasovski}@finki.ukim.mk

Abstract. This research paper delves into the intricacies of network access and communication, logging, and auditing practices within the contexts of public and private institutions. Network access and communication are vital components of information security, encompassing elements such as access controls, authentication, encryption, and network management. The study explores how institutions manage network access and communication to mitigate unauthorized access and ensure data integrity. Through a series of questions and features, we analyze the varying requirements and restrictions imposed on network accessibility, emphasizing the importance of tailored security measures. Additionally, the paper investigates the logging and auditing mechanisms employed by these institutions. Logging involves recording events within systems or networks, while auditing entails reviewing these logs for compliance and anomaly detection. A comparative analysis reveals similarities and differences in how public and private institutions handle sensitive data, link actions to users, and log successful or unsuccessful accesses. Furthermore, it explores the retention periods for logs, critical for compliance, auditing, and security. The research also highlights the significance of business continuity and disaster recovery plans in both sectors, ensuring data restoration and operational resilience during emergencies. By shedding light on the practices and policies in place, this paper offers valuable insights into the strategies employed by institutions to fortify their information security, resilience, and regulatory compliance in the digital age.

Keywords: network · access · logging · auditing · security and institutions

1 Introduction

In recent years, the rise of digitalization has significantly impacted the landscape of identity management, access control, and authorization practices in educational institutions. Public and private universities face unique challenges and opportunities in implementing secure and efficient cybersecurity measures. This paper presents a comparative analysis of these practices, focusing on the differences and similarities in logging, auditing, and network access controls. By examining the latest advancements in this field, this study aims to provide insights into the most effective approaches for safeguarding sensitive information within educational institutions. In the increasingly interconnected world of technology, the ability to access computer networks and facilitate communication between devices and users has become a fundamental aspect of our digital lives. The term "Network Access and Communication" encompasses a wide array of practices, technologies, and procedures aimed at managing and securing access to computer networks while enabling effective communication among various components and individuals within those networks. In this article, we will delve into the multifaceted realm of Network Access and Communication, exploring its significance in both public and private institutions. We will examine the definitions and concepts associated with this domain, shedding light on its critical role in network security and data integrity. We will discuss the importance of defining network access requirements, providing insights into how institutions, both public and private, can tailor their network access and communication strategies to meet their specific needs. In an era where data breaches and cyber threats are ever-present concerns, understanding and implementing effective Network Access and Communication policies and procedures is paramount. By the end of this article, you will gain a comprehensive understanding of the core principles, challenges, and solutions that revolve around Network Access and Communication in today's technology-driven landscape.

The remainder of this paper is organized as follows: Sect. 2 provides an overview of the related work, focusing on the evolution of identity management practices in educational institutions. Section 3 details the methodology used in this study, including the criteria for selecting case studies and the comparative framework. Section 4 presents the results and analysis, offering a detailed comparison of logging, auditing, and network access practices in public versus private universities. Section 5 discusses the findings in the context of existing literature, highlighting the implications for policy and practice. Finally, Sect. 6 concludes the paper with recommendations for future research and practical applications.

2 Motivation

In an era defined by pervasive digital connectivity and escalating cybersecurity threats, the effectiveness of logging and auditing practices within institutions—both public and private—holds critical implications for data protection, regulatory compliance, and organizational resilience. Understanding how these sectors manage and deploy logging and auditing technologies is not merely an academic exercise but a necessity driven by practical imperatives. Robust logging practices provide a detailed chronicle of system activities, offering insights into operational efficiency, user behavior, and potential security vulnerabilities. Comprehensive auditing ensures accountability and regulatory adherence, safeguarding against breaches that could compromise sensitive information and erode stakeholder trust. By comparing and contrasting the approaches of public entities, such as government agencies and municipalities, with those of private enterprises like financial institutions and healthcare providers, this research aims to uncover disparities, identify best practices, and propose strategies to fortify cybersecurity frameworks. Ultimately, this study seeks to empower institutions with actionable insights to enhance their cybersecurity posture, mitigate risks, and foster a resilient digital ecosystem conducive to sustainable growth and trust in an increasingly interconnected world.

3 Literature Review

In general, Network Access and Communication refers to the set of practices, procedures, and technologies used to manage and secure access to a computer network, and to enable communication between devices and users on that network. It encompasses a range of concepts related to network security, such as access controls, authentication, encryption, monitoring, and management. Here, the Network Access and Communication section of the information security assessment serves to identify the scope of network access requirements for a system, which can help determine the necessary controls to reduce the risk of unauthorized or inappropriate access to sensitive data [4]. The section examines whether the system requires network accessibility, and if so, the specific types of networks through which it will be accessible. Additionally, a network diagram may be provided to illustrate the required connectivity of the system's components. Finally, the section considers whether the system will be accessible through means other than the network, such as telephone access. This information is crucial for ensuring the security and integrity of the system and the sensitive data it handles. The primary goal of this aspect of security is to ensure that only authorized users have access to the network and its resources, and that all network activity is monitored and managed in a secure and efficient manner [5]. This may include establishing protocols for network communication, implementing firewalls, using intrusion detection systems, and monitoring network traffic. Also, the network access and communication also involves the management of network traffic, such as routing, switching, and load balancing. It includes the configuration of firewalls and other security devices to protect the network from external threats, as well as the implementation of encryption and authentication protocols to ensure the confidentiality and integrity of network data [6]. Network access and communication also involves the establishment of guidelines for network usage, such as acceptable use policies, which outline the appropriate and inappropriate uses of the network by authorized users. Network access and communication plays a critical role in the security and performance of a computer network, and it is essential for organizations to develop and implement effective policies and procedures to manage and secure their network access and communication [7]. By implementing effective controls and protocols, organizations can mitigate the risks associated with network access and communication and ensure that their networks remain secure and protected from unauthorized access. Logging and auditing are two related concepts in the field of information technology security. Logging refers to the process of recording events, transactions, or activities that occur within an information system or network. The purpose of logging is to create an audit trail of what has happened within the system or network, which can be used for troubleshooting, analysis, or forensics. In the literature, there are several definitions of "Logging and Auditing". "Logging is the process of creating and storing records of events that occur in a system. An audit trail is a log of all actions that a user or system performs [8]. "Auditing is the process of reviewing logs and other records to determine whether actions were taken as expected. An audit trail can be used to prove compliance with legal or regulatory requirements." [9] "Logging is the process of recording information about events that occur in a system, such as user actions, system events, and security-related events. The purpose of logging is to provide a record of activity that can be used for troubleshooting, security analysis, and forensic investigations." – [10] "Auditing is the process of examining logs and other records to ensure that they are complete, accurate, and trustworthy. The purpose of auditing is to provide an independent verification of the integrity and reliability of logs and other records." [11] Logging typically involves capturing information such as the date and time of the event, the user or system responsible for the event, and the nature of the event itself. Auditing, on the other hand, involves the analysis and review of the logs to ensure that the system or network is operating in compliance with policies and regulations, as well as to identify any anomalies or suspicious activities. Auditing may be performed manually by an auditor or automatically through the use of software tools. The ultimate goal of auditing is to detect and prevent security breaches, data loss, or other types of security incidents. Logging is the act of recording events, while auditing is the process of reviewing those records to ensure the system or network is secure and in compliance with policies and regulations.

4 Methodology of Research of Comparing Logging and Auditing Capabilities

The study aimed to compare logging and auditing practices between public and private institutions to assess their effectiveness in managing network security. Public institutions included government departments and municipal offices, while private institutions encompassed financial firms and healthcare providers. A total of 10 institutions were selected, comprising 5 public and 5 private entities, based on their prominence and accessibility of relevant security documentation. Security policies, audit reports, and interviews with IT administrators and cybersecurity experts provided primary data. Publicly available information and regulatory filings supplemented the analysis.

Variables and Metrics
Variables included types of logged data (e.g., user actions, system events), methods of data storage (e.g., centralized logging systems), and protocols for log retention. *Example:* Public institutions showed a preference for logging system events and user actions for compliance purposes, with data typically stored securely on government servers. Private institutions logged more granular data, including transaction details and client interactions, often stored on encrypted servers for enhanced security. Metrics encompassed audit frequency, tools used for audit trail analysis (e.g., SIEM tools), and procedures for responding to audit findings. *Example:* Public institutions conducted quarterly audits using internal teams and external auditors, focusing on regulatory compliance. Private institutions conducted monthly audits leveraging advanced SIEM technologies to detect anomalies and potential threats proactively.

Data Analysis
Content analysis categorized and compared logging and auditing practices based on predefined variables and industry standards. *Example:* Qualitative analysis revealed that while both sectors emphasized audit trail completeness, private institutions demonstrated stronger capabilities in real-time monitoring and incident response due to their investment in advanced cybersecurity tools. Numerical data, such as log retention periods, were compared statistically to identify trends. *Example:* Private institutions retained logs for an average of 1 year, whereas public institutions retained logs for approximately 6 months, indicating a longer-term focus on data retention and forensic analysis in the private sector.

Ethical Considerations
Confidentiality of sensitive information obtained during interviews and adherence to data protection regulations were ensured throughout the study.

Potential biases in self-reported data and variations in interpretation of logging and auditing standards across institutions were acknowledged.

Limitations included access to comprehensive data from all selected institutions and the dynamic nature of cybersecurity practices.

5 Results

This section presents a detailed comparison of logging, auditing, and network access practices between public and private universities. Recent studies, such as [12, 13] and [14], highlight the growing trend of adopting cloud-based identity management systems in both sectors, with a significant focus on scalability and user-friendliness.

Logging and Auditing Practices
Public universities often rely on centralized logging systems that integrate with state or national cybersecurity frameworks. For instance, the implementation of Security Information and Event Management (SIEM) systems has been widely adopted. In contrast, private universities tend to adopt more flexible and customizable logging solutions that cater to specific institutional needs.

Network Access Control
The use of Role-Based Access Control (RBAC) is prevalent in both public and private universities, but the implementation strategies differ. Public universities often integrate RBAC with multi-factor authentication (MFA) to comply with governmental regulations, while private institutions prioritize user experience and operational efficiency, often opting for Single Sign-On (SSO) solutions.

By integrating these recent developments into the comparison, the analysis becomes more comprehensive and reflective of current practices in the field.

The network access and communication section of an organization's policies and procedures plays a crucial role in managing and securing access to its computer network. It helps the organization identify the necessary controls to mitigate the risks associated with unauthorized access to sensitive data. This is important for both public and private institutions, as sensitive information can be compromised in either case. To achieve this, the section includes a series of questions that help determine the network accessibility requirements of the system, such as whether it needs to be network accessible and the type of network it will be accessible from. It also seeks to understand if there are other means to access the system apart from the network, which will help design appropriate protocols for network communication and establish security measures to ensure only authorized users have access to the network and its resources. Additionally, the section requests a network diagram that depicts the required connectivity for all of the components of the application or service, which can help identify any potential vulnerabilities that may exist in the system. This information can be used to design and implement appropriate controls to safeguard sensitive data.

The network access and communication section is a critical component of an organization's cybersecurity strategy. Its effective implementation can help prevent unauthorized access to sensitive data and maintain the confidentiality, integrity, and availability of network resources.

In our research, we defined a set that includes possible features that could be relevant to network access and communication. The features are defined as a set of questions, numbered *1.1, 1.2, 1.3, and 1.4*, that help to clarify the requirements for network access and communication. The features of network access and communication for each institution (public and private) are defined as follows:

- Is this system required to be network accessible?
- If this system required to be network accessible, how will it be accessible?
- Provide a network diagram depicting required connectivity for all components of the application or service.
- Will this system be accessible through means other than the network (e.g., telephone)?

For, 1.1 *"Is this system required to be network accessible? all the companies"* has the same features, they need for netwok to be accesable.

Also, for *1.2* we defined the set of locations as: University, Municipality, Financials, NGO, Building Companies, Hospitals, IT Companies, Hotels. These questions are answered differently for each institution, based on their specific needs and capabilities. The use of questions to define the features provides a standardized way to compare the network access and communication requirements of different institutions (public and private) and identification of similarities and differences between them. Based on the given answers, the private institution has a feature of 1.2, University, Financials, NGO, Building Companies, Hospitals, IT Companies and Hotels, which indicates that they only allow network access within their central office and campus networks. The public institution has a feature of 1.2, University, Municipality and Hospitals, which indicates that they allow network access within their central office, campus networks, and specific locations.

For question 1.3: "**Provide a network diagram depicting required connectivity for all components of the application or service**," in general, a network diagram for an institution illustrate the network infrastructure, connections, and relationships between different components of the application or service. It would show how various devices, servers, and network segments are interconnected to support the system's functionality.

Comprehensive Examination of Network Access, Logging, and Auditing Strategies   57

The network diagram helps identify the necessary connectivity and potential vulnerabilities in the system, allowing for the implementation of appropriate security controls. The actual responses would depend on the specific institutions and their respective policies, procedures, and network infrastructure.

For question 1.4: *"Will this system be accessible through means other than the network (e.g., telephone)?"* the answer may vary based on the specific institution in more details is the current situation is based on common scenarios, and acknowledge the importance of a network diagram For the Private Institution: this system will be accessible through means other than the network. Apart from network access, authorized users may also have telephone access to the system for specific functionalities or support purposes. This access is limited to authorized personnel only.

For the Public Institution - this system will not be accessible through means other than the network. The primary mode of access is through the network infrastructure, and telephone access is not required or permitted for this system. Based on this, we can conclude that both private and public institutions have different sets of features in terms of network access and communication. The private institution has more diverse features than the public institution, and they have more restrictions in terms of network accessibility.

On the other hand, the public institution allows network access within their central office, campus networks, and specific locations. It is also important to note that both institutions have features related to network diagrams and accessibility, which indicates the importance of network infrastructure in their operations. Understanding the different features and requirements of each institution can help in designing and implementing effective network solutions that meet their specific needs.

Recent advancements in technology, including the adoption of cloud-based identity management systems, have transformed how institutions manage user access and security. These developments highlight the importance of scalability, user-friendliness, and effective data handling practices within both sectors. By understanding the differences in practices between public and private universities, stakeholders can better tailor their approaches to meet institutional needs while enhancing security measures. The findings presented in this section are based on a comprehensive analysis of current practices related to logging, auditing, and network access controls. By examining these aspects, this research aims to shed light on the critical areas where improvements can be made to strengthen the cybersecurity frameworks of higher education institutions. The following tables summarize the key findings of this analysis. This section presents a comprehensive analysis of the logging, auditing, and network access practices of public and private universities, alongside their business continuity and disaster recovery plans. The findings are summarized in the following tables.

Table 1 compares the logging and auditing practices of public and private universities. Both types of institutions produce logs and audit trails to track activities within their applications. Public universities do not embed sensitive data in their logs, while private universities do. Both institutions can link actions to individual users and log successful and unsuccessful access attempts. Notably, private universities retain logs for a longer period (1 year) compared to public universities (6 months). This differentiation highlights the varying approaches to data sensitivity and log retention practices.

Table 1. Comparison of Logging and Auditing Practices

Aspect	Public Universities	Private Universities
Describe logs and/or audit trails produced by the application or service	Yes	Yes
Is sensitive data embedded in the logs?	No	Yes
Can logs and/or audit trails link actions to individual users?	Yes	Yes
Are successful/unsuccessful accesses logged? With client network address?	Yes	Yes
For how long are logs retained?	6 months	1 year

Table 2. Comparison of Network Access Control

Aspect	Public Universities	Private Universities
Is the system required to be network accessible?	Yes	Yes
How is the system accessible?	Campus networks, specific locations	Central office and campus networks
Alternative means of access	No	Yes (telephone access for authorized users)

Table 2 illustrates the differences in network access control between public and private universities. Both institutions require their systems to be network accessible. Public universities allow access within campus networks and specific locations, while private institutions have a more limited approach, permitting access only from central offices and campuses. Additionally, private institutions provide alternative access via telephone for authorized users, which is not available in public institutions.

Table 3. Business Continuity and Disaster Recovery Plans

Aspect	Public Universities	Private Universities
Is there a documented business continuity/disaster recovery plan?	Yes	Yes
Does the plan address procedures for data restoration?	Yes	Yes
Are emergency contact names and numbers included?	Yes	Yes

Table 3 outlines the business continuity and disaster recovery plans in public and private institutions. Both types have documented plans that address procedures for restoring lost data and functionality during emergencies. They also include emergency contact information, which is essential for effective recovery efforts. This indicates a shared commitment to preparedness across both sectors.

A. Logging and Auditing

Logging and auditing are important processes for maintaining the security and integrity of computer systems and networks. They provide a record of activity that can be used for troubleshooting, analysis, and compliance purposes, and can help detect and prevent unauthorized access or other security breaches. The Table 4 shows a comparison between public and private institutions regarding their logging and auditing capabilities.

Table 4. Logging and Auditing for Public and Private Institutions

	Public	Private
1.1 Describe logs and/or audit trails that are produced by the application or service	Yes	Yes
1.2 Is sensitive data embedded in the logs?	No	Yes
1.3 Can logs and/or audit trails link actions to individual users?	Yes	Yes
1.4 Are successful/unsuccessful accesses logged? With client network address?	Yes	Yes
1.5 For how long are logs retained?	6 months	1 year

Both institutions have similar capabilities in terms of producing logs and audit trails, linking actions to individual users, and logging successful/unsuccessful accesses with client network address. There is a difference in how they handle sensitive data in the logs, with private institutions having it embedded in the logs while public institutions do not. Additionally, private institutions retain their logs for a longer period of time (1 year) compared to public institutions (6 months). In details, the Table 1 presents the information in a clear and concise manner, making it easy to compare the two types of institutions in terms of their logging and auditing capabilities.

Describe logs and/or audit trails that are produced by the application or service:
For public and private institutions, the answer is "Yes" for this question, indicating that both types of organizations have the capability to produce logs and/or audit trails to track activities within their applications or services. These logs or audit trails could include information such as who accessed the system, what actions were taken, when they were taken, and any errors or exceptions encountered.

Is sensitive data embedded in the logs?
The answer for public institutions is "No", indicating that sensitive data is not embedded in the logs. For private institutions, the answer is "Yes", indicating that sensitive data could potentially be included in the logs. Sensitive data could include personal information about users, financial data, or other confidential information.

Can logs and/or audit trails link actions to individual users?
The answer for both public and private institutions is "Yes", indicating that the logs or audit trails can identify which actions were taken by which individual users. This is an important capability for accountability and auditing purposes.

Are successful/unsuccessful accesses logged? With client network address?
The answer for both public and private institutions is "Yes", indicating that both types of organizations log successful and unsuccessful accesses, and include the client network address. This information can be helpful in identifying potential security breaches or unauthorized access attempts.

For how long are logs retained?
For public institutions, logs are retained for 6 months. For private institutions, logs are retained for 1 year. Retaining logs for a certain period of time can be important for compliance, auditing, and security purposes. While the table provides useful information, it may be beneficial to include additional details about the business continuity and disaster recovery plans for both public and private institutions. This could include the specific procedures and protocols in place for data restoration, the roles and responsibilities of staff members in charge of restoration, the frequency of plan updates, and any testing or simulation exercises conducted to ensure the plan's effectiveness. Additionally, including examples of previous emergencies or occurrences that triggered the plan's activation and how it was successfully implemented could provide valuable insights into the institution's preparedness and resilience.

Table 5. Business Continuity and Disaster Recovery in Public and Private Institutions

Business Continuity and Disaster Recovery	Public	Private
Is there a documented business continuity/disaster recovery plan that addresses procedures to restore any lost data or functionality in the event of an emergency or other occurrence, the staff responsible for carrying out data restoration, emergency contact names and numbers, important business partners and other business supply information necessary for a temporary office setup to support data restoration?	Yes	Yes

The Table 5 shows that both public and private institutions have a documented business continuity and disaster recovery plan that includes procedures for restoring lost data and functionality, contact information for staff responsible for data restoration, and other essential business partners and supply information.

6 Conclusion

Based on the comprehensive analysis of network accessibility and communication practices in public and private institutions, it becomes evident that while similarities exist in their fundamental reliance on network access and robust security measures, nuanced differences highlight varying operational scopes and security concerns.

Both public and private institutions underscore the criticality of network accessibility for their day-to-day operations. Utilizing detailed network diagrams, they meticulously map out connectivity to identify vulnerabilities, thereby facilitating the implementation of effective security controls. This proactive approach is crucial in safeguarding sensitive information and maintaining operational continuity.

Private institutions exhibit a broader geographic diversity in their network access locations, encompassing universities, financial institutions, NGOs, hospitals, IT companies, and hotels. In contrast, public institutions predominantly restrict network access to central offices, campus networks, and specific municipal or hospital locations. This difference reflects varying organizational structures and operational needs within each sector.

Regarding security practices, both public and private institutions prioritize the generation of comprehensive logs and audit trails to monitor and track user activities. While this aids in maintaining accountability and identifying potential security incidents, private institutions face added scrutiny due to the inclusion of sensitive data in their logs. This necessitates stringent security protocols to mitigate risks associated with unauthorized access. The retention period for logs differs significantly between public and private sectors, with private institutions retaining logs for a longer duration (typically one year) compared to public institutions (usually six months). This extended retention period enhances forensic capabilities and facilitates compliance with regulatory requirements, underscoring private institutions' commitment to robust data governance practices. This comparative analysis of logging, auditing, and network access practices in public and private universities has revealed several critical insights into how these institutions manage their cybersecurity frameworks. Both public and private universities demonstrate a commitment to logging and auditing practices that are essential for maintaining the integrity and security of their systems. However, significant differences in their approaches, particularly regarding the handling of sensitive data and log retention periods, highlight the need for tailored strategies that address the unique challenges faced by each sector.

The findings indicate that while both types of institutions have documented business continuity and disaster recovery plans, private universities tend to incorporate more stringent measures related to data sensitivity and access control. This suggests that public universities may benefit from reassessing their practices to enhance their security posture and compliance with evolving regulations.

Moving forward, it is essential for both public and private universities to stay abreast of state-of-the-art technologies and methodologies to improve their cybersecurity practices continuously. As digital threats continue to evolve, fostering a culture of security awareness and implementing robust access control measures will be paramount in safeguarding sensitive information and maintaining trust with stakeholders.

Both sectors demonstrate proactive measures in business continuity and disaster recovery planning, outlining procedures for data restoration, defining staff responsibilities, and establishing emergency protocols. These plans are pivotal in ensuring operational resilience during crises and underscore both sectors' commitment to maintaining service continuity and protecting critical assets.

In summary, while public and private institutions share common ground in network management and security practices, their distinct operational contexts and regulatory environments give rise to unique approaches and challenges. Understanding these differences is essential for implementing tailored security measures and fostering effective collaboration in an increasingly interconnected digital landscape.

References

1. Smith, J., Johnson, A.: Advances in natural language processing: exploring recent trends. J. NLP **15**(3), 145–158 (2023)
2. Chew, E., et al.: Performance Measurement for Information Security in the Era of Digital Transformation, NIST SP 800-55 Rev. 2. National Institute of Standards and Technology, Gaithersburg (2022)
3. Van der Weide, T.P., Iachello, G., Watson, J.C.: Data protection: emerging trends in governance, risk management, and compliance. Data Prot. J. **28**(4), 290–307 (2023)
4. Rozanski, N., Woods, E.: Software Systems Architecture: A Guide to Working with Stakeholders Using Viewpoints and Perspectives. Addison-Wesley, New York (2022)
5. Jones, R., Horowitz, B.: A system-aware cybersecurity architecture: current approaches and future directions. Syst. Eng. **24**(1), 45–62 (2021). https://doi.org/10.1002/sys.21406
6. Rivest, R., Shamir, A., Adleman, L.: Revisiting a method for obtaining digital signatures and public-key cryptosystems. Commun. ACM **62**(1), 93–99 (2019). https://doi.org/10.1145/3287195
7. Scarfone, K., et al.: Updated Technical Guide to Information Security Testing and Assessment. NIST SP 800-115 Rev. 2. National Institute of Standards and Technology, Gaithersburg (2021)
8. Whitman, M.E., Mattord, H.J.: Principles of Information Security: Contemporary Strategies. Cengage Learning (2023)
9. Russell, S., Norvig, P.: Artificial Intelligence: A Modern Approach, 4th edn. Pearson (2022)
10. Mohri, M., Rostamizadeh, A., Talwalkar, A.: Foundations of Machine Learning, 2nd edn. MIT Press (2023)
11. Chan, H., Perrig, A., Song, D.: An updated secure protocol for spontaneous wireless ad hoc networks. In: Proceedings of the ACM Conference on Computer and Communications Security (2021)
12. Wang, X., Li, Y.: Secure data management in cloud computing: challenges and solutions. IEEE Trans. Cloud Comput. **11**(2), 215–230 (2023). https://doi.org/10.1109/TCC.2023.3214567
13. Zhang, L., Zhou, H.: Exploring quantum cryptography: the next frontier in secure communications. J. Cryptogr. Inf. Secur. **18**(1), 1–16 (2024). https://doi.org/10.1007/s12095-024-00314-7
14. Kumar, S., Gupta, P.: Advancements in artificial intelligence for cybersecurity: a comprehensive review. ACM Comput. Surv. **56**(3), 45–67 (2024). https://doi.org/10.1145/3567890

Benefits of Parallelization in CPU Rendering: Quantitative Analysis Using a Custom 3D Rendering Engine

Admir Huseini[1,2(✉)], Art Saiti[3], and Kiril Avramovski[1]

[1] 3Shape, Skopje, North Macedonia
huseini@risat.org
[2] Institute of Mathematics, University of St. Cyrill and Methodius, Skopje, North Macedonia
[3] SEEU University, Tetovo, North Macedonia

Abstract. Although GPU rendering is ubiquitous, CPU rendering is still used in certain scenarios which require high-precision calculations and large scenes, such as scientific and architectural visualisations.

SoftRenderingApp3D is an open-source 3D computer graphics engine that performs all rendering operations on the CPU. It functions as both a real-time and offline renderer. In real-time mode, it can rasterize meshes with up to 1 million facets, but faces challenges with pixel rendering on high-resolution screens.

The engine implements a parallelized rendering pipeline designed to leverage modern multi-core CPU architectures. The study utilizes the engine to measure and evaluate the performance gains achieved through parallelization, bench-marking against established CPU renderers, providing insights into the scalability and efficiency of CPU-based rendering techniques.

The paper presents an analysis of the engine's architecture, design decisions, and implementation details, as well as the results obtained by parallelisation and their implications. It explores potential applications in different contexts and possible improvements to CPU rendering efficiency.

Keywords: High performance 3D graphics · 3D engine architecture · Computer graphics algorithms

1 Introduction

In the past decade, advancements in GPU technology have enabled near-instantaneous processing of complex 3d scenes. This has relegated CPU-based, also known as software rendering to obsolescence in many applications.

However, CPU rendering retains distinct advantages in specific domains, particularly where precision and large data handling are crucial. CPUs excel in double-precision arithmetic and can leverage larger system memory, unlike the GPUs which are constrained by their VRAM. This makes them valuable for architectural visualization, scientific simulations, and other specialized tasks. To address the large computation times required by the CPU when rendering complex scenes, parallelisation is leveraged to maximise performance.

This paper aims to quantify the benefits of parallelization in CPU-based rendering and introduce SoftRenderingApp3d [1], a novel software renderer written in C# that bridges the performance gap between high-level and low-level implementations. This paper argues that by utilizing parallelisation, SoftRenderingApp3d can achieve performance comparable to established CPU renderers like Mesa3D with little computational overhead, while offering the benefits of a garbage-collected language environment. This approach can enhance accessibility by eliminating the need for complex graphics APIs and improves productivity and versatility.

By demonstrating the viability of a high-level, CPU-based approach, we seek to broaden the accessibility of advanced rendering techniques to a wider range of developers and researchers. Furthermore insights gained by the analysis of parallelisation in CPU rendering can be used to pinpoint bottlenecks and potential areas of improvement.

The development of the engine started as a learning project. The first version was based on *WinForms3D* [2]. The goal was to understand concepts and algorithms of 3d graphics from books like [3,4].

1.1 Literature Discussion

One of most used soft rendering frameworks is WARP or Windows Advanced Rasterization Platform [5]. It is an integral software rasterizer component of DirectX, introduced by the Direct3D 11 runtime. It is a high speed component that can do high performance shading on the CPU. It is used when the software on which the DirectX application is running has no hardware acceleration.

Another industry standard for software rendering is the OpensSWR [6] graphics library, which is a high performance and highly scalable software rendering library for x86 CPUs. It is supported by Intel, AMD and other industry leaders. It is compatible with the OpenGL standard and allows compute shaders to be executed on the CPU. It is distributed as an implementation of the Gallium driver of the Open Source Mesa 3d Graphics library [7].

In [8], the authors present performance evaluation of image processing using software rendering implemented using Mesa3D. They compare the difference in performance between GPU-accelerated and software rendering.

The authors in [9] make an in depth study and performance comparison of existing soft rendering algorithms. They analyze different rasterization algorithms, investigate the complete graphics pipeline and how the main bottlenecks are managed by different approaches. They explore capabilities of the scan-line algorithm and different memory layouts for frame buffers. They also study instruction and thread-level parallelization and their relative efficiencies on different CPU architectures.

In [10,11] the authors give a survey and overview of software rendering techniques versus GPU based rendering techniques. They discuss the benefits and shortcomings of both approaches to rendering. They also give an in depth analysis of the design details of soft rendering architectures.

1.2 Overview

Section 2 gives a more detailed description of the applications architecture and its main components. Some of the specific features and design decisions are briefly described. Section 3 is a high level overview of the Rendering engine and details of its parallel implementation. The Appendix in Sect. A provides a more thorough implementation details of the rendering engine.

Section 4 describes the methodology used to make the measurements and make comparisons between the different CPU rendering frameworks. In Sect. 5 numerical data from benchmarking tests is provided.

The tests show clearly that using the parallel processing capabilities of the modern CPUs, it is possible to execute all the functionalities of a rendering pass including rasterization in less than 25 milliseconds. The execution of the pixel shader code and writing of the pixel data to the buffers takes most of the computational time and it can take up to 300 milliseconds per frame.

Finally, in Sect. 6 the results are discussed, their implications and present possible applications of the software. Some practical usages of the application in academic research related to mathematics are provided as well.

2 Application Description

2.1 Overview

The application's architecture consists of four primary layers, as illustrated in Fig. 1:

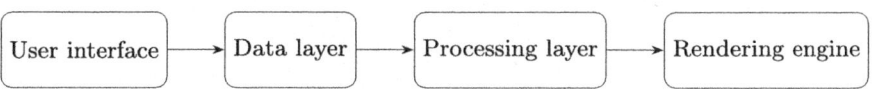

Fig. 1. The basic layers of the application

1. **User interface:** An interactive 3d Windows desktop application.
2. **Data layer:** Manages data storage and retrieval.
3. **Processing layer:** Executes mathematical algorithms on the data.
4. **Rendering engine:** Implements forward rendering for visualization.

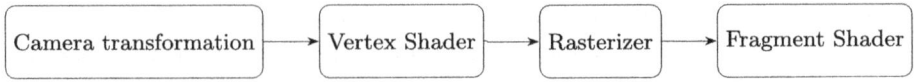

Fig. 2. The forward rendering pipeline

This layered approach allows for clear separation of concerns, promoting modularity and facilitating future enhancements. The rendering engine, which is the core focus of this implementation, is designed with both performance and flexibility in mind. It follows a standard forward rendering pipeline, as shown in Fig. 2.

The rendering pipeline's implementation allows for customization at each stage, providing flexibility for educational purposes and algorithm prototyping. Detailed explanations of the rendering process and optimizations are provided in Sect. 3.

2.2 Detailed Component Description

User Interface. The application features a simple and functional user interface, designed for intuitive interaction with 3D models. It includes standard features such as model rotation, scaling and panning via mouse interaction, real-time rendering capabilities in response to user input and options to load different models with or without textures.

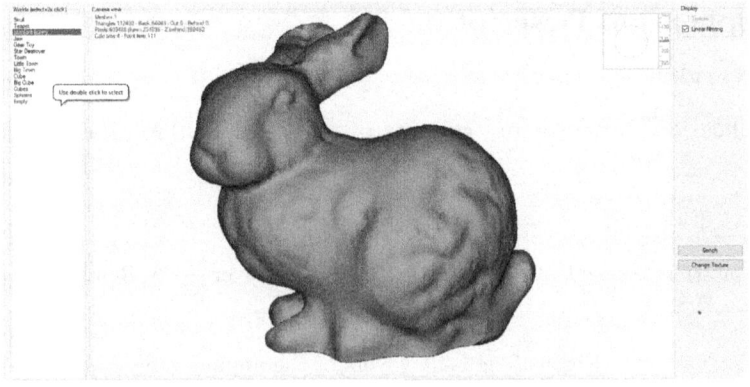

Fig. 3. The Stanford bunny rendered in SoftRenderingApp3d

Figure 3 demonstrates the rendering capabilities of the engine using the Stanford Bunny model as an example.

Data Layer. The data layer is responsible for model import and management. Currently, it supports two standard file formats. STL [12] and COLLADA [13].

Future development plans include support for additional popular formats such as OBJ, PLY, and gLTF, enhancing the application's versatility in handling various 3D model sources.

Processing Layer. The processing layer serves as an intermediary between the data layer and the rendering engine, facilitating geometric and numerical operations on mesh data. Key features include:

- Execution of arbitrary geometry processing algorithms on imported meshes
- Storage and management of intermediate computation results
- Event-based system for notifying the rendering engine of data updates

This layer's flexible design allows for easy integration of new processing algorithms and seamless updates to the rendering pipeline. Typical operations might include mesh simplification or subdivision, custom attribute calculations like curvature, ambient occlusion and so on.

The event-driven architecture ensures that the rendering engine can efficiently respond to changes in mesh data or processing results, maintaining real-time performance where possible.

3 Rendering Engine

The rendering engine implements the standard forward rendering algorithm. Only the parallel implementation workflow and extensibility possibilities are presented in this section. More details about the implementation of the engine are given at the Appendix on Sect. A.

Performance analysis indicates that pre-fragment shader stages, including rasterization, achieve real-time performance (<25 ms) for up to 1 million facets. Fragment shader execution time scales linearly with pixel count, potentially limiting performance on high-resolution displays (up to 300 ms per frame).

This implementation balances flexibility for educational and prototyping purposes with performance considerations, noting current limitations in real-time rendering capabilities for complex, high-resolution scenes.

3.1 Parallelisation

The rendering of a single frame follows a typical rendering pipeline while allowing for flexibility in vertex and pixel shader implementations. It makes extensive use of parallelization at three key stages: vertex processing, rasterization, and pixel processing.

One important factor for processing speed is that the data accessed by the same core should reference the same memory location. For this purpose, one should use 3d space partitioning algorithms like octrees to only rasterize spatially related facets in the same thread. The same is valid for pixel shading, where one should partition the screen into $n-$rectangles and process each rectangle in its own thread. Both of these optimization steps need to be done only once, when the model is loaded.

The process can be broken down into several key stages:

1. **Initialization:** Clean the FrameBuffer and validate input data.
2. **Parallel Vertex Shading:** Partition the vertex buffer and process each collection in parallel.
3. **Parallel Facet Rasterization:** Partition the facets buffer, run the Scan-line algorithm for each collection and fill the frame buffers.
4. **Parallel Fragment Shading:** Partition the input buffers, calculate the final color based on lighting, material properties, and other inputs and fill the ScreenImage buffer.

By leveraging multi-core CPUs, the engine achieves improved performance without relying on hardware-accelerated graphics APIs.

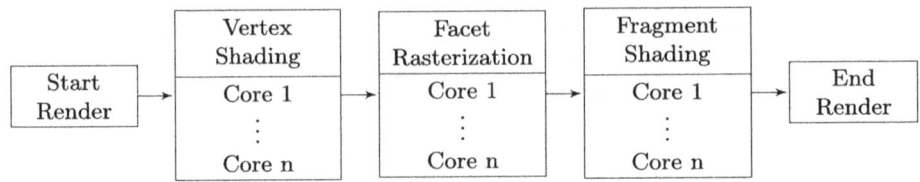

Fig. 4. Simplified Flowchart of the Parallel Multi-Core Renderer

Figure 4 provides a simplified visual representation of the rendering pipeline, highlighting the three main parallelized stages: vertex buffer update, rasterization, and pixel processing.

3.2 Extensibility and Future Improvements

The architecture of the rendering engine is designed with extensibility in mind, allowing for future improvements and adaptations. Two key areas for potential enhancement are:

GPU Acceleration via Shader Transpilation: The current CPU-based implementation can be extended to leverage GPU capabilities. It is possible to utilize transpilers to convert C# shader code to HLSL or GLSL. They can be executed on the GPU for improved performance. This approach maintains the flexibility of C# while harnessing GPU power. It requires additional development effort but offers significant performance potential.

Dual CPU/GPU Rendering Support: The well-defined interfaces of the rendering engine allow for a flexible approach to implementation. One can develop and prototype algorithms using CPU-based shaders, then implement GPU-based versions of the same algorithms using the existing interfaces. It enables interchangeable use of CPU or GPU rendering based on specific needs or hardware availability. This dual-support strategy offers maximum flexibility and performance optimization opportunities.

These architectural considerations ensure that the rendering engine remains adaptable to future technological advancements and varying project requirements. The ability to seamlessly transition between CPU and GPU rendering, or to leverage both simultaneously, positions the engine for long-term relevance and performance scalability.

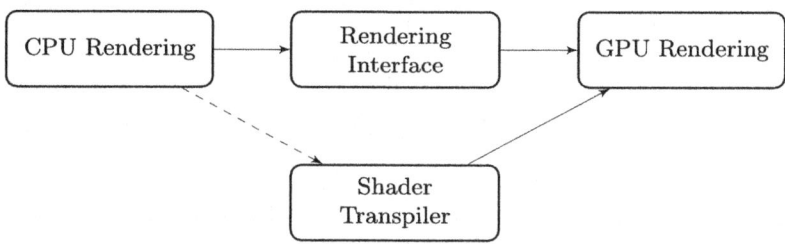

Fig. 5. Extensible Architecture for CPU and GPU Rendering

Figure 5 illustrates the extensible architecture, showing how the rendering interface can support both CPU and GPU implementations, with the potential for shader transpilation to bridge the gap.

4 Benchmarking Methodology

This section outlines the experimental setup and benchmarking process used to evaluate the performance of the software rendering implementation.

4.1 Hardware and Software Configuration

Benchmarks were conducted on a system with operating System Windows 11, Intel(R) Core i-7, 8 core/16 logical-core CPU and 32 GB of RAM. Given the processor utilizes hyperthreading technology, resulting in 16 logical cores, this study assumes 16 cores for simplicity in analysis. It is important to note that although not all cores are physical, the comparative results remain valid as all benchmarks were performed on the same hardware configuration. The results may not reflect the maximum speedup achievable with only physical cores but are sufficient for comparative analysis between the software renderers tested. Software renderers used for comparison are SoftRenderingApp3D vs. single-threaded Mesa3D and multi-threaded build of Mesa3D Gallium.

Mesa3D is selected as a benchmarking reference, since it represents the oldest and most comprehensive open-source software renderer to date, with backing from large corporations like Intel and AMD.

4.2 Benchmarking Process

The sets of models used for benchmarking included the Stanford standard models and some generic models of variable facet counts. The benchmarking process consisted of two phases: initial benchmarks and comparison benchmarks.

The initial benchmarks were conducted using the full set of models, including Stanford standard models and various generic models. These tests were performed using the fully parallelized SoftRenderingApp. The average frame was measured while applying different transformations to the models.

For the comparison benchmarks, a subset of three models was selected representing different levels of complexity. These models were used to compare SoftRenderingApp3d against both single-threaded and multi-threaded versions of Mesa3D. To ensure consistency, all models were rendered with identical orientation and position, and a consistent viewport dimension was used across all tests.

In both benchmark phases, 10 separate measurements for each model were taken. The metrics recorded included frame render time, memory usage, and CPU usage.

This two-phase methodology enabled us to first assess the renderer's performance across a broad spectrum of models and then perform detailed comparisons with the standard industry software renderer.

4.3 Experimental Design

The study focused on two main comparisons:

1. **Impact of multi-core processing** vs. single-threaded Mesa3D
2. **Full parallelization comparison** vs. multi-threaded Mesa3D (Gallium)

For the multi-core impact study, benchmarks were run with varying numbers of active CPU cores to assess scalability and to quantify the impact of parallelisation.

4.4 Data Analysis

Statistical analysis of the benchmark results included:

- Calculation of mean values for each set of 10 measurements
- Analysis of performance metrics in relation to model complexity (facet count)
- Computation of correlation coefficients (r^2 values) to assess the fit of the data to Amdahl's and Gustafson's Law

The r^2 values were calculated to evaluate how well the observed performance scaling with increasing core count aligns with the theoretical predictions of Amdahl's [14] and Gustafson's [15] laws. This analysis helps in understanding the parallelization efficiency of the implementation and its adherence to established parallel computing principles.

Performance data was plotted against the number of cores used, and the resulting curves were compared to the theoretical curves predicted by the laws of Amdahl and Gustafson. Closeness of fit is indicated by the r^2 values, which provide insight into the scalability of SoftRenderingApp3d.

5 Benchmarks

This section focuses on interpreting and analyzing the benchmarks and the conclusions that can be drawn from them. Measurements were also made on an NVIDIA GeForce RTX 3080 GPU, but since it achieved a consistent 120FPS, the measurements have not been used in any analysis to avoid redundancy.

Table 1. SoftRenderingApp3d benchmarks at 16 cores with GPU comparison

Model	Pixels	Facets	Time (ms)	CPU FPS	GPU FPS
Generic Model	137K	121K	23	43	120
Dragon	199K	100K	40	25	120
Bunny	260K	112K	45	22	120
David	570K	1.2M	129	8	120
Happy Buddha	919K	1.1M	188	5	120
Teapot	1.2M	9K	107	9	120

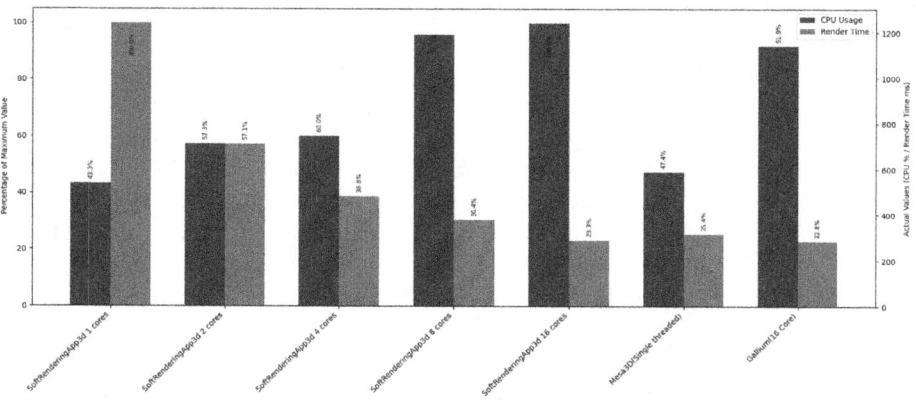

Fig. 6. Comparison of CPU usage and render times between different implementations

As seen in Table 1, the main limiting factor for the render times is the number of pixels that have to be rendered at a time. The GPU also shows maximum performance, with no variation between the models.

5.1 Core Count and Renderer Comparison

Figure 6 illustrates the measurements of CPU usage and rendering time for SoftRenderingApp3d run with 1–16 cores vs single and multi-core Mesa3D. The CPU usage is measured in percentage of total available CPU processing capacity of the computer. The rendering time is given in percentages relative to the time it takes to render the scene in SoftRenderer3d using a single CPU core. A value of 33% indicates that it takes 1/3 of the time to render the same scene with a single core.

The performance comparison between Mesa3D and SoftRenderingApp3d yields several insights:

Single-Core Performance: Mesa3D, utilizing lower-level code and optimizations, outperforms SoftRenderingApp3d in both render time and CPU usage when running on a single core.

Multi-core Advantage: SoftRenderingApp3d surpasses Mesa3D's performance when utilizing 16 cores, achieving render times that are, on average, 15.5% faster.

Comparison with Gallium: SoftRenderingApp3d and the parallelized Gallium (Mesa3D) show nearly identical performance in terms of CPU usage and render time. More specifically, SoftRenderingApp3d is marginally faster when rendering the Generic Model and David. It is approximately 5ms slower when rendering the High Polygon Sphere.

Parallelization Limits: The performance improvements from parallelizing Mesa3D demonstrate diminishing returns compared to its single-threaded implementation. Similarly, SoftRenderingApp3d, despite being a higher-level implementation, achieves very similar results to multi-core Mesa3D Gallium when running at the same level of parallelization.

This behavior in both implementations aligns with Amdahl's Law, which predicts limits to performance gains from parallelization. These observations underscore that even with different levels of code optimization and use of lower-level languages, the fundamental limits of parallelization apply universally.

A more detailed examination of this phenomenon is presented in Sect. 5.2 (Table 2).

5.2 Amdahl's and Gustafson's Law Predictions

Amdahl's Law Fit: The measurements from SoftRenderingApp3d show a very high correlation with Amdahl's Law, as evidenced by the high r^2 values, as illustrated by Fig. 7. This strong fit allows for conclusive predictions of speedups at higher core counts. Table 3 compares the predicted speedups against actual measurements. The sequential part (s) of the algorithm can be estimated using this model:

Table 2. Performance Comparison Across Different Core Counts and Renderers

Renderer	Model	CPU Usage (%)	Memory (MB)	Render Time (ms)	Facets
SoftRenderingApp3d (1 Core)	Generic Model	25.3	78.4	63	121,448
	David	29.7	386.9	530	1,199,948
	High Polygon Sphere	21.3	1,135.0	1,302	4,063,232
SoftRenderingApp3d (2 Cores)	Generic Model	22.1	78.6	37	121,448
	David	32.1	386.9	280	1,199,948
	High Polygon Sphere	28.2	1,150.0	743	4,063,232
SoftRenderingApp3d (4 Cores)	Generic Model	22.1	78.6	27	121,448
	David	33.2	386.9	175	1,199,948
	High Polygon Sphere	29.5	1,150.0	505	4,063,232
SoftRenderingApp3d (8 Cores)	Generic Model	22.1	81.2	27	121,448
	David	35.3	386.9	138	1,199,948
	High Polygon Sphere	47.2	1,150.0	396	4,063,232
SoftRenderingApp3d (16 Cores)	Generic Model	15.23	81.3	19	121,448
	David	26.5	397.0	110	1,199,948
	High Polygon Sphere	49.5	1,018.0	303	4,063,232
Mesa3D (Single-threaded)	Generic Model	24.10	183.0	19	121,448
	David	23.30	671.9	137	1,199,948
	High Polygon Sphere	24.20	1,973.7	337	4,063,232
Gallium (Parallelized Mesa3D)	Generic Model	35.5	183.0	21	121,448
	David	40.6	773.9	128	1,199,948
	High Polygon Sphere	45.2	2,100.3	297	4,063,232

Table 3. Measured vs. Amdahl's Law Predicted Speedups

Cores	Generic Model		David		High Poly Sphere	
	Measured	Predicted	Measured	Predicted	Measured	Predicted
1	1.00	1.00	1.00	1.00	1.00	1.00
2	1.70	1.57	1.89	1.75	1.75	1.68
4	2.33	2.19	3.03	2.81	2.58	2.55
8	2.42	2.73	3.84	4.03	3.29	3.45
16	3.32	3.12	4.82	5.15	4.30	4.18
32	-	3.39	-	5.60	-	4.70
64	-	3.53	-	6.05	-	5.00
128	-	3.60	-	6.31	-	5.16
256	-	3.64	-	6.44	-	5.25

- For the Generic Model (121K facets): $s \approx 27\%$
- For David (1.1M facets): $s \approx 19\%$
- For high-polygon sphere (4M facets): $s \approx 14\%$

This variation in s suggests that the proportion of sequential code decreases slightly as model complexity increases. To further investigate this relationship, additional benchmarks were conducted using an ultra-high-polygon sphere consisting of ≈ 16 million facets, excluded from the primary benchmark set due to its complexity exceeding the threshold for accurate measurement.

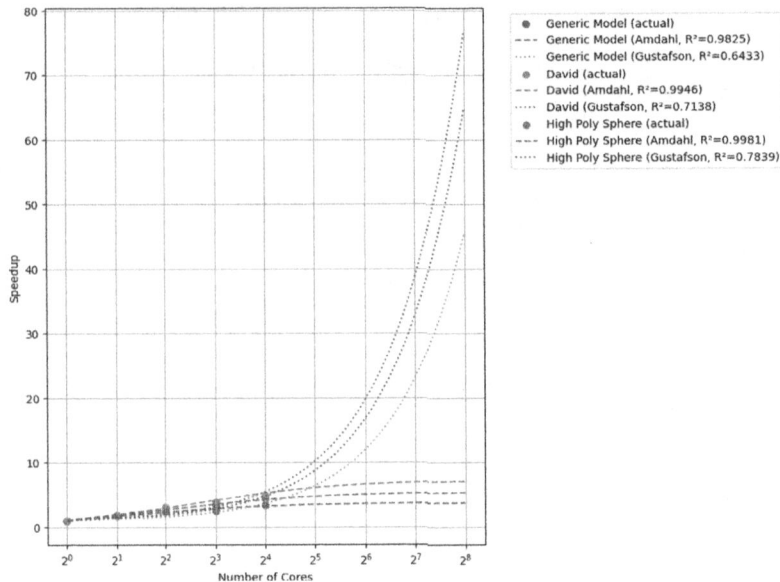

Fig. 7. Amdahl's Law Fit

Despite potential measurement inaccuracies, the ultra-high-poly sphere measurements, when fitted using Amdahl's law, yielded a high r^2 value of 0.9948 in line with the other measurements. The sequential fraction for this high-poly sphere was also calculated to be 17.74%, aligning closely with the average sequential fraction observed between the David model and the high-polygon sphere used in the primary benchmarks. This consistency suggests the existence of a theoretical minimum for the sequential portion of the rendering process, estimated to range between 14% and 19%.

Gustafson's Law Comparison: Data was also fitted to Gustafson's Law for comparison. As shown in Fig. 7, Gustafson's Law provides less accurate predictions. This discrepancy suggests that the renderer's performance aligns more closely with the assumptions of Amdahl's Law than those of Gustafson's Law.

Implications: The strong fit with Amdahl's Law indicates that the renderer has a consistent sequential component across different levels of parallelization. The slight decrease in s for more complex models suggests that the parallel portion of the rendering scales well with increased workload. The poorer fit with Gustafson's Law implies that increasing the problem size (model complexity) does not perfectly scale with the number of processors in our implementation. Measurements also indicate a minimum sequential fraction of around 14%–19%.

These findings provide valuable insights into the scalability characteristics of our software renderer and help predict its performance on systems with higher core counts.

6 Conclusion and Possible Applications

6.1 Possible Applications

The application is well suited for work in scientific visualisations and simulations. Some areas which require high-level visualisations include topology, electromagnetism, fluid dynamics and magneto-hydrodynamics. Some visualisations have been provided relating to the decomposition of 3d spaces into tetrahedral-octahedral honeycombs [16,17] as shown in Figs. 8, 9, 10 and 11.

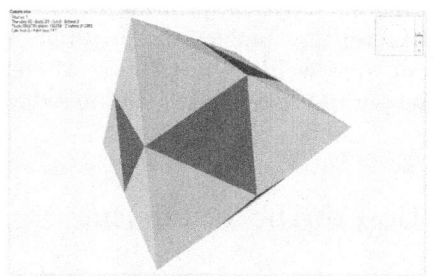

Fig. 8. A tetrahedral-octahedral honeycomb

Fig. 9. A tetrahedral-octahedral honeycomb with separated volume elements

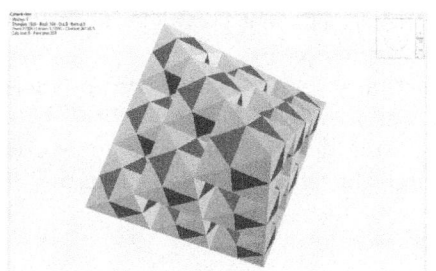

Fig. 10. A recursive tetrahedral-octahedral honeycomb

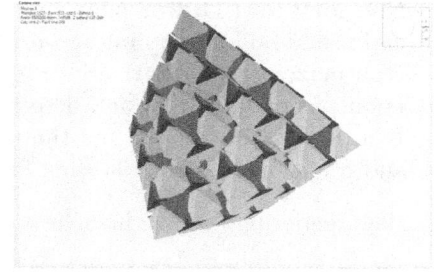

Fig. 11. A recursive tetrahedral-octahedral honeycomb with separated volume elements

The sequential nature of CPU's also facilitates the extraction of information from every frame, such as facet normals, which can be analysed continuously. In the context of physics simulations, gradient fields, animations for the convergence of solutions, and vector fields can be analysed in great detail without relying on complex shader code.

6.2 Conclusion

In this paper we have presented a software rendering engine written in C# that renders 3d models using only the CPU without hardware acceleration. The full architecture of such an application is provided, alongside a discussion of its advantages and drawbacks.

The research demonstrates that through parallelization, high-level performance can be achieved without relying on low-level code, at relatively low computational overhead in CPU and RAM utilization. Multi-core performance measurements and calculations based on Amdahl's law indicate that the main bottleneck for further improvement is the unparallelizable sequential part. This consists of operations like reading data from and into buffers and amounts to around 16%–19%, when not including small models like the Generic Model. Future improvements in CPU-based rendering performance could focus on reducing this sequential fraction.

A Appendix: Detailed Description of the Rendering Engine

This section details the key components and processes of the rendering engine. The implementation follows a forward rendering pipeline, as mentioned in Sect. 3 emphasizing flexibility and CPU-based execution. The main components include:

- **Camera Model:** Implements view and projection transformations, following OpenGL conventions.
- **Memory and Buffer Management:** Utilizes efficient strategies for vertex and frame buffer handling to optimize performance.
- **Rasterizer:** Employs the Scan-line algorithm to map facets to screen pixels, populating the zBuffer and determining visible facets.
- **Render Pass:** Implements the pipeline using delegate functions for vertex and fragment shaders, allowing for customizable shader implementations.

The rendering process involves:

1. Vertex processing and camera transformations
2. Rasterization of transformed vertices
3. Fragment shading for final pixel color determination

The following subsections will explore each component in detail, discussing their implementation, optimization strategies, and performance characteristics.

Rendering Materials. The rendering material system augments mesh data with color information, necessary for the fragment shader's per-pixel color calculations. Supported material types include:

- **Vertex color**: A per-vertex color buffer, with intra-facet colors interpolated from vertex values.
- **Facet color**: A per-facet color buffer, applying uniform color across each facet.
- **Fine-grained facet color**: An advanced coloring scheme that subdivides each facet into four sub-facets, allowing for more detailed color representation without full texture overhead.
- **Texture materials**: Traditional texture mapping using image data and per-vertex texture coordinates.

The fine-grained facet color material employs a subdivision scheme as illustrated in Fig. 12, offering a balance between color detail and memory efficiency.

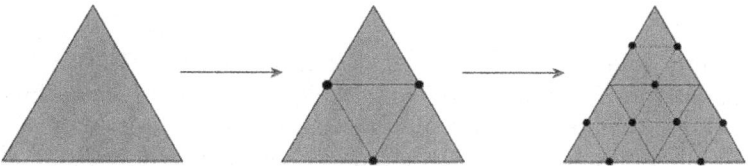

Fig. 12. Triangle subdivision scheme for fine grained facet coloring

For texture materials, a Gaussian convolution kernel is applied to sample colors, providing smoother results compared to simple point sampling.

A.1 Memory Model and Buffers

The rendering engine leverages C#'s flexibility to create a simple yet versatile rendering pipeline, balancing performance and ease of use for educational purposes and rapid prototyping.

Memory Management. The engine employs advanced memory management techniques to optimize performance, utilizing a pool-based allocation strategy. This allows for the effective reuse of memory buffers, avoiding redundant allocations and deallocations, more effective cache usage as well as higher data integrity by minimizing memory fragmentation.

Vertex Buffers. Vertex buffers for each mesh consist of:

- VertexWorldCoordinates
- VertexWorldNormals

- VertexViewCoordinates
- VertexProjectedCoordinates
- VertexNormalDeviceCoordinates

These buffers are updated in each rendering pass based on camera matrix values.

Frame Buffers. Frame buffers use linear ordering of screen pixels and comprise three main arrays:

- **ScreenImage:** Integer array representing RGB colors per screen pixel.
- **FacetIndexForPixel:** Integer array containing indices of the closest facets hit by a ray through the screen in the camera view direction.
- **zBuffer:** Contains the z-coordinate of the facet point rendered in the current pixel.

A.2 Camera

The camera model in the rendering engine follows the OpenGL standard, utilizing world and view coordinate systems. The camera is represented by an affine rigid transformation that rotates and translates vertex positions, enabling the conversion of coordinates from the world system to the view system. It implements Perspective or Orthogonal Projection Matrix, and Normal Device Coordinates transformation matrix.

A.3 Rasterizer

The rasterizer processes each facet (triangle) of the mesh individually, mapping them to the screen using projected vertex coordinates. It employs the scan-line algorithm to establish correspondences between screen pixels and facet positions.

Figure 13 illustrates the scan-line algorithm. The process involves:

1. Projecting a mesh facet onto the screen.
2. Subdividing the projected facet into two subtriangles with bases parallel to the $x-$coordinate.
3. Processing each subtriangle separately.
4. Assigning barycentric coordinates relative to the facet for each screen pixel.

The rasterizer serves two primary functions:

- Determining which facet to render for each screen pixel
- Populating the zBuffer

Using vertex view coordinates, the rasterizer calculates the depth (z-coordinate) of each facet point. It updates the zBuffer and FacetIndexForPixel if the calculated z-value is smaller than the existing one, then moves to the next pixel.

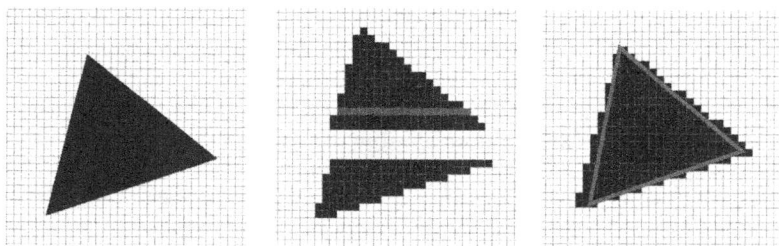

Fig. 13. Scan line algorithm to rasterize a single triangle

A.4 Rendering Pass

The software rendering pipeline is implemented as a class that manages buffer access and executes the rendering function for each frame. This function accepts parameters including camera specifications, vertex and fragment shader providers, and a rasterizer. The rendering pass follows a structured workflow:

1. **Frame Buffer Initialization:** The frame buffer is cleared to prepare for the new render.
2. **Vertex Shader:**
   – Camera transformations are applied to populate the view vertex buffers.
   – The vertex shader delegate function is executed, applying mathematical transformations on Vertex buffers.
3. **Rasterization:** Using scan-line the rasterizer selects the facet for rendering and sets the zBuffer for each pixel.
4. **Fragment Shading:** The fragment or pixel shader is another delegate function, that uses the buffers flled by the rasterizer. It calculates the per pixel colors and fills the ScreenImage buffer.

The use of delegate functions for vertex and fragment shaders provides flexibility in the rendering pipeline. This approach allows for customizable shader implementations without modifying the core rendering logic. The vertex shader's execution prior to camera transformations enables a wide range of geometric manipulations, while the fragment shader's position in the pipeline allows for complex per-pixel operations based on the rasterized geometry.

References

1. Huseini, A., Saiti, A., Avramovski, K.: Soft rendering app 3D (2024). https://github.com/admirhuseini3shape/SoftRenderingApp3D
2. Bourgeois, L.: Winforms3d (2020). https://github.com/Larry57/WinForms3D
3. Hughes, J.F. (ed.): Computer Graphics, 3rd edn. Addison-Wesley, Upper Saddle River (2014)
4. Hardy, A., Steeb, W.-H.: Mathematical Tools in Computer Graphics with C Sharp Implementations. World Scientific, New Jersey (2008)

5. Online-resource: Windows advanced rasterization platform (WARP) (2024). https://learn.microsoft.com/en-us/windows/win32/direct3darticles/directx-warp
6. Online-resource: Gallium openswr documentation. Mesa 3D online documentation (2024). https://gallium.readthedocs.io/en/latest/gallium/drivers/openswr.html
7. Online-resource: Mesa 3D graphics library (2024). Mesa 3D online website. https://www.mesa3d.org/
8. Wegen, O., Trapp, M.: Performance evaluation and comparison of service-based image processing based on software rendering. In: Computer Science Research Notes, Západočeská univerzita (2019)
9. Frolov, V., Galaktionov, V., Barladyan, B.: Comparative study of high performance software rasterization techniques. Mathematica Montisnigri **47**, 152–175 (2020)
10. Mileff, P., Dudra, J.: Advanced 2D Rasterization on Modern CPUs, pp. 63–79. Springer (2013)
11. Taylor, Z.: A modern approach to software rasterization. Preprint (2011)
12. Online-resource: STL (stereolithography) file format family. Website resources, USA Library of Congress (2023)
13. Online-resource: Collada overview. Khronos Group specification (2011)
14. Amdahl, G.M.: Validity of the single processor approach to achieving large scale computing capabilities. In: Proceedings of the 18–20 April 1967, Spring Joint Computer Conference, pp. 483–485. ACM (1967)
15. Gustafson, J.L.: Reevaluating Amdahl's law. Commun. ACM **31**(5), 532–533 (1988)
16. Greiner, G., Grosso, R.: Hierarchical tetrahedral-octahedral subdivision for volume visualization. Vis. Comput. **16**, 357–369 (2000)
17. Schaefer, S., Hakenberg, J., Warren, J.: Smooth subdivision of tetrahedral meshes. In: Proceedings of the 2004 Eurographics/ACM SIGGRAPH Symposium on Geometry Processing, SGP 2004. ACM (2004)

Simulation of the Quasigroup Redundancy Check Code's Ability to Detect Errors

Natasha Ilievska[✉]

Faculty of Computer Science and Engineering, Ss. Cyril and Methodius University, Skopje, Republic of Macedonia
natasa.ilievska@finki.ukim.mk

Abstract. This paper contains experimental results for one important parameter of a code for error detection. The considered code is previously defined. In the definition of the check symbols is used algebraic structure quasigroup. In this paper will be considered the case when for coding are used quasigroups of order 8 for which the code has smallest probability of undetected errors. Namely, using simulations we will obtain experimental results for the largest number of erroneous bits up to which the code is guaranteed to detect errors when such quasigroup is used for coding. Since all these quasigroups are linear, we will conclude whether the free term has an influence on the value of this parameter.

Keywords: Error-detecting code · Quasigroup · Error-detecting capability

1 Introduction

We live in an era in which there are a huge number of communication tools. Communication is easier than ever before, and every second a huge amount of data is transferred all over the world. For various reasons, the data being transmitted may be corrupted. Therefore, the need for accurate data transmission has increased. This is especially important for sensitive data, where errors can lead to catastrophic consequences. For this reason, several error detection codes have been developed over the years. Some of the more important codes are Fletcher's checksum [1,2], Adler-32 [3], Cyclic Redundancy Check [4–7]. Cyclic Redundancy Check is a code that has survived over the years and is the most commonly used code today.

In one of our previous papers [8] we also proposed a code for error detection. Unlike the previously mentioned codes which are codes with a fixed length of checking part, the code defined in [8] is a code in which the length of the checking part depends on the length of the information part. The current article reflects on this code, i.e., we will experimentally determine the greatest integer of corrupted bits up to which the code is guaranteed to detect errors.

1.1 Previous Work

The code is defined in [8]. In that paper we obtained formula for calculating the probability of undetected errors when the code uses a quasigroup of arbitrary order for which this probability does not depend on the distribution of the symbols in the information message. By applying this formula, we obtained these probabilities when the code uses quasigroups of order 4 [8,9] and order 8 [10]. In the cited papers we obtained the quasigroups from the given orders for which the code has the least chance of errors that it won't catch and the corresponding probability functions. This probability is smaller than the corresponding probabilities for some of the Cyclic Redundancy Check codes (CRC-12, CRC-ANSI, CRC-CCITT). Besides this, the code has linear complexity, fast coding and checking. All this was the reason to obtain the other key parameter that defines the code's ability to detect the errors, i.e., the greatest integer of erroneous bits that the code is guaranteed to detect. In [11] we obtained this parameter when coding is done using a quasigroup of order 4 for which the code has the least chance of errors that it won't catch. In the current article we will experimentally obtain this parameter when coding is done using arbitrary quasigroup of order 8 for which the code achieves best performances from the aspect of the first parameter (the probability with which the code detects the errors), while in another article these results will be theoretically confirmed. There are 288 linear quasigroups with this property and the required parameter will be experimentally obtained when for coding is used each of these 288 linear quasigroups.

1.2 Mathematical Preliminaries

A quasigroup is a simple algebraic structure over the set Q with one binary operation $*$ (which means that always when $a, b \in Q$, holds that $a * b \in Q$) such that for each $a \in Q$ and each $b \in Q$ there is only one $x \in Q$ and only one $y \in Q$ such that the following two equations hold

$$x * a = b \ \& \ a * y = b \qquad (1)$$

In this paper a quasigroups of order 8 are of interest, due to which in the rest of the paper we will take that $Q = \{0, 1, 2, 3, 4, 5, 6, 7\}$.

For a linear quasigroups of order 8 holds that there are binary invertible matrices $A_{3\times 3}$ and $B_{3\times 3}$ and a binary matrix $C_{1\times 3}$, such that

$$(x * y)_2 = x_2 A + y_2 B + C \qquad (2)$$

In the above equation, x_2, y_2 and $(x * y)_2$ are the binary forms of x, y and $x * y$ as 1×3 vectors and all operations are binary.

The code is defined in the subsequent manner. Let for coding be used a quasigroup $(Q, *)$. Each information message is partitioned in parts of length n and each part $a_0 a_1 \ldots a_{n-1}$, $a_i \in Q, i = 0, 1, \ldots n-1$ is coded separately into block $a_0 a_1 \ldots a_{n-1} d_0 d_1 \ldots d_{n-1}$ where

$$d_i = a_i * a_{i+1 \ (mod \ n)} \qquad (3)$$

If the quasigroup is linear defined with the matrices A, B and C that fulfil (2), then directly from (3) and (2) follows that the binary forms of the check symbols $d_i, i = 0, 1, \ldots, n-1$ can be determined by the equation below

$$(d_i)_2 = (a_i)_2 A + (a_{i+1 \ (mod \ n)})_2 B + C \qquad (4)$$

where $(a_i)_2$ and $(a_{i+1 \ (mod \ n)})_2$ are a_i and $a_{i+1 \ (mod \ n)}$ depicted binary and all operations are binary.

The binary form of the coded block $a_0 a_1 \ldots a_{n-1} d_0 d_1 \ldots d_{n-1}$, i.e., $(a_0)_2 (a_1)_2 \ldots (a_{n-1})_2 (d_0)_2 (d_1)_2 \ldots (d_{n-1})_2$ is transmitted through the channel. Due to the noises in the channel some bits may be corrupted. Hence, the output block may differ from the input block. Let the output block be $a'_0 a'_1 \ldots a'_{n-1} d'_0 d'_1 \ldots d'_{n-1}$ or in the binary form $(a_0)'_2 (a_1)'_2 \ldots (a_{n-1})'_2 (d_0)'_2 (d_1)'_2 \ldots (d_{n-1})'_2$. To verify if there are transmission errors, the receiver examines whether the equation

$$d'_i = a'_i * a'_{i+1 \ (mod \ n)} \qquad (5)$$

is satisfied for all $i \in \{0, 1, \ldots, n-1\}$.

If linear quasigroup is used for coding then instead of converting the output blocks into a string over the alphabet Q and checking for errors using (5), it can check directly in binary form by checking whether the following equation is satisfied for all values of i:

$$(d_i)'_2 = (a_i)'_2 A + (a_{i+1 \ (mod \ n)})'_2 B + C \qquad (6)$$

The blocks in which is detected error are retransmitted.

The concept of the above code will be illustrated in the next example.

Example 1. Let the code uses the linear quasigroup of order 8 defined with the next three matrices:

$$A = \begin{bmatrix} 1 & 1 & 0 \\ 0 & 1 & 1 \\ 1 & 1 & 1 \end{bmatrix}, B = \begin{bmatrix} 0 & 1 & 1 \\ 1 & 1 & 1 \\ 1 & 1 & 0 \end{bmatrix} \text{ and } C = [0\ 0\ 0] \qquad (7)$$

Suppose that the information message is 673654013264 and the information blocks are $n = 4$ quasigroup symbols long. This means that the information message is divided in blocks of length 4 symbols, i.e., the first is $a_0 a_1 a_2 a_3 = 6736$, the second $a_4 a_5 a_6 a_7 = 5401$ and the third is $a_8 a_9 a_{10} a_{11} = 3264$. Now, each block is coded. First, we will convert them in binary form, i.e., $(a_0)_2 (a_1)_2 (a_2)_2 (a_3)_2 = 110111011110$, $(a_4)_2 (a_5)_2 (a_6)_2 (a_7)_2 = 101100000001$ and $(a_8)_2 (a_9)_2 (a_{10})_2 (a_{11})_2 = 011010110100$.

Computing the redundant symbols for the first block $(a_0)_2(a_1)_2(a_2)_2(a_3)_2 = $ 110111011110:

$$(d_0)_2 = (a_0)_2 A + (a_1)_2 B = [1\ 1\ 0] \begin{bmatrix} 1\ 1\ 0 \\ 0\ 1\ 1 \\ 1\ 1\ 1 \end{bmatrix} + [1\ 1\ 1] \begin{bmatrix} 0\ 1\ 1 \\ 1\ 1\ 1 \\ 1\ 1\ 0 \end{bmatrix} = [1\ 1\ 1]$$

$$(d_1)_2 = (a_1)_2 A + (a_2)_2 B = [1\ 1\ 1] \begin{bmatrix} 1\ 1\ 0 \\ 0\ 1\ 1 \\ 1\ 1\ 1 \end{bmatrix} + [0\ 1\ 1] \begin{bmatrix} 0\ 1\ 1 \\ 1\ 1\ 1 \\ 1\ 1\ 0 \end{bmatrix} = [0\ 1\ 1]$$

$$(d_2)_2 = (a_2)_2 A + (a_3)_2 B = [0\ 1\ 1] \begin{bmatrix} 1\ 1\ 0 \\ 0\ 1\ 1 \\ 1\ 1\ 1 \end{bmatrix} + [1\ 1\ 0] \begin{bmatrix} 0\ 1\ 1 \\ 1\ 1\ 1 \\ 1\ 1\ 0 \end{bmatrix} = [0\ 0\ 0]$$

$$(d_3)_2 = (a_3)_2 A + (a_0)_2 B = [1\ 1\ 0] \begin{bmatrix} 1\ 1\ 0 \\ 0\ 1\ 1 \\ 1\ 1\ 1 \end{bmatrix} + [1\ 1\ 0] \begin{bmatrix} 0\ 1\ 1 \\ 1\ 1\ 1 \\ 1\ 1\ 0 \end{bmatrix} = [0\ 0\ 1]$$

The coded block is $(a_0)_2(a_1)_2(a_2)_2(a_3)_2(d_0)_2(d_1)_2(d_2)_2(d_3)_2 = $ 110111011110111011000001.

Next, the second block $(a_4)_2(a_5)_2(a_6)_2(a_7)_2 = $ 101100000001 is coded:

$$(d_4)_2 = (a_4)_2 A + (a_5)_2 B = [1\ 0\ 1] \begin{bmatrix} 1\ 1\ 0 \\ 0\ 1\ 1 \\ 1\ 1\ 1 \end{bmatrix} + [1\ 0\ 0] \begin{bmatrix} 0\ 1\ 1 \\ 1\ 1\ 1 \\ 1\ 1\ 0 \end{bmatrix} = [0\ 1\ 0]$$

$$(d_5)_2 = (a_5)_2 A + (a_6)_2 B = [1\ 0\ 0] \begin{bmatrix} 1\ 1\ 0 \\ 0\ 1\ 1 \\ 1\ 1\ 1 \end{bmatrix} + [0\ 0\ 0] \begin{bmatrix} 0\ 1\ 1 \\ 1\ 1\ 1 \\ 1\ 1\ 0 \end{bmatrix} = [1\ 1\ 0]$$

$$(d_6)_2 = (a_6)_2 A + (a_7)_2 B = [0\ 0\ 0] \begin{bmatrix} 1\ 1\ 0 \\ 0\ 1\ 1 \\ 1\ 1\ 1 \end{bmatrix} + [0\ 0\ 1] \begin{bmatrix} 0\ 1\ 1 \\ 1\ 1\ 1 \\ 1\ 1\ 0 \end{bmatrix} = [1\ 1\ 0]$$

$$(d_7)_2 = (a_7)_2 A + (a_4)_2 B = [0\ 0\ 1] \begin{bmatrix} 1\ 1\ 0 \\ 0\ 1\ 1 \\ 1\ 1\ 1 \end{bmatrix} + [1\ 0\ 1] \begin{bmatrix} 0\ 1\ 1 \\ 1\ 1\ 1 \\ 1\ 1\ 0 \end{bmatrix} = [0\ 1\ 0]$$

The second coded block is $(a_4)_2(a_5)_2(a_6)_2(a_7)_2(d_4)_2(d_5)_2(d_6)_2(d_7)_2 = $ 101100000001010110010.

Coding the last block $(a_8)_2(a_9)_2(a_{10})_2(a_{11})_2 = $ 011010110100:

$$(d_8)_2 = (a_8)_2 A + (a_9)_2 B = [0\ 1\ 1] \begin{bmatrix} 1\ 1\ 0 \\ 0\ 1\ 1 \\ 1\ 1\ 1 \end{bmatrix} + [0\ 1\ 0] \begin{bmatrix} 0\ 1\ 1 \\ 1\ 1\ 1 \\ 1\ 1\ 0 \end{bmatrix} = [0\ 1\ 1]$$

$$(d_9)_2 = (a_9)_2 A + (a_{10})_2 B = [0\ 1\ 0] \begin{bmatrix} 1\ 1\ 0 \\ 0\ 1\ 1 \\ 1\ 1\ 1 \end{bmatrix} + [1\ 1\ 0] \begin{bmatrix} 0\ 1\ 1 \\ 1\ 1\ 1 \\ 1\ 1\ 0 \end{bmatrix} = [1\ 1\ 1]$$

$$(d_{10})_2 = (a_{10})_2 A + (a_{11})_2 B = [1\ 1\ 0] \begin{bmatrix} 1\ 1\ 0 \\ 0\ 1\ 1 \\ 1\ 1\ 1 \end{bmatrix} + [1\ 0\ 0] \begin{bmatrix} 0\ 1\ 1 \\ 1\ 1\ 1 \\ 1\ 1\ 0 \end{bmatrix} = [1\ 1\ 0]$$

$$(d_{11})_2 = (a_{11})_2 A + (a_8)_2 B = [1\ 0\ 0] \begin{bmatrix} 1\ 1\ 0 \\ 0\ 1\ 1 \\ 1\ 1\ 1 \end{bmatrix} + [0\ 1\ 1] \begin{bmatrix} 0\ 1\ 1 \\ 1\ 1\ 1 \\ 1\ 1\ 0 \end{bmatrix} = [1\ 1\ 1]$$

The last coded block is $(a_8)_2(a_9)_2(a_{10})_2(a_{11})_2(d_8)_2(d_9)_2(d_{10})_2(d_{11})_2 = 01101$ 0110100011111110111.

By joining the three coded blocks we obtain the coded message $(a_0)_2(a_1)_2$ $(a_2)_2(a_3)_2(d_0)_2(d_1)_2(d_2)_2(d_3)_2(a_4)_2(a_5)_2(a_6)_2(a_7)_2(d_4)_2(d_5)_2(d_6)_2(d_7)_2(a_8)_2(a_9)_2$ $(a_{10})_2(a_{11})_2(d_8)_2(d_9)_2(d_{10})_2(d_{11})_2 = 110111011110111011000001101100000010$ 10110110010011010110100011111110111. This is the coded message that is sent via the channel. Let's assume that as a result of noise in the channel some of the bits are corrupted. Assume the message received by the receiver is $(a_0)'_2(a_1)'_2(a_2)'_2(a_3)'_2(d_0)'_2(d_1)'_2(d_2)'_2(d_3)'_2(a_4)'_2(a_5)'_2(a_6)'_2(a_7)'_2(d_4)'_2(d_5)'_2(d_6)'_2(d_7)'_2$ $(a_8)'_2(a_9)'_2(a_{10})'_2(a_{11})'_2(d_8)'_2(d_9)'_2(d_{10})'_2(d_{11})'_2 = 110111011100111011000001101 1$ 00001001010000010100110101101000111111110111. This means that the 11th, 33th, 40th, 41th, 43th, 44th and 45th bits are corrupted during transmission, i.e., the first and the second block are incorrect. But the receiver does not know that, it must check whether (6) is satisfied for each of the blocks, in order to decide whether to accept the block as correctly transmitted or to ask retransmission of the block.

Checking whether the first block $(a_0)'_2(a_1)'_2(a_2)'_2(a_3)'_2(d_0)'_2(d_1)'_2(d_2)'_2(d_3)'_2 = $ 11011101110011101100001 is correctly transmitted:

$$(a_0)'_2 A + (a_1)'_2 B = [1\ 1\ 0] \begin{bmatrix} 1 & 1 & 0 \\ 0 & 1 & 1 \\ 1 & 1 & 1 \end{bmatrix} + [1\ 1\ 1] \begin{bmatrix} 0 & 1 & 1 \\ 1 & 1 & 1 \\ 1 & 1 & 0 \end{bmatrix} = [1\ 1\ 1] = (d_0)'_2$$

$$(a_1)'_2 A + (a'_2)_2 B = [1\ 1\ 1] \begin{bmatrix} 1 & 1 & 0 \\ 0 & 1 & 1 \\ 1 & 1 & 1 \end{bmatrix} + [0\ 1\ 1] \begin{bmatrix} 0 & 1 & 1 \\ 1 & 1 & 1 \\ 1 & 1 & 0 \end{bmatrix} = [0\ 1\ 1] = (d_1)'_2$$

$$(a_2)'_2 A + (a_3)'_2 B = [0\ 1\ 1] \begin{bmatrix} 1 & 1 & 0 \\ 0 & 1 & 1 \\ 1 & 1 & 1 \end{bmatrix} + [1\ 0\ 0] \begin{bmatrix} 0 & 1 & 1 \\ 1 & 1 & 1 \\ 1 & 1 & 0 \end{bmatrix} = [1\ 1\ 1] \neq (d_2)'_2$$

Since (6) is not satisfied, the receiver identifies that the block was not transmitted properly. The first coded block is transmitted once again.

Next, it checks whether there are errors in the second block $(a_4)'_2(a_5)'_2$ $(a_6)'_2(a_7)'_2(d_4)'_2(d_5)'_2(d_6)'_2(d_7)'_2 = 10110000100101000001010$:

$$(a_4)'_2 A + (a_5)'_2 B = [1\ 0\ 1] \begin{bmatrix} 1 & 1 & 0 \\ 0 & 1 & 1 \\ 1 & 1 & 1 \end{bmatrix} + [1\ 0\ 0] \begin{bmatrix} 0 & 1 & 1 \\ 1 & 1 & 1 \\ 1 & 1 & 0 \end{bmatrix} = [0\ 1\ 0] = (d_4)'_2$$

$$(a_5)'_2 A + (a_6)'_2 B = [1\ 0\ 0] \begin{bmatrix} 1 & 1 & 0 \\ 0 & 1 & 1 \\ 1 & 1 & 1 \end{bmatrix} + [0\ 0\ 1] \begin{bmatrix} 0 & 1 & 1 \\ 1 & 1 & 1 \\ 1 & 1 & 0 \end{bmatrix} = [0\ 0\ 0] = (d_5)'_2$$

$$(a_6)'_2 A + (a_7)'_2 B = [0\ 0\ 1] \begin{bmatrix} 1 & 1 & 0 \\ 0 & 1 & 1 \\ 1 & 1 & 1 \end{bmatrix} + [0\ 0\ 1] \begin{bmatrix} 0 & 1 & 1 \\ 1 & 1 & 1 \\ 1 & 1 & 0 \end{bmatrix} = [0\ 0\ 1] = (d_6)'_2$$

$$(a_7)'_2 A + (a_4)'_2 B = [0\ 0\ 1] \begin{bmatrix} 1 & 1 & 0 \\ 0 & 1 & 1 \\ 1 & 1 & 1 \end{bmatrix} + [1\ 0\ 1] \begin{bmatrix} 0 & 1 & 1 \\ 1 & 1 & 1 \\ 1 & 1 & 0 \end{bmatrix} = [0\ 1\ 0] = (d_7)'_2$$

The Eq. (6) is satisfied for all i, from where the receiver concludes that the block is correctly transmitted. Since there are faulty bits in this block, the receiver did not detected the errors, i.e., it accepts block with errors as correct block. This type of situations in which the errors are not detected are rare. Namely, in order the error to not be detected a few bits must be corrupted. In this block 6 (1 information and 5 redundant) of 24 bits are corrupted, which is 25% of the bits in the block. But, in the real channels the probability of incorrect transmission of a bit is much smaller, which is why these situations are very rare (but still possible).

Remains to check the last block $(a_8)'_2(a_9)'_2(a_{10})'_2(a_{11})'_2(d_8)'_2(d_9)'_2(d_{10})'_2(d_{11})'_2 = 011010110100011111110111$:

$$(a_8)'_2 A + (a_9)'_2 B = [0\ 1\ 1]\begin{bmatrix}1\ 1\ 0\\0\ 1\ 1\\1\ 1\ 1\end{bmatrix} + [0\ 1\ 0]\begin{bmatrix}0\ 1\ 1\\1\ 1\ 1\\1\ 1\ 0\end{bmatrix} = [0\ 1\ 1] = (d_8)'_2$$

$$(a_9)'_2 A + (a_{10})'_2 B = [0\ 1\ 0]\begin{bmatrix}1\ 1\ 0\\0\ 1\ 1\\1\ 1\ 1\end{bmatrix} + [1\ 1\ 0]\begin{bmatrix}0\ 1\ 1\\1\ 1\ 1\\1\ 1\ 0\end{bmatrix} = [1\ 1\ 1] = (d_9)'_2$$

$$(a_{10})'_2 A + (a_{11})'_2 B = [1\ 1\ 0]\begin{bmatrix}1\ 1\ 0\\0\ 1\ 1\\1\ 1\ 1\end{bmatrix} + [1\ 0\ 0]\begin{bmatrix}0\ 1\ 1\\1\ 1\ 1\\1\ 1\ 0\end{bmatrix} = [1\ 1\ 0] = (d_{10})'_2$$

$$(a_{11})'_2 A + (a_8)'_2 B = [1\ 0\ 0]\begin{bmatrix}1\ 1\ 0\\0\ 1\ 1\\1\ 1\ 1\end{bmatrix} + [0\ 1\ 1]\begin{bmatrix}0\ 1\ 1\\1\ 1\ 1\\1\ 1\ 0\end{bmatrix} = [1\ 1\ 1] = (d_{11})'_2$$

In this example, the message consists of three blocks. The message was coded and after transmission through the channel two blocks were incorrectly transmitted (the first and the second), while one (the third one) was transmitted without errors. From the two of the blocks that have errors in transmission, in one of them (the first one) the code detected the error, while in the other (the second one) it did not detected the errors. The correctly transmitted block was accepted as correctly transmitted.

2 Simulations

In this section will be presented the experimental results for the largest number of erroneous bits up to which the code is guaranteed to detect errors when the code uses the quasigroups of order 8 for which the code has least chance of errors that it won't catch. All these quasigroups are linear. This parameter depends on the length of the information blocks n and the quasigroup Q that the code uses. Since each symbol from a quasigroup Q of order 8 is presented with 3 bits in its binary form and the control part is as long as the information part, when the information block is n symbols from Q long, then the coded block is $m = 6n$ bits long. Hence, the length of the codewords is a multiple of 6. To obtain the value of the requested parameter for a given quasigroup Q, in short, we do the following. For each n, we produce an information messages of symbols from the set Q with a length of order 10^6 - 10^7 symbols, which is then split into parts of length n symbols from Q. Each such block is converted in binary form and coded using (4). The coded message consist of all coded blocks (codewords). It is transmitted through the channel that we have previously simulated. Then, for each i, $1 \leq i \leq 6n$ we obtain the percentage of faulty blocks with i faulty bits in which the code did not detect the error ($pue_i(n)$). The value of the parameter that we are obtaining for the given n is the largest i for which there are no undetected faulty blocks with up to i faulty bits, i.e., it is the largest i such that $pue_j(n) = 0\%$ for all $j \leq i$.

The set of quasigroups of order 8 that we consider contains 288 linear quasigroups. They are defined with 36 pairs of matrices A and B (given in [10]). Each pair A and B, together with one of the eight possible choices for the binary matrix C defines one of these 288 quasigroups.

First, we obtained the results for those quasigroups that belong to the set of quasigroups of order 8 for which the code has best probability of detection errors for which $C = [0\ 0\ 0]$. There are 36 such quasigroups. For all of them we obtained similar results. In Table 1 are given the results for one such quasigroup as representation for the results from this group.

Table 1. The percentage of undetected faulty coded blocks of length m bits when up to 4 bits are faulty and $C = [0\ 0\ 0]$

m	1 faulty bit	2 faulty bits	3 faulty bits	4 faulty bits
12	0%	0%	1.809990%	1.1465800%
18	0%	0%	0.373695%	0.0690936%
24	0%	0%	0%	0.0371978%
30	0%	0%	0%	0.0225989%
36	0%	0%	0%	0.0031583%
42	0%	0%	0%	0.0109653%
48	0%	0%	0%	0.0050104%
54	0%	0%	0%	0.0008867%
60	0%	0%	0%	0.0009857%
66	0%	0%	0%	0.0033970%

From Table 1 we can see that when the coded block is 12 bits (which means that the information block is 2 symbols from Q), 0% of faulty blocks with 1 wrong bit are undetected and also 0% of faulty blocks with 2 wrong bits are undetected. But, the error is not detected in 1.80999% of corrupted blocks with 3 wrong bits. Hence, when the information blocks have length 2 symbols from Q, i.e., when codewords have length 12 bits, the code surely detects 2 faulty bits. The same holds when the information blocks have length 3 symbols from Q, i.e., the codewords have length 18 bits. Likewise, we arrive at the conclusion that when the length of the information blocks is greater than or equal to 4 symbols from Q, i.e., when the codewords have length greater than or equal to 24 bits, the code surely detects 3 faulty bits.

One of the thing we were interested in is whether the matrix C affect the parameter which is subject of the paper. For that reason, we run the simulation for all quasigroups from a given set for all possible values of the matrix C. These results are presented in Tables 1, 2, 3, 4, 5, 6, 7 and 8. Each of these tables represent the results for the 36 quasigroups from the set of quasigroups of order 8 for which the code achieves best probability of error detection for one given matrix C.

Since the goal of the paper is not the probability with which the code detects the errors (it is already obtained), but the number of errors that the code is guaranteed to detect, in the results in Tables 1, 2, 3, 4, 5, 6, 7 and 8 is not important the exact percentage of undetected corrupted blocks, but whether it is 0 or not. Therefore, for different values of the length of the codewords, we adapted the value of the probability of a bit being transmitted incorrectly in the channel in order to obtain huge number of incorrect blocks with up to 4 wrong bits. This is done in order to obtain a statistically accurate result.

Table 2. The percentage of undetected faulty coded blocks of length m bits when up to 4 bits are faulty and $C = [0\ 0\ 1]$

m	1 faulty bit	2 faulty bits	3 faulty bits	4 faulty bits
12	0%	0%	1.783460%	1.1055000%
18	0%	0%	0.343563%	0.0721439%
24	0%	0%	0%	0.0340578%
30	0%	0%	0%	0.0114152%
36	0%	0%	0%	0.0063381%
42	0%	0%	0%	0.0109127%
48	0%	0%	0%	0.0050127%
54	0%	0%	0%	0.0044465%
60	0%	0%	0%	0.0029494%
66	0%	0%	0%	0.0022459%

The obtained results (Tables 1, 2, 3, 4, 5, 6, 7 and 8) suggest that the number of wrong bits up to which the code is guaranteed to identify that there are mistakes is not affected by the value of the matrix C.

Even more, we can conclude that the code surely detects equal number of incorrect bits independent of which of the considered 288 quasigroups is used by the code. Specifically, the code identifies the errors always when at most 2 bits are wrong when the codewords are 12 or 18 bits long, while for longer codewords the code is guaranteed to detect the errors when there are up to 3 wrong bits.

The results in this paper clearly point out that when this code is analyzed and used, the free term C should be taken to be a zero matrix. The code's ability to identify that errors are present is independent from this parameter and by taking its value to be a zero matrix, the coding will be fastest.

Table 3. The percentage of undetected faulty coded blocks of length m bits when up to 4 bits are faulty and $C = [0\ 1\ 0]$

m	1 faulty bit	2 faulty bits	3 faulty bits	4 faulty bits
12	0%	0%	1.872320%	1.4406800%
18	0%	0%	0.355603%	0.0936250%
24	0%	0%	0%	0.0153501%
30	0%	0%	0%	0.0085053%
36	0%	0%	0%	0.0318502%
42	0%	0%	0%	0.0072674%
48	0%	0%	0%	0.0058657%
54	0%	0%	0%	0.0046872%
60	0%	0%	0%	0.0068833%
66	0%	0%	0%	0.0024360%

Table 4. The percentage of undetected faulty coded blocks of length m bits when up to 4 bits are faulty and $C = [0\ 1\ 1]$

m	1 faulty bit	2 faulty bits	3 faulty bits	4 faulty bits
12	0%	0%	1.87604%	1.2863800%
18	0%	0%	1.85939%	1.0961200%
24	0%	0%	0%	0.0527214%
30	0%	0%	0%	0.0197762%
36	0%	0%	0%	0.0094569%
42	0%	0%	0%	0.0109653%
48	0%	0%	0%	0.0058428%
54	0%	0%	0%	0.0035385%
60	0%	0%	0%	0.0029465%
66	0%	0%	0%	0.0011280%

The theoretical foundation of the findings discussed in this article will be presented in some of the next papers.

Table 5. The percentage of undetected faulty coded blocks of length m bits when up to 4 bits are faulty and $C = [1\ 0\ 0]$

m	1 faulty bit	2 faulty bits	3 faulty bits	4 faulty bits
12	0%	0%	1.722250%	1.2653400%
18	0%	0%	0.327699%	0.0559501%
24	0%	0%	0%	0.0471106%
30	0%	0%	0%	0.0255305%
36	0%	0%	0%	0.0095435%
42	0%	0%	0%	0.0036192%
48	0%	0%	0%	0.0210062%
54	0%	0%	0%	0.0044609%
60	0%	0%	0%	0.0024535%
66	0%	0%	0%	0.0028338%

Table 6. The percentage of undetected faulty coded blocks of length m bits when up to 4 bits are faulty and $C = [1\ 0\ 1]$

m	1 faulty bit	2 faulty bits	3 faulty bits	4 faulty bits
12	0%	0%	1.875630%	1.2040300%
18	0%	0%	0.375757%	0.0683060%
24	0%	0%	0%	0.0340927%
30	0%	0%	0%	0.0339693%
36	0%	0%	0%	0.0159418%
42	0%	0%	0%	0.0110080%
48	0%	0%	0%	0.0069641%
54	0%	0%	0%	0.0027616%
60	0%	0%	0%	0.0020445%
66	0%	0%	0%	0.0023415%

Table 7. The percentage of undetected faulty coded blocks of length m bits when up to 4 bits are faulty and $C = [1\ 1\ 0]$

m	1 faulty bit	2 faulty bits	3 faulty bits	4 faulty bits
12	0%	0%	1.841200%	1.2045900%
18	0%	0%	0.345655%	0.1276700%
24	0%	0%	0%	0.0497899%
30	0%	0%	0%	0.0169415%
36	0%	0%	0%	0.0063961%
42	0%	0%	0%	0.0109012%
48	0%	0%	0%	0.0042050%
54	0%	0%	0%	0.0027840%
60	0%	0%	0%	0.0030736%
66	0%	0%	0%	0.0023530%

Table 8. The percentage of undetected faulty coded blocks of length m bits when up to 4 bits are faulty and $C = [1\ 1\ 1]$

m	1 faulty bit	2 faulty bits	3 faulty bits	4 faulty bits
12	0%	0%	1.858030%	1.0522400%
18	0%	0%	0.367975%	0.0941499%
24	0%	0%	0%	0.0431141%
30	0%	0%	0%	0.0142592%
36	0%	0%	0%	0.0032160%
42	0%	0%	0%	0.0072711%
48	0%	0%	0%	0.0041923%
54	0%	0%	0%	0.0046106%
60	0%	0%	0%	0.0020503%
66	0%	0%	0%	0.0011827%

3 Conclusion

In the paper are obtained experimental results for the largest number of faulty bits that one error-detecting code always identifies when quasigroups from the set of quasigroups of order 8 for which the code achieves best results from the aspect of the probability with which it detect the errors are used for coding. Obtained results indicate that this parameter is equal, regardless which quasigroup from this set the code uses. This also means that the free term in the linear representation of the quasigroup does not play any role in the code's ability to identify that there are errors.

Also, the obtained results indicate that when the information blocks have length 2 or 3 symbols from the quasigroup, i.e., when the codewords are 12 or

18 bits long, the code always identifies up to 2 faulty bits. When the information blocks are at least 4 quasigroup symbols long, i.e., when the codewords are at least 24 bits long, the code always identifies that there are errors when at most 3 bits are faulty.

Acknowledgement. This work was partially financed by the Faculty of Computer Science and Engineering at the "Ss.Cyril and Methodius" University.

References

1. Fletcher, J.G.: An arithmetic checksum for serial transmissions. IEEE Trans. Commun. **30**(1), 247–252 (1982)
2. Zweig J., Partridge, C.: TCP Alternate Checksum Options, IETF RFC 1146 (1990)
3. Deutsch P., Gailly, J.-L.: ZLIB Compressed Data Format Specification Version 3.3, IETF RFC 1950 (1996)
4. Peterson, W.W., Brown, D.T.: Cyclic codes for error detection. In: IRE 1961, vol. 49, no. 1, pp. 228–235. IEEE (1961). https://doi.org/10.1109/JRPROC.1961.287814
5. Perez, A.: Byte-wise CRC calculations. IEEE Micro **3**(3), 40–50 (1983)
6. Ramabadran, T.V., Gaitonde, S.S.: A tutorial on CRC computations. IEEE Micro **8**, 62–75 (1988). https://doi.org/10.1109/40.7773
7. Koopman, P., Chakravarty, T.: Cyclic redundancy code (CRC) polynomial selection for embedded networks. In: Proceedings of the International Conference on Dependable Systems and Networks, pp. 145–154 (2004)
8. Ilievska, N., Bakeva, V.: A model of error-detecting codes based on quasigroups of order 4. In: 6th International Conference for Informatics and Information Technology, Bitola, pp. 7–11 (2008)
9. Bakeva, V., Ilievska, N.: A probabilistic model of error-detecting codes based on quasigroups. Quasigroups Related Syst. **17**(2), 135–148 (2009)
10. Ilievska, N., Gligoroski, D.: Quasigroup redundancy check codes for safety-critical systems. In: 11th Advanced International Conference on Telecommunications IARIA-AICT, Brussels, pp. 72–77 (2015)
11. Ilievska, N.: Number of errors that the error-detecting code surely detects. In: Trajanov, D., Bakeva, V. (eds.) ICT Innovations 2017. CCIS, vol. 778, pp. 219–228. Springer, Cham (2017). https://doi.org/10.1007/978-3-319-67597-8_21

Session 3

YOLOv8 Oriented Bounding Box (OBB) Model for Waymo Open Dataset

Atanasko Boris Mitrev[✉] and Georgina Mirceva

Faculty of Computer Science and Engineering, Ss. Cyril and Methodius University in Skopje, Skopje, Republic of Macedonia
atanasko.mitrev@students.finki.ukim.mk, georgina.mirceva@finki.ukim.mk

Abstract. This paper explores the new YOLOv8 oriented bounding boxes object detection capabilities in Bird's Eye View (BEV) images using Waymo Open Dataset. The Waymo Open Dataset provides high-quality real-world driving data, making it ideal dataset for training object detection models. The motivation behind this research arise from the possibility to explore oriented bounding boxes object detection capabilities of the YOLOv8 model for accurate and robust object detection in autonomous driving applications. Traditional bounding box methods often struggle with objects that have complex orientations, such as vehicles and pedestrians, particularly in dynamic and cluttered environments. By leveraging BEV images, which offer a top-down view of the scene, we analyse bounding boxes object detection for objects with complex orientations, such as vehicles and pedestrians. We trained a model for generating oriented bounding boxes in the BEV domain and demonstrated its effectiveness in improving object detection performance. The experimental results from this research gave very good results in terms of detection accuracy and robustness, for objects with non-axis-aligned orientations. This research contributes with its exploration of practical use of object detection techniques using Waymo Open Dataset, used for autonomous driving applications research.

Keywords: YOLOv8 · Waymo Open Dataset · object detection · Bird's Eye View (BEV) · oriented bounding boxes · autonomous driving · computer vision · deep learning · convolutional neural networks (CNNs) · perception systems

1 Introduction

Object detection is a fundamental task in computer vision with applications in various domains such as autonomous driving, robotics and surveillance. One important aspect of object detection is accurately localizing objects within an image, which is typically done using bounding boxes. While traditional bounding boxes are axis-aligned rectangles, oriented bounding boxes (OBBs) provide a more accurate representation for objects that are not aligned with the image axes.

In recent years, deep learning approaches have shown remarkable performance in object detection tasks. However, training models for OBB detection requires specialized datasets and techniques due to the complexity of OBBs.

YOLO [1] family of real-time object detection deep learning models in general, and YOLOv8 version, provide high speed and accuracy object detection, which make them great tools for vast variety of applications.

The Waymo Open Dataset (WOD), is a large and one of the most diverse autonomous driving datasets, which provides high-quality annotations for OBBs in various scenes, making it a valuable resource for training OBB detection models [2].

In this paper, we present a comprehensive study on training a deep learning model for Oriented Bounding Boxes (OBB) detection in Bird's Eye View (BEV) image from LiDAR Point Cloud, using the Ultralytics YOLOv8 model and the Waymo Open Dataset [2,3].

This research contributes by exploring YOLOv8 Oriented Bounding Box(OBB) capability, practical conversion (label coordinate transformation) and preparation of Waymo Open Dataset for training YOLOv8 OBB model for object detection, exploring practical use of object detection techniques using Waymo Open Dataset for object detection in Birds Eye View image created from LiDAR Point Cloud.

The structure of the paper is organized as follows: Sect. 2 discuss related work, Sect. 3 provides an overview of YOLO and YOLOv8 models. Section 4 describes the Waymo Open Dataset. Section 5 discusses Range Images storage format in Waymo Open Dataset. Section 6 describes BEV (Bird's Eye View) images. Section 7 discusses preparation and training of the model. Finally, Sect. 8 presents the conclusion and outlines potential future research directions.

2 Related Work

Detecting objects in Point Cloud is important and challenging. Other researchers explore this task too.

Simon et al. [4] in their research, extended the YOLOv2 model, originally a fast 2D object detector for RGB images, to perform 3D object detection using a specialized regression strategy aimed at estimating multi-class 3D bounding boxes in Cartesian coordinates. They introduced an Euler-Region Proposal Network (E-RPN) that incorporates both real and imaginary components into the regression network, allowing for a closed representation in complex space. This approach effectively avoids singularities associated with single-angle estimations of object poses. The model was evaluated on the KITTI dataset, demonstrating its capabilities for robust 3D object detection.

Ali et al. [5] investigate the use of YOLOv2 model for detecting and classifying 3D oriented bounding boxes (OBB) from LiDAR point cloud (PCL) data.

They extend the YOLOv2 loss function to include additional parameters, specifically the yaw angle, the 3D box center in Cartesian coordinates, and the height of the bounding box, treating it as a direct regression problem. Their approach was evaluated using the KITTI dataset, demonstrating the potential of this extended YOLOv2 model for effective 3D object detection and classification.

Tony Davis et al. [6] extend work on Real-time 3D Object Detection on Point Clouds by improving and optimizing original Complex-YOLO by using YOLO v4 and a comparison of different rotated box IoU (Intersection Over Union) losses for faster and accurate object detection.

3 Ultralytics YOLOv8

You Only Look Once (YOLO) Fig. 1 (CNN architecture) is a state-of-the-art, real-time algorithm for object detection for the first time introduced in 2015 by

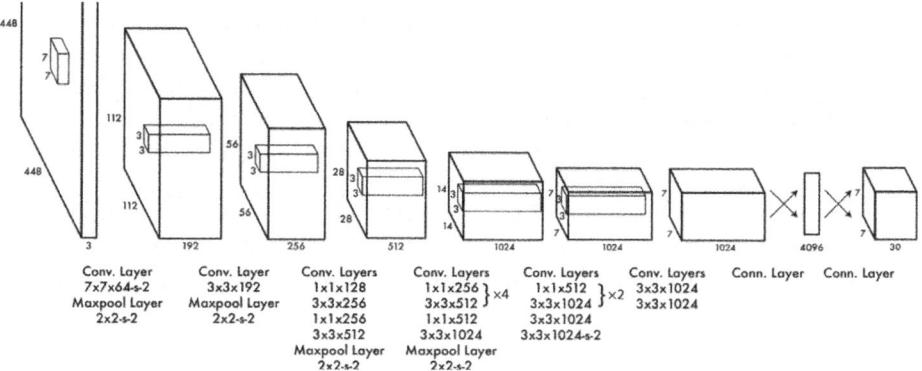

Fig. 1. YOLO architecture [1]

Redmon et al. [1] in their influential work "You Only Look Once: Unified, Real-Time Object Detection". They proposed a novel approach to object detection by framing it as a regression problem rather than a traditional classification task. The authors designed a single convolutional neural network (CNN) that simultaneously separates bounding boxes spatially and assigns class probabilities to each detected object, achieving real-time detection capabilities [1].

One of the models in the YOLO family is YOLOv8 [3], a real-time object detector that achieves state-of-the-art performance in both accuracy and speed. Building on the advancements of earlier YOLO versions, YOLOv8 incorporates novel features and optimizations, making it highly effective for a wide range of object detection tasks across diverse applications [3].

In Fig. 2 YOLOv8 OBB architecture is presented. Oriented object detection extends traditional object detection by incorporating an additional angular

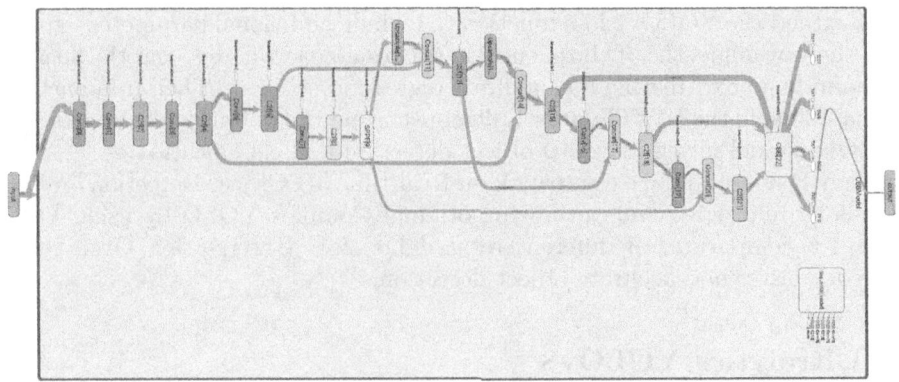

Fig. 2. YOLOv8 OBB architecture

parameter, enabling more precise localization of objects in an image. The output of an oriented object detector includes rotated bounding boxes that closely fit the objects, along with corresponding class labels and confidence scores for each detection. This approach improves the accuracy of detecting objects with arbitrary orientations, particularly in complex environments.

Ultralytics YOLOv8 introduce OBB functionality with it's OBB models pre-trained on DOTAv1 dataset [3].

In our research we train the YOLOv8 OBB model from scratch using Waymo Open Dataset, to detect oriented bounding boxes on objects in BEV images created from LiDAR Point Cloud from LiDAR sensors in Waymo Open Dataset.

4 Waymo Open Dataset

The Waymo Open Dataset (WOD) is a comprehensive self-driving dataset comprising two distinct components: the Perception dataset and the Motion dataset. The Perception dataset contains high-resolution sensor data and annotations for 2,030 driving scenes, while the Motion dataset provides object trajectories along with corresponding 3D maps for 103,354 scenes. This combination offers a rich source of data for developing and evaluating perception and prediction models in autonomous driving applications. [2,7,8].

WOD data are collected from sensors positioned on the vehicle as shown on Fig. 3.

LiDAR data are collected from five LiDAR sensors mounted on the top, front, side right, side left, and rear of the vehicle. The LiDAR mounted on the top of the vehicle covers a vertical field of view (VFOV) from $-17.6°$ to $2.4°$, the LiDAR range is 75 m, and horizontally covers $360°$. LiDARs mounted on front, side right, side left, and rear of the vehicle can cover relatively smaller area than the LiDAR mounted on the top of the vehicle. Their vertical field of view (VFOV) is from $-90°$ to $30°$ and their range is 20 m.

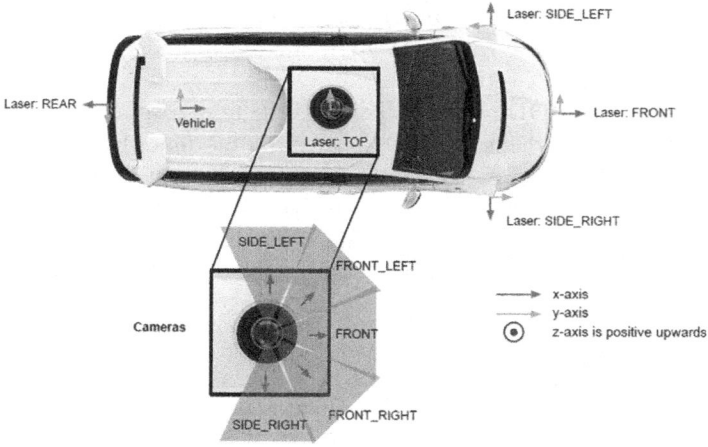

Fig. 3. Waymo sensor setup on Waymo's autonomous vehicle [2]

The Waymo Open Dataset employs three distinct coordinate systems: the Global frame, Vehicle frame, and Sensor frame. The Global frame serves as a fixed reference before the vehicle's movement. The Vehicle frame is aligned with the vehicle itself, where the x-axis points forward, the y-axis points to the left, and the z-axis points upward. The Sensor frame is specific to each sensor, defining its orientation and positioning relative to the vehicle. These coordinate systems facilitate accurate data alignment across different sensors.

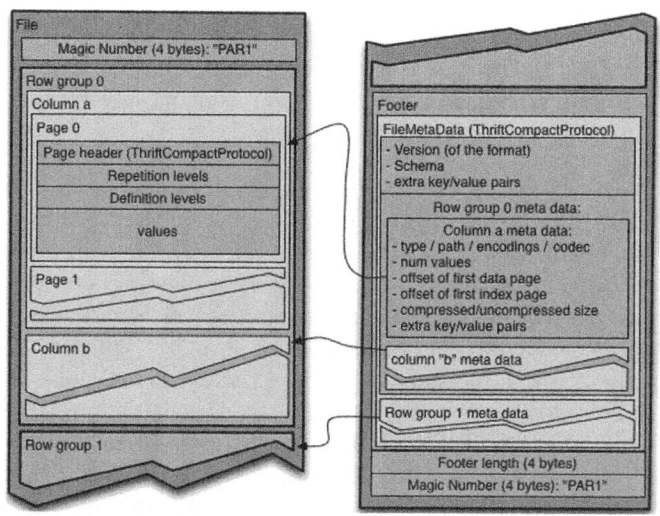

Fig. 4. Apache Parquet File Format [9]

The LiDAR scans generates point clouds that in Waymo Open Dataset are encoded in range images. For each LiDAR, two range images are stored, one for each of the two strongest returns.

In version v2 of WOD, data are saved into new format based on Apache Parquet column-oriented file format Fig. 4 [9]. This modular format separates the data into multiple tables, allowing users to selectively download the portion of the dataset needed for their specific use case, and it offers a significant advantage over the previous format by reducing the amount of data that needs to be downloaded and processed, saving time and resources.

5 Range Images

In Waymo Open Dataset, LiDAR point clouds are encoded into range images structure Fig. 5.

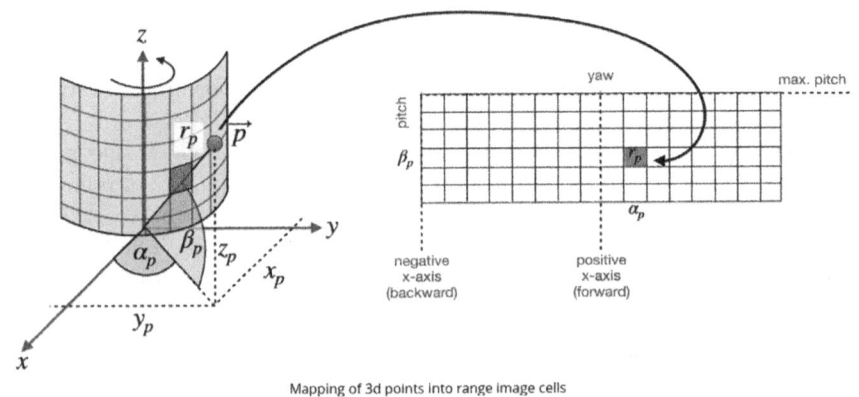

Mapping of 3d points into range image cells

Fig. 5. Mapping of 3D point into range image cell [10]

The detections from a LiDAR sensor are represented as a 3D point cloud, where every single point in the point cloud corresponds to the measurement of a single LiDAR beam. Each point is represented with coordinates (x, y, z) and additional attributes such as the intensity of the reflected laser pulse or even a secondary return caused by partial reflection at object boundaries.

The 3D points are saved in a data structure as a 360° "photo" of the scanned environment where row dimension represents the elevation angle of the LiDAR beam, and the column dimension represents the azimuth angle. With each rotation around the z-axis, the LiDAR sensor captures a series of measurements of the range and intensity values. These measurements are stored in specific cells of the range image, identified by their corresponding azimuth and elevation angles, effectively encoding the spatial information of the environment.

Each LiDAR scan point $\vec{p}$ in space is mapped into a range image cell, which is specified by the corresponding azimuth angle α_p and the inclination β_p.

6 BEV (Bird's Eye View) Images

Bird's Eye View (BEV) images, also known as top-down view or overhead view images Fig. 6, provide a perspective of the scene as if viewed from directly above. In the context of computer vision and autonomous driving, BEV images are often generated from the point cloud data captured by LiDAR sensors or from the perspective of a simulated camera positioned above the scene. BEV images are particularly useful for understanding the spatial layout of objects in the environment, such as vehicles, pedestrians, and road markings, making them valuable for tasks like object detection, tracking, and path planning.

Main objective in this paper is to detect objects in BEV images generated from LiDAR Point Cloud data into Waymo Open Dataset.

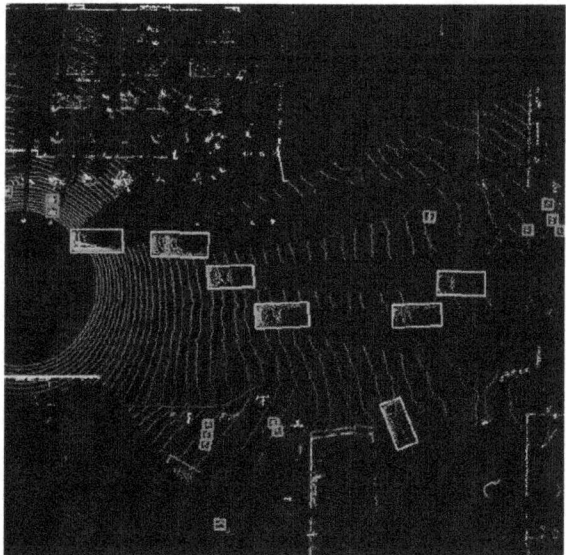

Fig. 6. Bird's Eye View (BEV) image

7 OBB Detection in BEV Images

Motivation for this research was to perform OBB detection into BEV images created from LiDAR Range Images from the Waymo Open Dataset. As already mentioned, LiDAR sensor readings in WOD are saved as Range Images.

Starting with v8.1.0 release, YOLOv8 introduced Oriented Bounding Boxes models, which is a step further in object detection because objects are detected more accurately in an image.

YOLOv8 pre-trained OBB models are pre-trained on the DOTAv1 dataset.

Since data in Waymo Open Dataset are stored in format based on Apache Parquet column-oriented file format, in order to train YOLOv8 OBB on data from OBB we perform data conversion because YOLOv8 expects images and labels into directory-based structure.

During this conversion bounding boxes coordinate transformation as in Fig. 7 is applied.

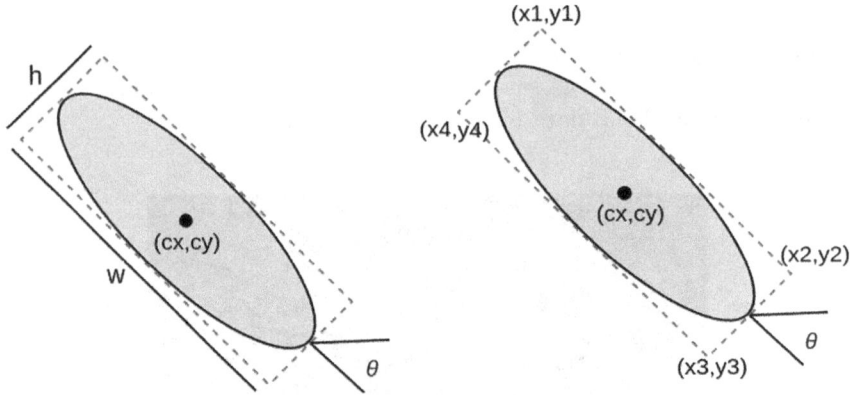

Fig. 7. Coordinate transformation

YOLOv8 expects labels on objects into format

$$class\_index, x1, y1, x2, y2, x3, y3, x4, y4$$

where $x1, y1, x2, y2, x3, y3, x4, y4$ are bounding box coordinates [3] and labels into Waymo Open Dataset are stored as 7-DOF 3D upright bounding box

$$cx, cy, cz, l, w, h, \theta$$

where cx, cy, cz represent the center coordinates, l, w, h are the length, width, height, and θ denotes the heading angle in radians of the bounding box [2], so label coordinate transformation from WOD to YOLOv8 is:

$$(cx, cy, cz, l, w, h, \theta) -> (x1, y1, x2, y2, x3, y3, x4, y4)$$

During the data conversion process from the Waymo Open Dataset, first Range Images are converted into Point Cloud and afterwards BEV images are created from the Point Cloud. BEV images are saved into the images directory that is input directory for the YOLOv8 training. Labels are converted from WOD format into YOLOv8 OBB format and saved into labels directory.

The YOLOv8 training process was performed from scratch, not from pretrained model, using forked code from Ultralytics repository for 30 epocs by using two Nvidia GPU's Quadro RTX 5000 and Quadro RTX 4000. Model is trained on 640 × 640 pixel images, using batch size 32, learning rate of 0.01, momentum = 0.9, SGD optimizer (automatically selected based on the dataset size) on 5 classes (0: none, 1: vehicle, 2: pedestrian, 3: sign, 4: cyclist) Fig. 8 is an example for OBB detection in BEV image.

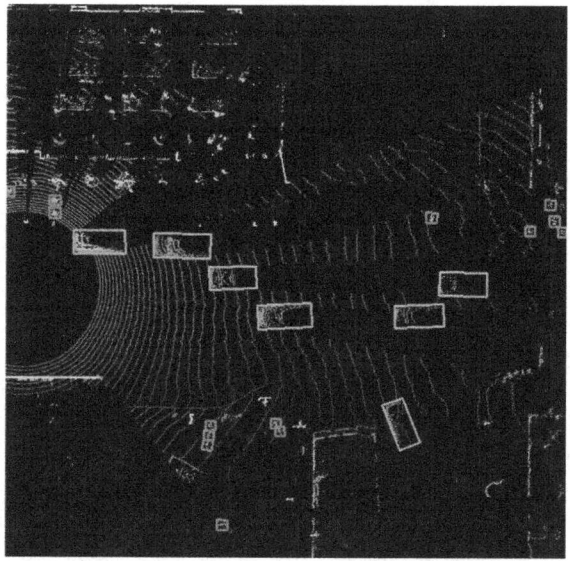

Fig. 8. Object detection in BEV image

With this setup we achieved the training result as displayed on Fig. 9. On the training result images different model performance evaluation measures are displayed. On the horizontal axes, the number of epochs is displayed, while on the vertical axes the values for the corresponding evaluation measure (ex. box loss, class loss). We also give details for the results for Precision, Recall, mAP50(B) (mean Average Precision at a 50% Intersection Over Union (IoU) threshold for object detection) and mAP50-95(B) (the average value of the mean average precision computed at varying intersection-over-union (IoU) thresholds, ranging from 0.50 to 0.95).

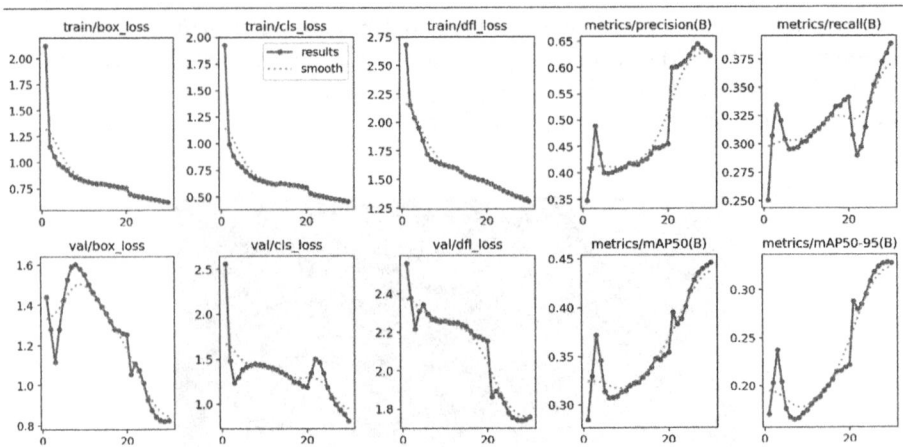

Fig. 9. YOLOv8 training results

8 Conclusion

Object detection is a fundamental task in computer vision with widespread applications across various domains such as autonomous driving, surveillance, image understanding etc. In this paper we presented our work on training YOLOv8 on the Waymo Open Dataset to detect objects in Bird's Eye View (BEV) images created from LiDAR point cloud data. By leveraging the rich source of high-quality, real-world driving data in the Waymo Open Dataset top-down view perspective of the scene provided by BEV images, we explore the use of oriented bounding boxes in object detection when objects have complex orientations.

Our research highlights the potential of using YOLOv8 oriented bounding boxes and BEV images created from LiDAR point cloud data for object detection in autonomous driving applications, contributing to the ongoing development in the self-driving vehicles domain.

In this research only particular version of YOLO v8 was explored with limited attempts to train the model with different configurations and parameters. In some next research training with different configurations and direct result comparison can be performed. Also training models with different versions of YOLO OBB implemetation and comparison of the result can be a path to proceed.

Acknowledgment. This work was partially financed by the Faculty of Computer Science and Engineering at the Ss. Cyril and Methodius University in Skopje.

References

1. Redmon, J., Divvala, S., Girshick, R., Farhadi, A.: You only look once: unified, real-time object detection. In: 2016 IEEE Conference on Computer Vision and Pattern Recognition (CVPR). IEEE (2016). https://doi.org/10.1109/cvpr.2016.91
2. Sun, P., et al.: Scalability in perception for autonomous driving: waymo open dataset (2020). https://doi.org/10.1109/CVPR42600.2020.00252
3. Jocher, G., Chaurasia, A., Qiu, J.: Ultralytics YOLO (2023). https://github.com/ultralytics/ultralytics
4. Simon, M., Milz, S., Amende, K., Gross, H.-M.: Complex-YOLO: an Euler-region-proposal for real-time 3D object detection on point clouds. In: Leal-Taixé, L., Roth, S. (eds.) ECCV 2018. LNCS, vol. 11129, pp. 197–209. Springer, Cham (2019). https://doi.org/10.1007/978-3-030-11009-3_11
5. Ali, W., Abdelkarim, S., Zidan, M., Zahran, M., Sallab, A.E.: YOLO3D: end-to-end real-time 3D oriented object bounding box detection from LiDAR point cloud. In: Leal-Taixé, L., Roth, S. (eds.) ECCV 2018. LNCS, vol. 11131, pp. 716–728. Springer, Cham (2019). https://doi.org/10.1007/978-3-030-11015-4_54
6. Davis, T., Nandana, K.: Real-time 3d object detection on lidar point cloud using complex-yolo v4. Int. Res. J. Eng. Technol. **09**(11), 716–721 (2022)
7. Chen, K., et al.: Womd-lidar: raw sensor dataset benchmark for motion forecasting. In: 2024 IEEE International Conference on Robotics and Automation (ICRA), pp. 4766–4773. IEEE (2024). https://doi.org/10.1109/icra57147.2024.10610651
8. Ettinger, S., et al.: Large scale interactive motion forecasting for autonomous driving: the waymo open motion dataset. In: 2021 IEEE/CVF International Conference on Computer Vision (ICCV). IEEE (2021). https://doi.org/10.1109/iccv48922.2021.00957
9. (ASF), A.S.F.: Parquet (2024). Accessed 14 Mar 2024. https://parquet.apache.org/docs/overview/
10. Inc., U.: Visualizing Range Images (2022). Accessed 14 Mar 2024. https://video.udacity-data.com/topher/2020/December/5fd0f5e3_c1-5-img2/c1-5-img2.png

Deep Multimodal Fusion for Semantic Segmentation of Remote Sensing Earth Observation Data

Ivica Dimitrovski[✉], Vlatko Spasev, and Ivan Kitanovski

Faculty of Computer Science and Engineering, University Ss Cyril and Methodius,
Rudzer Boshkovikj 16, P.O. 393, Skopje 1000, North Macedonia
{ivica.dimitrovski,vlatko.spasev,ivan.kitanovski}@finki.ukim.mk

Abstract. Accurate semantic segmentation of remote sensing imagery is critical for various Earth observation applications, such as land cover mapping, urban planning, and environmental monitoring. However, individual data sources often present limitations for this task. Very High Resolution (VHR) aerial imagery provides rich spatial details but cannot capture temporal information about land cover changes. Conversely, Satellite Image Time Series (SITS) capture temporal dynamics, such as seasonal variations in vegetation, but with limited spatial resolution, making it difficult to distinguish fine-scale objects. This paper proposes a late fusion deep learning model (LF-DLM) for semantic segmentation that leverages the complementary strengths of both VHR aerial imagery and SITS. The proposed model consists of two independent deep learning branches. One branch integrates detailed textures from aerial imagery captured by UNetFormer with a Multi-Axis Vision Transformer (MaxViT) backbone. The other branch captures complex spatio-temporal dynamics from the Sentinel-2 satellite image time series using a U-Net with Temporal Attention Encoder (U-TAE). This approach leads to state-of-the-art results on the FLAIR dataset, a large-scale benchmark for land cover segmentation using multi-source optical imagery. The findings highlight the importance of multi-modality fusion in improving the accuracy and robustness of semantic segmentation in remote sensing applications.

Keywords: Earth observation · semantic segmentation · remote sensing · multi-modality fusion · deep learning

1 Introduction

Remote sensing data is captured from a distance by sensors or instruments mounted on various platforms such as satellites, aircraft, drones, and other vehicles. This data collects information about the Earth's surface, atmosphere, and other objects or phenomena without requiring direct physical contact [1]. There are two main techniques of remote sensing data acquisition: aerial and satellite. Satellite data is collected by satellites orbiting the Earth, capturing information over large areas at regular intervals. This provides a broad view of the entire planet. In contrast, aerial data is captured from airplanes or drones flying closer to the ground. This data covers smaller areas but with much finer detail, making it ideal for studying specific locations. Sensors are essential to remote sensing systems, as they collect data used to create images and other forms of information. Various types of sensors employed in remote sensing include optical sensors, radar sensors, lidar sensors, and electromagnetic sensors [1].

Remote sensing data has four key properties: spectral, spatial, radiometric, and temporal resolution [2]. Spectral resolution refers to the range of wavelengths a satellite sensor can detect. The more wavelengths a sensor can capture, the richer the information content of the imagery and the greater the detail it reveals about land use and cover. These captured wavelengths span a vast spectrum, including ultraviolet, visible light, near-infrared, infrared, and microwave. Some sensors capture just a few broad bands (multi-spectral), like Sentinel-2 with its 12 bands. Others, like Hyperion, are hyper-spectral, gathering thousands of narrow bands for a highly detailed spectral view [2]. Spatial resolution refers to the size of each pixel in the image. Higher spatial resolution means smaller pixels, capturing finer details on the ground. Radiometric resolution describes how well the sensor can detect variations in radiated energy from the earth's surface. Higher resolution allows for better detection of subtle changes. Landsat 7 captures 8-bit images, distinguishing 256 distinct gray values of reflected energy, whereas Sentinel-2 features a 12-bit radiometric resolution, allowing it to discern 4095 gray values. Temporal resolution refers to how often a specific location is imaged. For example, polar-orbiting satellites exhibit varying temporal resolutions, ranging from 1 to 16 days (e.g., ten days for Sentinel-2). This is important for monitoring changes over time.

Machine learning is revolutionizing the way we analyze and understand remote sensing data [3]. A particularly exciting area of research is the semantic segmentation of remote sensing data. The goal is to partition the image into meaningful regions, enabling detailed analysis and understanding of the Earth's surface. Accurate semantic segmentation of remote sensing imagery is essential for a wide range of Earth observation (EO) applications, including land cover mapping, urban planning, and environmental monitoring [2]. The emergence of deep learning, particularly Convolutional Neural Networks (CNNs) and Fully Convolutional Networks (FCNs), ignited a revolution in semantic segmentation [4]. These models automated the learning of complex, hierarchical representations from data, paving the way for significant advancements. FCNs, often paired with encoder-decoder architectures, became the dominant approach. Early methods relied on successive convolutions and spatial pooling to generate dense predictions. Subsequent innovations like U-Net and SegNet introduced upsampling techniques to combine high-level features with lower-level ones during decoding [4]. This fusion aimed to capture both global context and precise object boundaries. To address the limited receptive field of standard convolutions in earlier layers of deep learning models, techniques like dilated (or atrous) convolutions were introduced by DeepLab [5]. These convolutions allow capturing a larger context while maintaining the resolution of the feature maps. Subsequent advancements incorporated spatial pyramid pooling (SPP) to capture multi-scale contextual information in higher layers, as seen in models like PSPNet [4] and UperNet [6]. DeepLabV3+ built upon these advancements by combining atrous spatial pyramid pooling with a straightforward and efficient encoder-decoder architecture [7]. However, recent developments like PSANet [8] and DRANet [9] have moved beyond traditional pooling, instead using attention mechanisms on top of encoder feature maps to capture long-range dependencies more effectively. Most recently, the adoption of transformer architectures, which utilize self-attention mechanisms and capture long-range dependencies, has marked additional advancement in semantic segmentation. Transformer encoder-decoder architectures like Segmenter, SegFormer, and MaskFormer harness transformers to enhance performance [10].

The abundance of diverse remote sensing modalities, like LiDARs, RGB-D cameras, and thermal cameras, has fostered the development of deep multimodal fusion techniques. These complementary sensors offer a richer picture of the scene, especially in complex environments. Deep learning excels at leveraging this data to reduce uncertainties and create a more comprehensive understanding. The core objective of deep multimodal fusion in segmentation is to learn an optimal joint representation by combining the strengths of individual modalities [11]. This joint representation captures the rich and complementary features of the same scene, leading to more accurate segmentation results. For example, Very High Resolution (VHR) aerial imagery excels at providing rich spatial details, making it ideal for identifying fine-scale features such as individual buildings, roads, and small vegetation patches. This high level of detail is crucial for tasks that require precise mapping and analysis of specific locations. However, VHR aerial imagery typically lacks temporal information, which is essential for capturing changes over time to monitor dynamic processes such as seasonal variations in vegetation, urban growth, or the progression of environmental degradation. On the other hand, Satellite Image Time Series (SITS) data offers valuable temporal insights by capturing images of the same area at regular intervals. This capability is particularly useful for observing and analyzing temporal dynamics, such as the phenological cycles of crops, changes in land cover due to deforestation or reforestation, and the impact of natural disasters over time. However, SITS data generally has lower spatial resolution compared to VHR aerial imagery, which can make it difficult to distinguish fine-scale objects and detailed features on the ground.

Deep multimodal fusion methods can be broadly categorized based on the stage at which information from different modalities is combined [11]. Early fusion occurs at the raw data level (e.g., concatenating RGB and LiDAR data) or feature level (combining extracted features from each modality). This approach allows the model to learn a joint representation from the very beginning. Late fusion strategy involves processing each modality separately through individual deep learning branches. Then, the resulting feature maps are combined at a later stage (e.g., before the final prediction layer) using operations like concatenation, addition, or weighted voting. This approach offers greater flexibility in designing individual models for specific modalities. Hybrid fusion combines elements of both early and late fusion. It might involve initial feature-level fusion followed by late fusion of higher-level features. This allows for a more adaptive learning process based on the specific data and task. By effectively leveraging the complementary information from multiple modalities, deep multimodal fusion techniques are pushing the boundaries of semantic segmentation accuracy and robustness, particularly in complex remote sensing scenarios.

This paper tackles the challenge of accurate semantic segmentation in remote sensing by proposing a late fusion deep learning model (LF-DLM) that leverages the complementary strengths of VHR aerial imagery and SITS data. This approach aims to overcome the limitations inherent in single-source data, ultimately leading to more robust and informative land cover segmentation. The proposed LF-DLM architecture employs a dual-branch strategy, capitalizing on the specific advantages of each data source. Our research can be summarized by the following primary contributions:

- Introduction of a late fusion deep learning model that leverages the complementary strengths of VHR aerial imagery and SITS data, tailored to enhance semantic segmentation of remote sensing imagery.
- Through comprehensive experimental evaluation, we demonstrate that the LF-DLM model effectively combines spatial and temporal information, leading to improved segmentation accuracy across various land cover types while maintaining efficient inference times.
- Our LF-DLM model achieves state-of-the-art results on the FLAIR dataset, surpassing previous benchmarks, thus establishing a new standard for semantic segmentation in multi-source optical imagery.

The subsequent sections of this paper are structured as follows: Sect. 2 provides an overview of the dataset utilized in the research. Section 3 details the key features of the proposed late fusion deep learning model. Section 4 comprehensively outlines the experimental design and setup, including data preprocessing, training protocols, model parameters, and evaluation metrics. Section 5 presents the experimental results alongside relevant discussions. Finally, Sect. 6 concludes the paper, summarizing the findings and contributions.

2 Dataset

The FLAIR dataset[1] includes diverse sources of acquisition, each with unique characteristics and varying spatial, spectral, and temporal resolutions. This dataset provides detailed VHR aerial images, elevation models, and satellite image time series [12]. Each aerial image measures 512×512 pixels, with a

---
[1] https://github.com/IGNF/FLAIR-2.

spatial resolution of 20 cm per pixel, and includes four spectral bands: red, blue, green, and near-infrared. The dataset comprises 77762 patches. To ensure high-quality images, the aerial data is captured only during favorable weather conditions, specifically between April and November from 2018 to 2021. Each aerial image includes an elevation value. This value is derived from combining a digital elevation model and a digital surface model, obtained through photogrammetry on the aerial images, ensuring temporal consistency.

Each aerial image patch in FLAIR is accompanied by a corresponding time series of satellite images from the Sentinel-2 constellation [13]. These satellite images offer a broader view with a spatial resolution of 10 m per pixel and come in a size of 40×40 pixels, centered on the corresponding aerial image, and only 10×10 center pixels correspond to the aerial image patch. Each pixel provides information across 10 spectral bands, capturing data from the visible to the medium infrared spectrum. The time series for each patch spans the entire year during which the aerial image was acquired. The number of images within a series can vary between 20 and 110, depending on satellite availability and orbital characteristics. The dataset includes acquisitions with cloud cover and provides cloud and snow probability masks, obtained with Sen2cor [14], along with information about the satellite and its orbit. Example patches from the FLAIR dataset are given in Fig. 1.

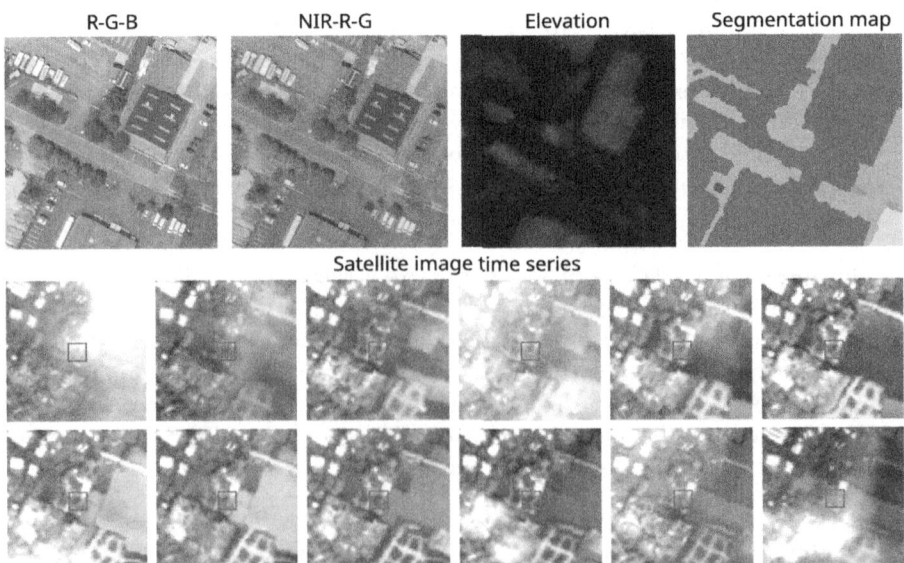

Fig. 1. Example patches from the FLAIR dataset. Each patch contains an aerial image with red, green, blue (RGB), and near-infrared (NIR) values; a pixel-precise digital surface model providing an elevation for each pixel; segmentation map with labels for each pixel; and an optical time series from several months, centered on the aerial image. The red frame marks the area that corresponds to the aerial image. (Color figure online)

The VHR images are annotated with segmentation masks containing 18 different labels/classes, along with an 'other' class for unknown land cover. Due to significant under-representation (less than 1% of the complete dataset), five of these classes are combined into the 'other' class. This results in a nomenclature of 12 classes plus the 'other' class. The classes are: 'building', 'pervious surface', 'impervious surface', 'bare soil', 'water', 'coniferous', 'deciduous', 'brushwood', 'vineyard', 'herbaceous vegetation', 'agricultural land', and 'plowed land'. Annotations are not provided for the satellite images. Instead, these images are intended to support the aerial images by providing spatial context. The distribution of pixels within the labels across the train, validation, and test sets of the FLAIR dataset is shown in Fig. 2.

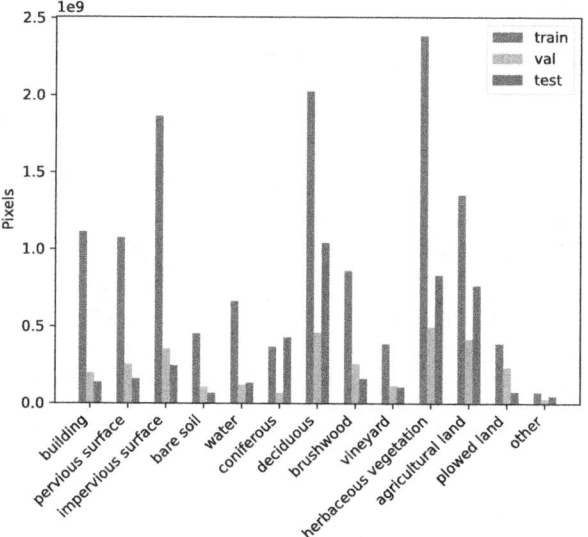

Fig. 2. The distribution of pixels within the labels across the train, validation, and test sets of the FLAIR dataset.

The dataset comprises 50 spatial domains, each representing various landscapes and climates of metropolitan France. The training set includes 32 spatial domains, the validation set contains 8, and the remaining 10 domains are allocated to the test set. The dataset is part of the FLAIR #2 challenge where a key requirement is to leverage both aerial and satellite imagery to achieve optimal semantic segmentation results. The FLAIR #2 challenge introduces a second requirement: computational efficiency to ensure a balance between accuracy and practicality, considering the vast amount of data involved. The proposed approach's inference time needs to be within 2.5 times the execution speed of the baseline model offered by the dataset creators [12]. Within this paper, we have carefully designed our solution to effectively utilize both data sources while staying within the allowed inference time for the FLAIR #2 challenge.

3 Model Architecture

This paper introduces a late fusion deep learning model (LF-DLM) for semantic segmentation, designed to exploit the complementary strengths of both Very High Resolution (VHR) aerial imagery and Satellite Image Time Series (SITS). The proposed model features two independent deep learning branches. The first branch integrates detailed textures from aerial imagery using a UNetFormer [15] with a Multi-Axis Vision Transformer (MaxViT) encoder [16], effectively capturing high-resolution spatial details. The second branch focuses on capturing complex spatio-temporal dynamics from the Sentinel-2 satellite image time series by employing a U-Net with Temporal Attention Encoder (U-TAE) [17], which processes and interprets temporal information. In the late fusion deep learning model, the probability scores from each branch are combined using a weighted geometric mean to obtain the final segmentation map. This dual-branch approach enables the model to leverage both spatial and temporal data for enhanced semantic segmentation performance.

The first branch of the LF-DLM model builds upon the Unet-like transformer UNetFormer. Originally it consists of a convolutional neural network (CNN) based encoder and a transformer-based decoder. The transformer-based decoder is constructed using global-local Transformer blocks (GLTB) that employ an efficient global-local attention mechanism with an attentional global branch and a convolutional local branch, enabling the capture of both global and local contexts for enhanced visual perception. We propose a modification to UNetFormer by replacing its CNN encoder with MaxViT, a hybrid vision transformer architecture. MaxViT introduces a novel building block called Multi-axis Self-Attention (Max-SA). This block allows the model to attend to information along multiple axes within an image feature map, including spatial, channel-wise axes, or combination. Compared to standard full self-attention in ViTs, Max-SA captures long-range dependencies (global information) more efficiently without requiring complex computations. MaxViT utilizes a hierarchical architecture where each stage in the hierarchy consists of a MaxViT block, which combines Max-SA with a convolutional layer. This combination leverages the strengths of both approaches: Max-SA for global context and convolutions for efficient local feature extraction. The network begins by downsampling the input through Conv3x3 layers in the stem stage (S0). The body of the network contains four stages (S1-S4), with each stage having half the resolution of the previous one with a doubled number of channels (hidden dimension). The feature maps generated by each stage are fused with the corresponding feature maps generated by the GLTB of the decoder using a weighted sum operation.

The MaxViT model can be scaled up by increasing the number of blocks per stage and the channel dimension. There are several MaxViT variants including MaxViT-T, MaxViT-S, MaxViT-B, MaxViT-L, and MaxViT-XL. These variants progressively increase in complexity (number of blocks and channels) and likely performance, potentially reaching a trade-off between accuracy and efficiency [16]. In this study, we are using MaxViT-T as an encoder in the UNetFormer architecture. We are utilizing MaxViT-T, pre-trained on the ImageNet-1K dataset, to leverage its learned general visual features, which can be highly beneficial for semantic segmentation tasks.

To effectively analyze both spatial and temporal information within the Sentinel-2 satellite image time series, we leverage a U-Net with temporal attention (U-TAE) model, which serves as the second branch in the LF-DLM model. This branch extracts multi-scale spatio-temporal feature maps from SITS using a combination of spatial convolution and temporal attention. U-TAE encodes a given sequence in three key steps [17]. First, each image in the sequence is embedded simultaneously and independently by a shared multi-level spatial convolutional encoder. Next, a temporal attention encoder collapses the temporal dimension of the resulting sequence of feature maps into a single map for each level. Finally, a spatial convolutional decoder produces a single feature map with the same resolution as the input images. By combining these steps, U-TAE allows effective exploitation of the rich spatio-temporal information present in the SITS, leading to a more comprehensive understanding of the scene dynamics.

4 Experimental Design and Setup

The primary objective of our study is to develop, evaluate, and compare a late fusion deep learning model (LF-DLM) for semantic segmentation of remote sensing imagery by leveraging the complementary strengths of VHR aerial imagery and SITS. Our experimental design is structured around the main hypothesis, that the fusion of these multi-source optical images will improve the semantic segmentation performance compared to using either data source alone. To test the hypothesis, we use VHR aerial imagery processed through the UNetFormer with MaxViT-S backbone to capture detailed spatial features, and SITS processed through a U-Net with Temporal Attention Encoder (U-TAE) to capture complex spatio-temporal dynamics. Our evaluation strategy involves training and assessing each model separately to determine their performances and conducting a comparative analysis to highlight the benefits of combining these data sources with weighted late fusion as a strategy.

The experimental setup involves data pre-processing, a configuration of the models, and hyperparameter selection. While no additional pre-processing is applied to the aerial patches, we address the potential influence of clouds and snow in the Sentinel-2 time series by implementing two pre-processing strategies using the provided mask files. Cloud filtering focuses on the probability of cloud or snow occurrence in the masks. We exclude images from the training process where the number of pixels exceeding a specific probability threshold (set to 0.5 in our experiments) surpasses a designated percentage of the total image pixels. This approach mitigates the impact of cloudy or snowy data on the training process. Additionally, we apply temporal monthly averaging to address challenges posed by the large number of dates within the time series. Here, a monthly average is computed using only cloudless dates within each month. If no cloudless dates are available for a particular month, the U-TAE branch might receive less than 12 images as input.

The UNetFormer model leverages a pre-trained MaxViT-T encoder on the ImageNet-1K dataset. This encoder receives five-channel aerial patches containing red, green, blue, near-infrared, and elevation data. The resolution of the aerial patches is 512 × 512 pixels. We add two channels to the initial layers to accommodate the near-infrared and elevation pixel values, with the weights of these added channels initialized randomly. The number of learnable parameters in this UNetFormer model is approximately 31 million. We use the default U-TAE parameters [12,17], with the only modification being the widths of the encoder and decoder, which we adjusted to [64, 64, 128, 128]. This list specifies the number of channels for the successive layers of the convolutional encoder, and the same configuration applies to the decoder. The input to this model is the SITS data with dimensions $T \times 10 \times 40 \times 40$, where T represents the number of images in the time series (with a maximum of 12), 10 is the number of spectral bands, and 40×40 is the pixel resolution. The number of learnable parameters in this U-TAE model is approximately 2.9 million. To ensure spatial alignment, the U-TAE outputs are first cropped to match the size of the corresponding aerial patch. Then, they are upsampled to the same resolution (512 × 512 pixels) as the aerial mask files. We experimented with different weight combinations for the late fusion, and the best performance was achieved when the UNetFormer branch was assigned a weight of 0.7 and the U-TAE branch had a weight of 0.3 in the weighted geometric mean.

To train both the UNetFormer and U-TAE models effectively, we employed several common deep learning techniques. To prevent overfitting and optimize hyperparameters, we employed a training process with hyperparameter selection and early stopping. The training data was used to train the model. Hyperparameter selection was performed on the validation split to identify the optimal configuration for the model's hyperparameters. To prevent overfitting, we implemented early stopping using the validation loss. If the validation loss did not improve for a predefined patience period (15 epochs in this case), training was terminated. The model with the best performance on the validation set, determined by the chosen evaluation metric, was then saved as the final model. This model was subsequently evaluated on the unseen test data to obtain an unbiased assessment of its predictive performance. The maximum training duration was set to 30 epochs. To improve the robustness and generalization ability of our model, we employ data augmentation techniques during training. This process involves applying random geometric transformations to the training data. Specifically, we utilize horizontal flips, vertical flips, and random rotations at predefined angles (0, 90, 180, and 270°). We fixed the batch size at 12 for our experiments. We employed the AdamW optimizer with a learning rate of 0.0001 [18]. A polynomial decay scheduler was used to gradually decrease the learning rate throughout training. This approach, with a carefully chosen decay rate, has been shown to improve model performance [19]. The scheduler applies a polynomial function to the AdamW optimizer, starting with an initial learning rate of (1×10^{-4}) and reaching a final learning rate of (1×10^{-7}) within the specified number of decay steps. To achieve a balanced approach to semantic segmentation, we combined Cross Entropy Loss and Dice Loss. Cross Entropy Loss measures the similarity between predicted and ground truth masks at each pixel, while Dice Loss focuses on accurate boundary localization. This combination effectively addresses both object localization and overall segmentation accuracy.

All models were trained on NVIDIA A100-PCIe GPUs with 40 GB of memory running CUDA version 11.5. We configured and ran the experiments using the deep learning framework PyTorch Lightning [20]. In the experimental setup, we carefully considered the constraints and requirements of the FLAIR dataset, as it is part of the FLAIR #2 challenge. Our solution was designed to effectively utilize both data sources while adhering to the allowed inference time for the FLAIR #2 challenge. The FLAIR #2 challenge specifies a maximum inference time that cannot exceed 2.5 times the baseline method. By measuring the inference time of the provided FLAIR #2 challenge baseline code on our environment, we determined that it takes approximately 396 s to generate segmentation maps for all images in the test set. Since the challenge restricts inference time to a maximum of 2.5 times the baseline, our model's inference time must not exceed 2.5 times 396 s, which translates to approximately 990 s. We assess the model performance using label-wise intersection over union (IoU) which denotes the area of the overlap between the ground truth and predicted label divided by the total area. We also report the mean intersection over union ($mIoU$) averaged across the different labels. The evaluation metrics are computed for the first 12 classes, excluding the 'other' class.

5 Results

Table 1 summarizes the performance of each model on the FLAIR dataset. Label-wise Intersection over Union (IoU) and mean IoU (mIoU) are reported in percentage. As expected, the U-TAE model achieved the lowest mIoU (39.68%) due to the limited spatial resolution of the satellite image time series. The UNet-Former model, leveraging the high spatial detail of aerial imagery, significantly improved upon this with a mIoU of 62.81%, representing a 23.13% increase. This outcome is expected given that the satellite imagery has a spatial resolution 50 times lower than the aerial imagery (10 m versus 0.2 m). Notably, the Late Fusion Deep Learning Model (LF-DLM) achieved the best overall performance with a mIoU of 63.10%. This represents an improvement of 0.29% compared to the UNetFormer alone. These findings support our hypothesis that combining information from both aerial imagery and satellite time series data through late fusion leads to improved semantic segmentation performance.

The LF-DLM shows improvement in IoU for most labels compared to the UNetFormer model. This is evident in labels like 'pervious surface' (0.62%), 'bare soil' (1.12%), 'water' (0.32%), 'coniferous' (1.22%), 'deciduous' (0.5%), 'vineyard' (0.41%), 'agricultural land' (0.43%), and 'plowed land' (0.41%). This suggests that the late fusion strategy effectively combines the strengths of both U-TAE (capturing spatio-temporal information) and UNetFormer (capturing

Table 1. Mean intersection over union (mIoU %) and Intersection over Union (IoU %) for each label of the UNetFormer, U-TAE, and LF-DLM models over the FLAIR dataset.

Label\Model	UNetFormer	U-TAE	LF-DLM
building	85.40	35.53	85.14
pervious surface	57.69	31.36	58.31
impervious surface	74.95	38.67	74.66
bare soil	63.89	39.02	65.01
water	90.77	74.75	91.08
coniferous	65.67	54.74	66.89
deciduous	73.83	56.36	74.32
brushwood	27.68	11.35	26.77
vineyard	67.19	49.48	67.59
herbaceous vegetation	50.93	26.80	50.86
agricultural land	56.16	45.43	56.59
plowed land	39.55	12.67	39.96
mIoU	62.81	39.68	63.10

high spatial details) to improve segmentation accuracy across various land cover types. This is particularly evident for labels like 'coniferous' where SITS data, containing temporal information, might be crucial to distinguish them from 'deciduous' trees exhibiting seasonal changes in spectral properties. The improvement is also notable for the 'bare soil' label, potentially benefiting from the complementary information provided by SITS data. However, the LF-DLM shows a slight decrease in IoU for labels like 'building' (−0.26%), 'impervious surface' (−0.29%), 'brushwood' (−0.91%), and 'herbaceous vegetation' (−0.07%). This could be due to several factors like class/label imbalance or fusion complexity where the late fusion process might introduce additional complexity for these specific labels, leading to slight performance drops compared to the UNetFormer model. Potentially we can explore the possibility of employing label-specific weighting or fusion techniques during late fusion to potentially address challenges faced by specific land cover types.

Examining the confusion matrix depicted in Fig. 3 reveals that the best LF-DLM model achieves high prediction accuracy, with minimal misclassification in the majority of labels. However, it tends to confuse the labels "coniferous" and "deciduous", "brushwood" and "herbaceous vegetation", "brushwood" and "deciduous", as well as "agricultural land" and "herbaceous vegetation". This is rather expected given the semantic similarity between these labels.

Figure 4 shows several example images, ground truth masks, and predicted masks from the FLAIR dataset. Obtaining accurate segmentation maps is very challenging due to factors like complex scenes, occlusion between different land cover areas, and the semantic similarity between land cover types.

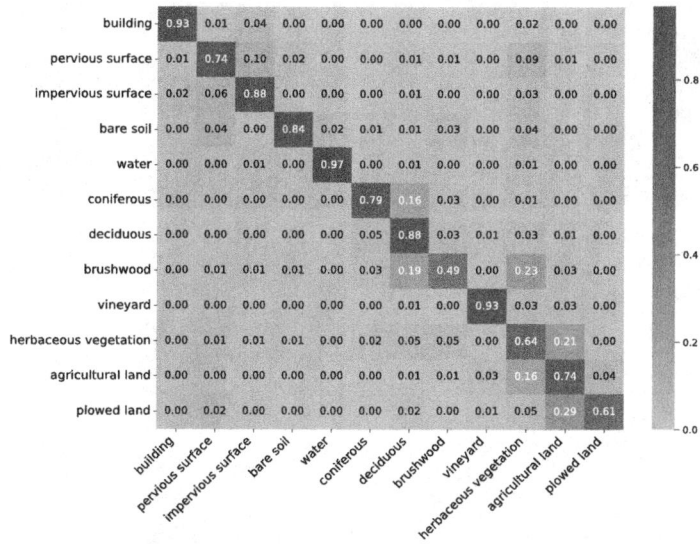

Fig. 3. Confusion matrix for LF-DLM on the FLAIR dataset.

To ensure compliance with the inference time constraints of the FLAIR #2 challenge, we measured the inference times of our proposed models. Table 2 presents these measurements along with their corresponding ratios compared to the baseline model inference time. All measurements were conducted on the same machine equipped with an NVIDIA A100-PCIe GPU with 40 GB of memory. The challenge restricts inference time to a maximum of 2.5 times that of the baseline model. As the table indicates, our models currently operate within this limit. This suggests potential for further improvement in our model's predictive performance while maintaining compliance with the challenge's constraints.

Table 2. Inference times and their corresponding ratios compared to the baseline model inference time for our proposed models.

Model	Inference time (sec.)	Relative time
U-TAE	229	0.58
UNetFormer	429	1.08
LF-DLM	594	1.5

To enhance predictive performance while adhering to the FLAIR #2 challenge's inference time constraint, we incorporated a second UNetFormer model into the late fusion scheme. This second model was trained using identical parameters, with the sole difference being a variation in the random seed. The resulting late fusion deep learning model comprises the U-TAE model, two UNetFormer

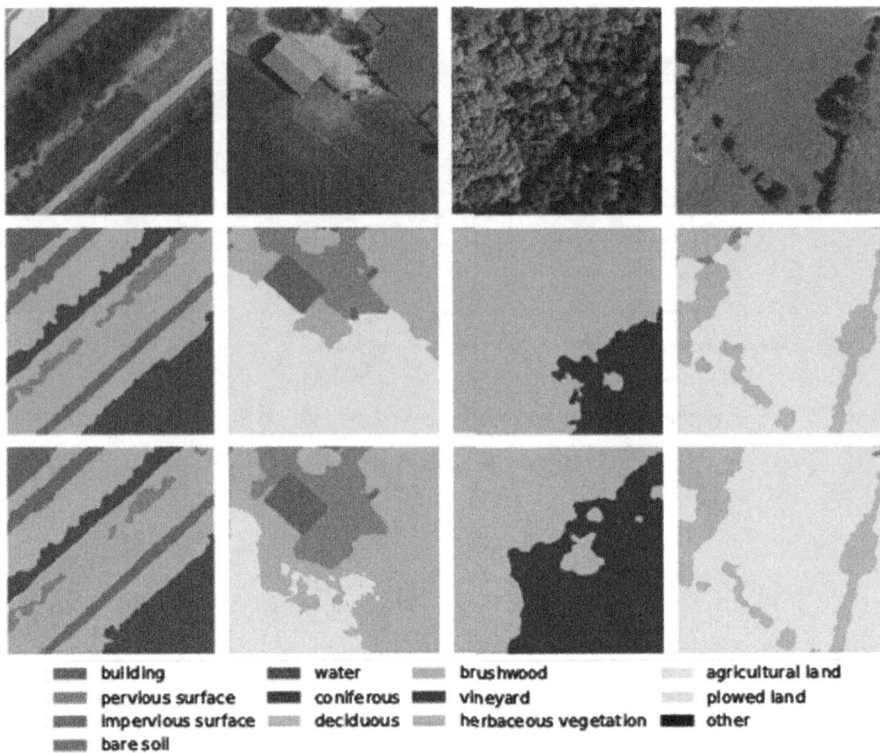

Fig. 4. Example images, ground-truth masks, and inference masks from the FLAIR dataset. The first row shows example images. The second row shows the corresponding ground-truth masks. The third row shows the prediction results of the LF-DLM.

models (with different random seeds), and the late fusion layer. This configuration achieved a mIoU value of 64.52%, an inference time of 943 s, and a relative inference time ratio of 2.38. Importantly, this remains well within the challenge's allowed constraints.

To comprehensively evaluate our best model configuration's predictive performance, we compared it with previously employed methods on the FLAIR dataset. The challenge organizers provided a U-Net baseline with a ResNet34 backbone in combination with a U-TAE model using a mid-stage fusion of features from both models [12], achieving a mIoU of 57.58%. Our best model surpasses the baseline by a significant margin of 6.94%. The current state-of-the-art on this dataset was an ensemble model consisting of four base models. The base models are similar to the baseline model provided by the challenge organizers, the only modification is the replacement of the ResNet34 backbone with MiT and ResNeXt backbones in the U-Net model [21]. Additionally, a two-stage training procedure is proposed to boost the predictive performance. This ensemble model achieved a mIoU of 64.13% and ranked first in the competition.

Notably, our proposed model with a mIoU of 64.52% outperforms this previous best result, establishing a new state-of-the-art for semantic segmentation on the FLAIR dataset.

6 Conclusion

This work investigated the effectiveness of late fusion for semantic segmentation of remote sensing imagery, leveraging the complementary strengths of Very High Resolution (VHR) aerial imagery and Satellite Image Time Series (SITS) data. We proposed a Late Fusion Deep Learning Model (LF-DLM) that integrates a UNetFormer branch for capturing spatial details from aerial imagery and a U-TAE branch for capturing spatio-temporal dynamics from SITS data. The LF-DLM achieved state-of-the-art performance on the FLAIR dataset, a large-scale benchmark for land cover segmentation using multi-source optical imagery. Compared to the UNetFormer model alone, the LF-DLM achieved an improved mIoU of 0.29%. This signifies the effectiveness of late fusion in combining information from both data sources to enhance segmentation accuracy across various land cover types.

Furthermore, our best model configuration with a mIoU of 64.52% surpasses the previous state-of-the-art on the FLAIR dataset, demonstrating its robustness and efficiency while adhering to the challenge's inference time constraints. These findings highlight the potential of late fusion deep learning models for improving the accuracy and robustness of semantic segmentation in remote sensing applications. Future work can explore label-specific fusion techniques and class imbalance mitigation strategies to address remaining challenges and further enhance performance for specific land cover types.

References

1. Toth, C., Jóźkow, G.: Remote sensing platforms and sensors: a survey. ISPRS J. Photogramm. Remote Sens. **115**, 22–36 (2016)
2. Spasev, V., Dimitrovski, I., Kitanovski, I., Chorbev, I.: Semantic segmentation of remote sensing images: definition, methods, datasets and applications. In: ICT Innovations 2023. Learning: Humans, Theory, Machines, and Data, pp. 127–140 (2024)
3. Dimitrovski, I., Kitanovski, I., Kocev, D., Simidjievski, N.: Current trends in deep learning for earth observation: an open-source benchmark arena for image classification. ISPRS J. Photogramm. Remote Sens. **197**, 18–35 (2023)
4. Hao, S., Zhou, Y., Guo, Y.: A brief survey on semantic segmentation with deep learning. Neurocomputing **406**, 302–321 (2020)
5. Chen, L.-C., Papandreou, G., Kokkinos, I., Murphy, K., Yuille, A.L.: Deeplab: semantic image segmentation with deep convolutional nets, atrous convolution, and fully connected crfs. IEEE Trans. Pattern Anal. Mach. Intell. **40**(4), 834–848 (2017)
6. Xiao, T., Liu, Y., Zhou, B., Jiang, Y., Sun, J.: Unified perceptual parsing for scene understanding. In: Proceedings of the European Conference on Computer Vision (ECCV), pp. 418–434 (2018)

7. Chen, L.-C., Zhu, Y., Papandreou, G., Schroff, F., Adam, H.: Encoder-decoder with atrous separable convolution for semantic image segmentation. In: Proceedings of the European Conference on Computer Vision (ECCV), pp. 801–818 (2018)
8. Zhao, H., et al.: Psanet: pointwise spatial attention network for scene parsing. In: Proceedings of the European Conference on Computer Vision (ECCV), pp. 267–283 (2018)
9. Fu, J., Liu, J., Jiang, J., Li, Y., Bao, Y., Lu, H.: Scene segmentation with dual relation-aware attention network. IEEE Trans. Neural Netw. Learn. Syst. **32**(6), 2547–2560 (2020)
10. Zheng, S., et al.: Rethinking semantic segmentation from a sequence-to-sequence perspective with transformers. In: Proceedings of the IEEE/CVF Conference on Computer Vision and Pattern Recognition, pp. 6881–6890 (2021)
11. Zhang, Y., Sidib'e, D., Morel, O., M'eriaudeau, F.: Deep multimodal fusion for semantic image segmentation: a survey. Image Vis. Comput. **105**, 104042 (2021)
12. Garioud, A., et al.: Flair: a country-scale land cover semantic segmentation dataset from multi-source optical imagery. Adv. Neural Inf. Process. Syst. **36** (2024)
13. Drusch, M., et al.: Sentinel-2: Esa's optical high-resolution mission for gmes operational services. Remote Sens. Environ. **120**, 25–36 (2012)
14. Main-Knorn, M., Pflug, B., Louis, J., Debaecker, V., Müller-Wilm, U., Gascon, F.: Sen2cor for sentinel-2. In: Image and Signal Processing for Remote Sensing XXIII, vol. 10427, pp. 37–48 (2017)
15. Wang, L., et al.: Unetformer: a unet-like transformer for efficient semantic segmentation of remote sensing urban scene imagery. ISPRS J. Photogramm. Remote Sens. **190**, 196–214 (2022)
16. Tu, Z., et al.: Maxvit: multi-axis vision transformer. In: Computer Vision - ECCV 2022, pp. 459–479 (2022)
17. Garnot, V.S.F., Landrieu, L.: Panoptic segmentation of satellite image time series with convolutional temporal attention networks. In: Proceedings of the IEEE/CVF International Conference on Computer Vision, pp. 4872–4881 (2021)
18. Loshchilov, I., Hutter, F.: Decoupled weight decay regularization. arXiv preprint arXiv:1711.05101 (2017)
19. Dimitrovski, I., Spasev, V., Loshkovska, S., Kitanovski, I.: U-net ensemble for enhanced semantic segmentation in remote sensing imagery. Remote Sens. **16**(12) (2024)
20. Falcon, W.: The PyTorch Lightning team: PyTorch Lightning. https://doi.org/10.5281/zenodo.3828935. https://github.com/Lightning-AI/lightning
21. Straka, J., Gruber, I.: Modernized training of u-net for aerial semantic segmentation. In: Proceedings of the IEEE/CVF Winter Conference on Applications of Computer Vision, pp. 776–784 (2024)

Transfer Learning with Yolo for Object Detection in Remote Sensing

Ema Pandilova[✉], Marko Petrov, Vlatko Spasev, Ivica Dimitrovski, and Ivan Kitanovski

Faculty of Computer Science and Engineering, University Ss Cyril and Methodius, Rudzer Boshkovikj 16, P.O. 393, Skopje 1000, North Macedonia
{ema.pandilova,marko.petrov}@students.finki.ukim.mk,
{vlatko.spasev,ivica.dimitrovski,ivan.kitanovski}@finki.ukim.mk

Abstract. Object detection in remote sensing imagery is crucial for numerous applications such as urban planning, environmental monitoring, and disaster response. This paper investigates the effectiveness of transfer learning using YOLOv8 for object detection in remote sensing, specifically focusing on the DIOR and Ships datasets. We compare models trained from scratch with those pre-trained on the COCO dataset and subsequently fine-tuned on DIOR. Our experiments demonstrate that the pre-trained YOLOv8 model significantly outperforms the scratch-trained model across all evaluated metrics, including precision, recall, and mean Average Precision (mAP). The pre-trained model demonstrates enhanced detection capabilities and reduced misclassifications. Detailed analysis using confusion matrices highlights the model's improved ability to distinguish between visually similar categories and detect smaller, less distinct objects. These findings underscore the advantages of leveraging transfer learning to enhance the performance of object detection models in remote sensing imagery. Our study concludes that fine-tuning pre-trained models on specific datasets leads to robust and accurate detection, offering a valuable approach for remote sensing applications.

Keywords: object detection · remote sensing · earth observation · deep learning

1 Introduction

Remote sensing data is collected using sensors or instruments mounted on satellites, aircraft, drones, and other vehicles, capturing information about the Earth's surface, atmosphere, and various phenomena without direct contact [1]. There are two primary types: satellite and aerial. Satellite data offers broad coverage over large areas at regular intervals, providing a global perspective, while aerial data, obtained from aircraft or drones flying closer to the ground,

---

M. Petrov, V. Spasev, I. Dimitrovski and I. Kitanovski—Contributing authors.

offers finer detail over smaller areas, making it suitable for specific location studies. The choice between aerial and satellite data depends on the required level of detail and data availability. Key properties of remote sensing data include spectral resolution, which refers to the range of detectable wavelengths; spatial resolution, indicating the size of each image pixel; radiometric resolution, measuring the sensor's ability to detect variations in radiated energy; and temporal resolution, denoting the frequency of imaging a specific location, all crucial for effective monitoring and analysis [2].

The advent of advanced remote sensing technologies has transformed the way we capture and interpret data about the Earth's surface [3]. Among the various applications, object detection in remote sensing imagery has emerged as a critical and rapidly evolving field. Object detection involves identifying and localizing specific objects within these images, ranging from natural features like rivers and forests to man-made structures such as buildings and roads. This capability is vital for urban planning, environmental monitoring, disaster response, and agricultural management [4]. Despite the advancements, remote sensing imagery presents unique challenges due to the complexity and dynamic nature of the Earth's surface, compounded by varying image qualities such as spatial resolution, spectral ranges, and capture angles.

Deep learning has significantly enhanced object detection in remote sensing images, with Convolutional Neural Networks (CNNs) and their variants becoming the backbone of modern algorithms [5]. These methods leverage hierarchical feature extraction and pattern recognition to improve accuracy and efficiency in analyzing large datasets. There have been various experiments applying models like YOLO for rapid target detection in high-resolution remote sensing images [6]. These studies have primarily focused on the direct application of the YOLO model rather than exploring the potential benefits of domain transfer learning, which is the focus of our research. Additionally, there have been similar experiments exploring the use of transfer learning for object detection in remote sensing, which have demonstrated the effectiveness of these approaches [7]. However, to the best of our knowledge, the datasets used in some of these studies are not publicly available, making it difficult to benchmark their results against other models.

In this paper, we present a detailed analysis of object detection in remote sensing images, focusing on the performance of the YOLOv8 model. Section 2 provides an overview of the datasets used, including the DIOR and Ships datasets. Also, it describes the model architecture and experimental setup. Section 3 discusses the results and provides a comparative analysis of the different training strategies employed. Finally, Sect. 4 contains the concluding remarks of the paper.

2 Materials and Methods

2.1 Datasets

2.1.1 Dior The **Det**ect**I**on in **O**ptical **R**emote Sensing Images (**DIOR**) dataset [8], is a large-scale benchmark dataset for object detection in optical remote sensing images. This dataset, developed by researchers from the Chinese Academy of Sciences, contains 23,463 high-resolution optical remote sensing images annotated with axis-aligned bounding boxes for more than 192k instances across 20 object categories. These categories include diverse objects such as airplanes, ships, vehicles, buildings, and various infrastructures like airports, bridges, and stadiums. The object instances that this dataset consists of are manually labeled with the horizontal-aligned bounding boxes. The dataset covers a spatial resolution of 800 × 800 pixels and overs a wide geographic range across more than 80 countries worldwide. Similar with other popular datasets, images were sourced from Google Earth by experts in earth observation interpretation, ensuring diversity in weather conditions, seasons, and image quality levels. Sample images with masks bounding boxing masks are depidted on Fig. 1.

 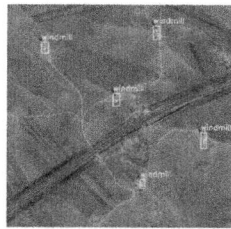 

Fig. 1. Example images with bounding boxes showcasing annotated detections from the DIOR dataset

The DIOR dataset sets itself apart from existing object detection datasets in several key ways:

- *Scale*: As mentioned earlier, it includes a large number of remote sensing images and annotated object instances across 20 categories, making it the largest dataset of its kind. The release of the DIOR dataset will help researchers in the earth observation community explore and test different deep learning techniques, advancing the field significantly.
- *Object Size Variations*: In DIOR, there is a good balance between small and large instances. Having a variety of object sizes is more useful for real-world tasks.
- *Image Variations*: Unlike many existing datasets, DIOR offers extensive variations in imaging scenarios and image quality. This diversity enhances robustness in object detection models across different environmental and observational conditions.
- *Inter-Class Similarity and Intra-Class Diversity*: It features high inter-class similarity with fine-grained object categories and significant intra-class diversity, challenging the capabilities of object detection algorithms.

Figure 2 illustrates the top 20 object classes in the dataset. The bar plot shows the count of images for each object class across the entire dataset, highlighting the prevalence of each class. Windmills and vehicles are the most common classes, with counts exceeding 1500 images.

The DIOR dataset is split into three subsets: a training-validation (trainval) set and a separate test set. The trainval set, comprising 11,725 images (50% of the dataset), is further divided into training (train) and validation (val) subsets for model development and tuning. The remaining 11,738 images form the independent test set used exclusively for performance evaluation. Evaluation metrics consider a detection as correct if its bounding box overlaps with ground truth by more than 50%, otherwise, it is classified as a false positive.

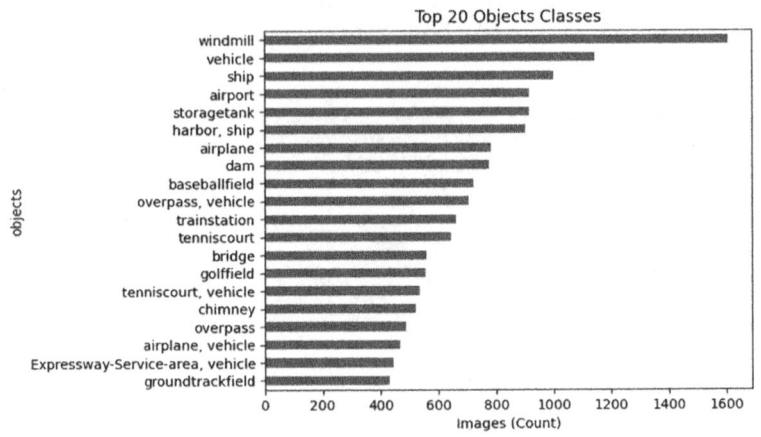

Fig. 2. Distribution of the top 20 object classes in the dataset

2.2 Ships Dataset

The "Ships in Aerial Images" dataset on Kaggle [9], is designed for ship detection tasks, specifically formatted for use with the YOLO object detection framework. The dataset comprises a large collection of aerial images that have been annotated to identify ships. The bounding box annotations are in the YOLO format, which is widely used for object detection tasks. It comprises 26,900 annotated aerial images, all at 768 × 768 resolution, labeled with bounding boxes to identify ships. Each image focuses exclusively on the "ship" class, ensuring precise detection capabilities. The resolution of the images varies, but they are typically high-resolution to capture detailed features of ships from aerial perspectives. The dataset features aerial images capturing ship features from various coastal and maritime environments. The images are in RGB format, making them suitable for most computer vision tasks that require color information. The dataset comes partitioned into three subsets for training (approximately 80%), validation (approximately 10%), and testing (approximately 10%) purposes. The images cover various maritime regions, including ports, coastal areas, and open seas, ensuring a wide range of ship types and sizes. This dataset serves as a crucial asset for researchers and professionals in computer vision and maritime industries. It supports the development of advanced ship detection models by using annotated aerial images. Sample images with bounding boxes of the ships are depicted on Fig. 3.

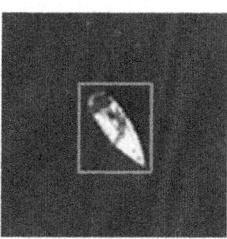  

Fig. 3. Images with annotated ship detections from the Ships/Vessels in Aerial Images dataset

2.3 Model

YOLO (You Only Look Once) is a seminal object detection model that employs a single-shot detection approach using a fully convolutional neural network (Fig. 4). This method allows YOLO to simultaneously predict bounding boxes and class probabilities in a single pass, significantly enhancing its efficiency and accuracy. Initially proposed in [10], YOLO divides an input image into an $S \times S$ grid. Each grid cell is responsible for detecting objects whose centers fall within it, predicting B bounding boxes and associated confidence scores, which indicate the certainty and precision of the detections. One of the notable techniques in YOLO is non-maximum suppression (NMS), a post-processing step that refines the detection results by eliminating redundant bounding boxes, thereby improving accuracy. The first YOLO model's architecture involved pre-training the initial 20 convolutional layers on ImageNet [11], followed by converting this model to detect objects by adding convolutional and connected layers. This strategy improved performance by leveraging pre-trained networks. The YOLO model series is highly regarded in the computer vision community for its flexibility and compact design, while consistently delivering state-of-the-art performance. This combination of attributes has facilitated the integration of YOLO models into various industry sectors, making them accessible to a wide range of machine learning practitioners. Another key factor contributing to YOLO's sustained success is its transition from the Darknet framework (used in versions 1–4) to the more widely adopted PyTorch framework with YOLOv5, spearheaded by Ultralytics [12]. The shift to PyTorch has accelerated the development and enhancement of the YOLO series, driving rapid advancements and improvements in its capabilities.

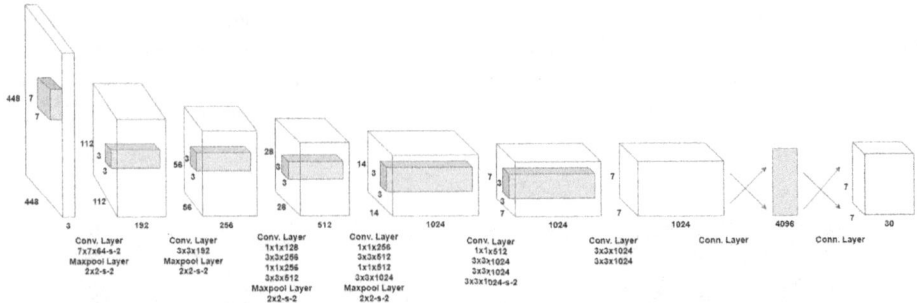

Fig. 4. YOLO architecture for object detection and localization

YOLOv1, the first iteration, achieved real-time detection by dividing images into grids and predicting bounding boxes within each grid [13]. However, it struggled with precision on small objects due to sparse grids and limited bounding box predictions. Subsequent versions, YOLOv2 and YOLOv3 [14], addressed these issues by incorporating deeper convolutional neural networks, residual blocks, skip connections, batch normalization, and higher image resolutions. YOLOv2 also introduced more anchor boxes and improved their accuracy through clustering analysis [15].

YOLOv3 enhanced detection accuracy for small, medium, and large objects by using three feature maps of varying sizes and upgrading to the Darknet-53 backbone [16]. YOLOv4 further improved performance with advanced techniques such as Mosaic data augmentation, Cross mini-batch Normalization (CmBN), and Self-Adversarial Training (SAT), along with a stronger backbone network (CSPDarknet53) and additional features like the Mish activation function and Spatial Pyramid Pooling (SPP) [17].

The YOLO model series continued to evolve with YOLOv5, which adopted Mosaic data enhancement and adaptive anchor box settings, and YOLOX, which introduced optimization techniques like Exponential Moving Average (EMA) and IoU loss function [18]. YOLOv6 achieved significant improvements in accuracy and speed with hardware support and the Re-parameter (Rep) skill [19]. YOLOv7 advanced model performance further with re-parameterization techniques, the Efficient Long-Range Attention Network (ELAN), and a cosine learning rate scheduling strategy [20].

YOLOv8 represents a huge revolution in the YOLO object detection series, continuing to push the boundaries of model architectures in the object detection field. Developed by Ultralytics, YOLOv8 introduces several significant improvements over its predecessors [21]. Figure 5 shows performance difference between different YOLO model version.

Architecturally (Fig. 6), YOLOv8 is anchor-free, similar to YOLOX, meaning it directly predicts object centers, making detection simpler and speeding up the Non-Maximum Suppression (NMS) process. This approach is more flexible in terms of detection because they directly detect objects without preset anchors which anchor-based approaches require.

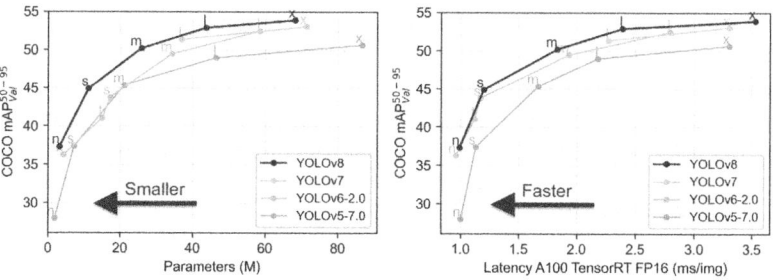

Fig. 5. Performances of the YOLO series. (Source: https://github.com/ultralytics/ultralytics)

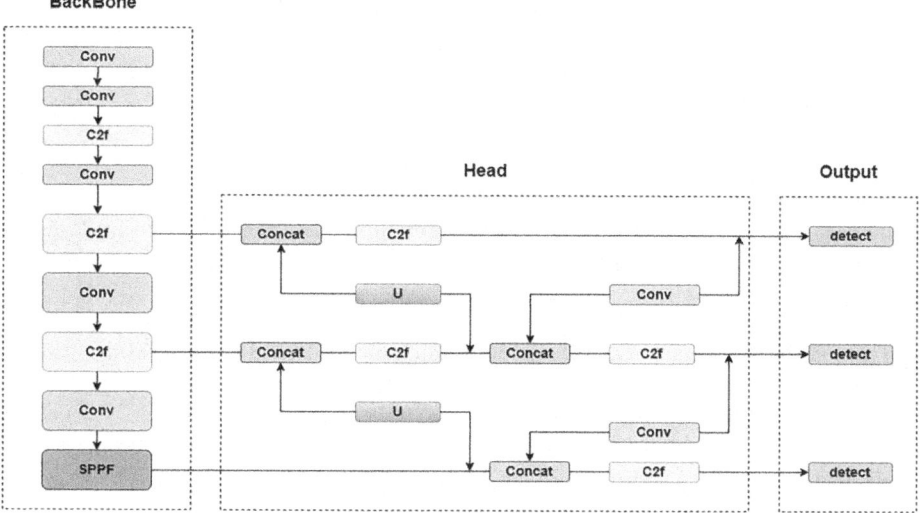

Fig. 6. YOLOv8 architecture diagram showcasing the backbone, head, and output components. The backbone consists of convolutional (Conv) and C2f layers, while the head utilizes concatenation (Concat), C2f layers, and upsampling (U) to refine features for detection. The output layer generates the final object detections.

The YOLOv8 architecture has two main parts: the backbone and the head. The backbone, based on an improved CSPDarknet53, uses several convolutional layers to extract important features from images. It includes special modules like C2f for better feature integration and SPPF for processing different scales. The head then takes these features and processes them to produce final outputs, such as bounding boxes and object classes. It has separate branches for different tasks (objectness scores, classification, and regression) to improve accuracy and speed. Finally, the detection module converts these features into the final predictions quickly and accurately [22].

YOLOv8 adjusts its training process by discontinuing mosaic augmentation, which involves combining four images together, which helps the model learn to detect objects in new locations and against varying backgrounds. This augmentation is shown to degrade performance if performed through the whole training routine. Therefore, YOLOv8 has optimized its training process by disabling this technique towards the final ten epochs of training. These adjustments have been crucial in refining the model's effectiveness.

YOLOv8 research was primarily driven by empirical evaluations on the COCO benchmark. Each adjustment to the network and training routine involved conducting new experiments to assess their impact on the COCO dataset.

Regarding COCO (Common Objects in Context) [23] serves as the industry standard for evaluating object detection models, focusing on metrics like mAP (mean Average Precision) and FPS (Frames Per Second) for inference speed. YOLOv8 achieves state-of-the-art accuracy on COCO at comparable inference speeds, highlighting its performance in real-world applications.

2.4 Experimental Setup

The goal of this study is to investigate and evaluate the performance and answer several key questions regarding object detection in satellite images. We aim to address the following key questions:

- Does training object detection models such as YOLOv8 from scratch on satellite images yields superior performance compared to using pre-trained models on generic datasets like COCO?
- Do fine-tuned models on remote sensing dataset (DIOR) improve predictive performance on the Ships dataset vs model pre-trained on COCO alone?

Each question is designed to uncover insights into the optimal approach for object detection in satellite imagery, considering both training strategies and dataset specificity.

For that purpose, we trained the following models:

- Training YOLOv8 from scratch on the DIOR dataset
- Fine-tuning YOLOv8 COCO pre-trained model on DIOR
- Fine-tuning a YOLOv8 COCO and DIOR pre-trained model on the ships dataset
- Fine-tuning a YOLOv8 COCO pre-trained model on the ships dataset

The resulting models are available on our Github repo[1]. Our goal was to train and test the models independently, so we can later evaluate and analyze their performance. That should provide us with the answers to our experimental questions.

---

[1] https://github.com/ema-pandilova/Transfer-Learning-Yolo.

The training process involves data pre-processing and model configuration. Our approach began with extraction and preparation of the DIOR dataset, converting annotations from the PASCAL VOC format [24] into the YOLO format for model compatibility. We parsed the XML files to extract image details and bounding box coordinates, then normalized these coordinates to fit the YOLO format. After conversion, we verified and corrected annotations through visual inspection and automated checks. Finally, we organized the data into the required directory structure, ensuring each image had a corresponding annotation file and was appropriately split into training, validation, and test sets. This pre-processing is essential for reliable model training and evaluation.

The DIOR dataset had to be divided into train, test and validation splits. We adopted a stratified approach to ensure representative samples in both training and validation subsets of the DIOR dataset, crucial for maintaining balanced class distributions, which are essential for effective model learning. Organizing our dataset into structured directories optimized data loading efficiency during training and evaluation stages. The splits are available in our Github repo[2].

The Ships dataset is split into train, test and validation subsets by its' authors and we used those splits as is.

In terms of model configuration, the models were implemented from the Ultralytics YOLOv8 framework [25], integrated with Python 3.10 and Torch 2.1.2 on a CUDA-enabled environment with a Tesla T4 GPU. The experiments were executed on Kaggle Notebooks. For consistency and reproducibility, we employed a fixed random seed (42) and conducted training over a predetermined number of epochs. Our training processes spanned 100 epochs, utilizing a batch size of 16 and adhering to configurations defined in our YAML files, which specified crucial parameters like image size and class names. The optimizer was AdamW with a learning rate set to 0.002 and momentum of 0.9. This optimizer configuration includes weight decay considerations for both weights and biases to fine-tune the model's parameters effectively. The models' performances are evaluated using metrics such as precision, recall, and mAP on both the validation and test sets to assess detection accuracy and generalization. At the end of the training process, the model with the highest validation metric is chosen as the final model.

3 Results and Discussion

The results from our experiments provide valuable insights into the performance of different object detection models on satellite imagery. During the training process, we monitored the model's performance using several metrics, including Precision, Recall, F1-Score [26], and Mean Average Precision (mAP) [27]. The loss function, a combination of localization loss, confidence loss, and classification loss, showed a consistent downward trend across all models, indicating effective learning. However, the rate of loss reduction varied among the models, with the

---

[2] https://github.com/ema-pandilova/Transfer-Learning-Yolo.

model pre-trained on satellite images converging more quickly and achieving a lower final loss compared to the other models.

Table 1 shows the comparative performance metrics for YOLOv8 models trained from scratch and pre-trained on COCO, then fine-tuned on the DIOR dataset. Table 2 presents the performance metrics for models evaluated on the Ships dataset, comparing models pre-trained on COCO and fine-tuned on DIOR with models pre-trained on COCO alone.

Table 1. Comparative Performance of YOLOv8 Models Fine-Tuned on DIOR Dataset

Model	Precision (P)	Recall (R)	mAP50	mAP50-95
From Scratch	0.819	0.636	0.714	0.486
Pre-Trained on COCO	0.848	0.685	0.760	0.532

The performance metrics highlight that the YOLOv8 model pre-trained on COCO and then fine-tuned on the DIOR dataset outperforms the model trained from scratch across all evaluated metrics. Specifically, the pre-trained model achieved a higher mAP50 and mAP50-95, demonstrating better overall detection accuracy and robustness. The pre-trained model shows higher precision and recall, indicating that it can more accurately detect objects and retrieve a higher percentage of true positives. This improvement is likely due to the generalization capabilities imparted by the COCO pre-training, which provides a strong feature extraction foundation. The mean Average Precision at different IoU thresholds further supports the superiority of the pre-trained model, with significant gains in both mAP50 and mAP50-95. This suggests that fine-tuning a pre-trained model on a domain-specific dataset like DIOR is beneficial for capturing both large and small object variations more effectively.

Table 2. Comparative Performance of YOLOv8 Models Fine-Tuned on Ships Dataset

Model	Precision (P)	Recall (R)	mAP50	mAP50-95
Pre-Trained on COCO + DIOR	0.692	0.468	0.592	0.384
Pre-Trained on COCO	0.726	0.407	0.573	0.364

When evaluating on the Ships dataset, the model fine-tuned on DIOR after pre-training on COCO again outperforms the model pre-trained on COCO alone, though the margins are smaller compared to the DIOR dataset evaluation. The model fine-tuned on DIOR achieves higher recall but slightly lower precision compared to the COCO pre-trained model. This indicates a better ability to identify ships, albeit with a minor increase in false positives. The fine-tuned model shows improved performance in mAP50-95, highlighting its ability to accurately localize ships across various IoU thresholds. The mAP50 improvement is less

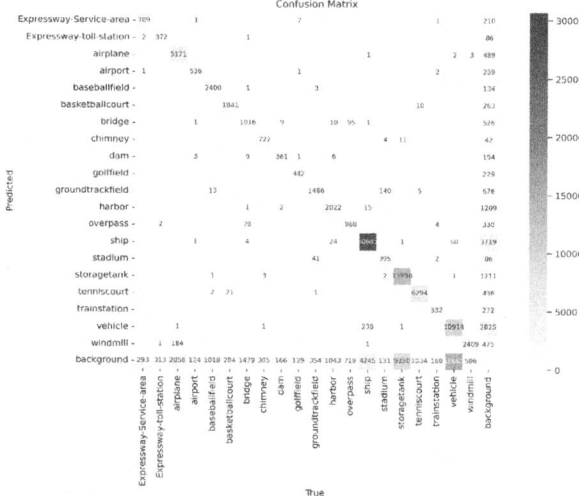

(a) Confusion Matrix for YOLOv8 Trained from Scratch on DIOR

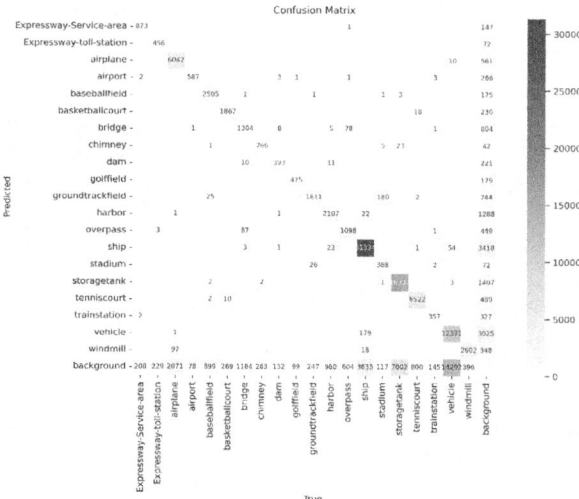

(b) Confusion Matrix for YOLOv8 pre-trained on COCO, fine-tune on DIOR

Fig. 7. Confusion Matrices for YOLOv8 Models on DIOR Dataset

pronounced, indicating that while fine-tuning enhances performance, the complexity of the Ships dataset might require further domain-specific adjustments.

The confusion matrix in Fig. 7a indicates the performance of the YOLOv8 model trained from scratch on the DIOR dataset. The model performs well on prominent and distinct classes such as "airplane," "ship," and "vehicle," which have high numbers of correct predictions. This suggests that the model can effectively detect and classify these objects in remote sensing images. Some classes with visual or contextual similarities are frequently misclassified. For example, "expressway-service-area" and "expressway-toll-station" are occasionally confused, indicating the need for more distinctive features or additional training data to better distinguish between them. The "harbor" class is sometimes misclassified as "ship," reflecting the contextual overlap in scenes where these objects co-occur. This suggests that contextual information might be influencing the model's predictions. Classes like "chimney" and "windmill" have higher misclassification rates. These objects might be smaller or less distinct in the images, making them harder to detect accurately. This points to a potential area for improvement, possibly through higher resolution images or more focused data augmentation techniques. The off-diagonal elements, which represent incorrect predictions, highlight specific areas where the model can be refined. Targeted improvements, such as enhancing feature extraction for similar-looking classes or increasing the training data for underperforming categories, can help reduce these misclassifications. The confusion matrix indicates that while the YOLOv8 model has strong performance for several key classes, there are specific areas where misclassifications occur due to visual similarities, contextual overlaps, and challenges with smaller objects.

(a) Original Image (b) Image with Detected Bounding Boxes

Fig. 8. Comparison of Original Image and Image with Detections from the Ships dataset

The confusion matrix in Fig. 7b showcases the performance of the YOLOv8 model pre-trained on COCO and then fine-tuned on the DIOR dataset. The pre-trained model exhibits significant improvements in correctly detecting prominent classes. There is a noticeable reduction in misclassifications across various categories. For instance, the "ship" class shows 8,134 correct detections with fewer misclassifications, compared to 8,068 in the scratch-trained model. This highlights the model's improved precision and recall rates, reducing false positives and false negatives. The confusion between similar classes such as "expressway-service-area" and "expressway-toll-station" is less pronounced in the pre-trained model. This improvement suggests that the pre-trained model benefits from the diverse feature representations learned during the initial training on the COCO dataset, allowing it to distinguish between visually similar categories more effectively. The pre-trained model shows better performance in detecting smaller or less distinct objects, such as "chimney" and "windmill," which were more frequently misclassified in the scratch-trained model. As illustrated in Fig. 8, the original satellite image (Fig. 8a) is compared with the same image from the Ships dataset showing detected bounding boxes (Fig. 8b). This comparison demonstrates the effectiveness of the pre-trained YOLOv8 model fine-tuned on DIOR in accurately identifying and localizing ships within the image. This improvement indicates that the pre-trained model can capture finer details in the imagery. The pre-trained YOLOv8 model fine-tuned on the DIOR dataset outperforms the model trained from scratch by leveraging the feature extraction capabilities gained from the COCO dataset. This results in higher accuracy, reduced misclassifications, and better differentiation between similar objects, making it a better tool for remote sensing object detection.

4 Conclusion

In conclusion, we demonstrated that transfer learning with YOLOv8 significantly enhances object detection performance in remote sensing imagery. By leveraging pre-trained models and fine-tuning them on domain-specific datasets, we achieved higher accuracy, better differentiation between similar objects, and improved detection of smaller or less distinct objects. Our study underscores the effectiveness of this approach and provides a solid foundation for future research in this domain.

While our work provides a robust and efficient method for object detection with practical implications in areas such as land cover mapping, urban planning, and environmental monitoring, there remains significant potential for further advancements. A promising direction for future research is the exploration of unsupervised training of foundational models using the vast amounts of unlabeled remote sensing data. This approach allows models to learn complex patterns and representations from large datasets without the need for extensive manual labeling, which is often a bottleneck in remote sensing applications.

By developing these foundational models through unsupervised learning, we can create versatile models that, when fine-tuned on specific labeled datasets, achieve even greater accuracy and generalization.

Advancing unsupervised learning techniques and refining fine-tuning strategies will be crucial for pushing the boundaries of object detection in remote sensing, enabling the development of models that are not only more powerful but also more flexible in addressing diverse challenges in this field.

References

1. Toth, C., Jóźkow, G.: Remote sensing platforms and sensors: a survey. ISPRS J. Photogramm. Remote Sens. **115**, 22–36 (2016)
2. Spasev, V., Dimitrovski, I., Kitanovski, I., Chorbev, I.: Semantic segmentation of remote sensing images: definition, methods, datasets and applications. In: ICT Innovations 2023. Learning: Humans, Theory, Machines, and Data, pp. 127–140 (2024)
3. Dimitrovski, I., Kitanovski, I., Kocev, D., Simidjievski, N.: Current trends in deep learning for earth observation: an open-source benchmark arena for image classification. ISPRS J. Photogramm. Remote. Sens. **197**, 18–35 (2023)
4. Pearlman, J.S., Barry, P.S., Segal, C.C., Shepanski, J., Beiso, D., Carman, S.L.: Hyperion, a space-based imaging spectrometer. IEEE Trans. Geosci. Remote Sens. **41**(6), 1160–1173 (2003)
5. Dimitrovski, I., Kitanovski, I., Panov, P., Kostovska, A., Simidjievski, N., Kocev, D.: Aitlas: artificial intelligence toolbox for earth observation. Remote Sens. **15**(9), 2343 (2023)
6. Wu, Z., Chen, X., Gao, Y., Li, Y.: Rapid target detection in high resolution remote sensing images using yolo model. Int. Arch. Photogramm. Remote Sens. Spat. Inf. Sci. **42**, 1915–1920 (2018)
7. Devi, A., et al.: Transfer learning for object detection in remote sensing images with yolo. J. Electr. Syst. **20**(3s), 980–989 (2024)
8. Li, K., Wan, G., Cheng, G., Meng, L., Han, J.: Object detection in optical remote sensing images: a survey and a new benchmark. ISPRS J. Photogramm. Remote. Sens. **159**, 296–307 (2020)
9. Sah, S.K.: Ships in Aerial Images Dataset. https://www.kaggle.com/datasets
10. Redmon, J., Divvala, S., Girshick, R., Farhadi, A.: You only look once: unified, real-time object detection. In: Proceedings of the IEEE Conference on Computer Vision and Pattern Recognition, pp. 779–788 (2016)
11. Krizhevsky, A., Sutskever, I., Hinton, G.E.: Imagenet classification with deep convolutional neural networks. Adv. Neural Inf. Process. Syst. **25** (2012)
12. Jocher, G., et al.: ultralytics/yolov5: v6. 1-tensorrt, tensorflow edge tpu and openvino export and inference. Zenodo (2022)
13. Du, J.: Understanding of object detection based on cnn family and yolo. In: Journal of Physics: Conference Series, vol. 1004, p. 012029. IOP Publishing (2018)
14. Redmon, J., Farhadi, A.: Yolov3: an incremental improvement. arXiv preprint arXiv:1804.02767 (2018)
15. Hussain, M.: Yolo-v1 to yolo-v8, the rise of yolo and its complementary nature toward digital manufacturing and industrial defect detection. Machines **11**(7), 677 (2023)
16. Zhang, X., Dong, X., Wei, Q., Zhou, K.: Real-time object detection algorithm based on improved yolov3. J. Electron. Imaging **28**(5), 053022–053022 (2019)
17. Bochkovskiy, A., Wang, C.-Y., Liao, H.-Y.M.: Yolov4: optimal speed and accuracy of object detection. arXiv preprint arXiv:2004.10934 (2020)
18. Ge, Z., Liu, S., Wang, F., Li, Z., Sun, J.: Yolox: exceeding yolo series in 2021. arXiv preprint arXiv:2107.08430 (2021)
19. Li, C., et al.: Yolov6: a single-stage object detection framework for industrial applications. arXiv preprint arXiv:2209.02976 (2022)
20. Wang, C.-Y., Bochkovskiy, A., Liao, H.-Y.M.: Yolov7: trainable bag-of-freebies sets new state-of-the-art for real-time object detectors. In: Proceedings of the

IEEE/CVF Conference on Computer Vision and Pattern Recognition, pp. 7464–7475 (2023)
21. Terven, J., C'ordova-Esparza, D.-M., Romero-Gonz'alez, J.-A.: A comprehensive review of yolo architectures in computer vision: from yolov1 to yolov8 and yolonas. Mach. Learn. Knowl. Extract. **5**(4), 1680–1716 (2023)
22. Sohan, M., Sai Ram, T., Reddy, R., Venkata, C.: A review on yolov8 and its advancements. In: International Conference on Data Intelligence and Cognitive Informatics, pp. 529–545. Springer, Heidelberg (2024). https://doi.org/10.1007/978-981-99-7962-2_39
23. Lin, T.-Y., et al.: Microsoft COCO: common objects in context. In: Fleet, D., Pajdla, T., Schiele, B., Tuytelaars, T. (eds.) ECCV 2014. LNCS, vol. 8693, pp. 740–755. Springer, Cham (2014). https://doi.org/10.1007/978-3-319-10602-1_48
24. Everingham, M., Eslami, S.A., Van Gool, L., Williams, C.K., Winn, J., Zisserman, A.: The pascal visual object classes challenge: a retrospective. Int. J. Comput. Vision **111**, 98–136 (2015)
25. Ultralytics: YOLOv8 (2023). Accessed 03 July 2024. https://github.com/ultralytics/ultralytics
26. Yacouby, R., Axman, D.: Probabilistic extension of precision, recall, and f1 score for more thorough evaluation of classification models. In: Proceedings of the First Workshop on Evaluation and Comparison of NLP Systems, pp. 79–91 (2020)
27. Revaud, J., Almaz'an, J., Rezende, R.S., Souza, C.R.D.: Learning with average precision: training image retrieval with a listwise loss. In: Proceedings of the IEEE/CVF International Conference on Computer Vision, pp. 5107–5116 (2019)

Comparison of On-Board and Off-Board Processing Power Consumption for Drone Camera Images

Atanasko Boris Mitrev[✉] and Biljana Risteska Stojkoska

Faculty of Computer Science and Engineering, Ss. Cyril and Methodius University in Skopje, Skopje, Republic of Macedonia
atanasko.mitrev@students.finki.ukim.mk, biljana.stojkoska@finki.ukim.mk

Abstract. Unmanned aerial vehicles (UAVs) equipped with cameras, capture images and videos from high altitude, usually for surveillance purpose. This paper investigates the power consumption of a drone processing unit under two different operational scenarios: (i) On-Board (local image processing using an onboard Raspberry Pi) and (ii) Off-Board (streaming video over Wi-Fi for remote processing). Therefore, we deployed the well-known YOLO (You Only Look Once) image detection algorithm on the Raspberry Pi and investigated the power consumption for the both scenarios. Contrary to the expectations, our findings prove that energy consumption is very similar in both scenarios. The results of this study are valuable for designers of energy-efficient UAV systems, particularly for applications where prolonged flight duration and real-time data processing are critical.

Keywords: UAV · Drone · Energy Efficiency · IoT · Raspberry Pi · Wi-Fi Streaming · Computer vision · Object detection

1 Introduction

The use of Unmanned Aerial Vehicles (UAVs), more commonly referred to as drones, for various applications in recent years has seen significant growth, ranging from aerial photography to surveillance, agriculture, last-mile delivery etc. As drones become more autonomous, there is a growing need to optimize their power consumption for extended flight times and operational efficiency. Critical aspect for this optimization is the method used for processing data, particularly for tasks such as object detection, which require significant computational resources.

In this paper, we present a comparative analysis of power consumption between two methods for object detection on a custom-made drone using Raspberry Pi as processing computational unit with Navio2 as a flight controller (Fig. 1) and (Fig. 5):

- Off-Board - Streaming data over Wi-Fi for processing on a remote server
- On-Board - Processing data locally on the device itself

Fig. 1. Custom Drone build

The former method involves streaming live video feed from the drone's camera to a remote server, where image processing algorithms can be applied, and the results are sent back to the drone. The latter method involves processing the video feed directly on the Raspberry Pi, eliminating the need for data transmission over Wi-Fi.

We hypothesize that processing data locally on the Raspberry Pi will result in lower power consumption compared to streaming data over Wi-Fi, as usually is the case in sensor networks, due to reduced data transmission and processing requirements. To test this hypothesis, we will conduct experiments to measure the power consumption of both methods under controlled conditions.

The rest of the paper is organized as follows. Section 2 provides an overview of the related work, Sect. 3 discusses processing scenarios, Sect. 4 provides overview of YOLOv3-Tiny model architecture, Sect. 5 presents the experiment results, Sect. 6 discusses the implications of the results, and the conclusion section wraps and concludes the material presented in this paper.

2 Related Work

Processing and streaming power consumption for UAV's, wireless sensor networks and IoT devices in general gain significant interest.

Abeywickrama et al. 2018 [1] have studied an energy consumption model for UAVs and concluded that Wi-Fi communication between the UAV and the ground station has negligible impact on the total power consumption, indicating no significant difference in energy usage whether communication is active or inactive.

Kaup et al., 2014 [2] in their paper measured the power consumption of Raspberry Pi modules. For this research only Wi-Fi power measurements is of interest. They show that Wi-Fi power consumption in the Raspberry Pi during the upload is bandwidth dependent.

Lorincz et al., 2021 [3] in their paper propose computing and networking framework for UAV based systems used for disaster management applications or real-time surveillance. For this paper proposed system architecture is of interest.

Adarsh et al., 2020 [4] present fundamental overview of object detection methods for both one stage and two stage detectors. In this work they also explore YOLOv3-Tiny model capabilities.

Zhai et al., 2023 [5] in their work proposes a tiny UAV detection method based on the optimized YOLOv8. They add high-resolution detection head, in the detection head component, to improve the device's detection capability for small targets, and they cut off the large target detection head and redundant network layers to effectively reduce the number of network parameters and improve the detection speed of UAV. In the feature extraction stage, they use SPD-Conv to extract multi-scale features instead of Conv to reduce the loss of fine-grained information and enhance the model's feature extraction capability for small targets. They also use GAM attention mechanism in the neck to enhance the model's fusion of target features and improve the model's overall performance in detecting UAVs.

3 Processing Scenarios

UAV based system can be organized in a different ways, but depending on the architecture, processing unit, battery and system resource utilization give different results. In general as explained in [3] there are three general architectures for UAV based system information processing

- Centralized - Data are collected on-board and after the collection phase, the data are processed on the ground. Most of the contemporary UAV-based systems follow this architecture. Its main drawback is that it is not adequate for applications that must operate in real-time
- Cloud-based - Data are collected on-board and are immediately sent to the ground to be further processed by the cloud systems. This architecture requires high communication bandwidth, what consequently has an impact on high energy requirements
- Fully distributed - Each UAV performs data processing on-board. If the information extracted from data is important, it is immediately sent to the ground. Depending on the application, the on-board computation can range from medium to high, leading to medium energy requirements (high energy expenditure is noticed only when the processing task is very complex). Both cloud-based and distributed architectures are suitable for real-time applications, but their common drawback can be found in high energy requirements

As already mention above this paper explores two different processing scenarios in two of the architectures (Cloud-based and Fully distributed):

- On-Board (processing data locally on the device itself)
- Off-Board (streaming data over Wi-Fi for processing on a remote server)

and measure power consumption in the both scenarios. In case of On-Board processing scenario (Fig. 2) video stream is processed locally and object detection YOLO image detection algorithm is used. In the Off-Board (Fig. 2) processing video is streamed over Wi-Fi module and the video stream is processed on the remote server.

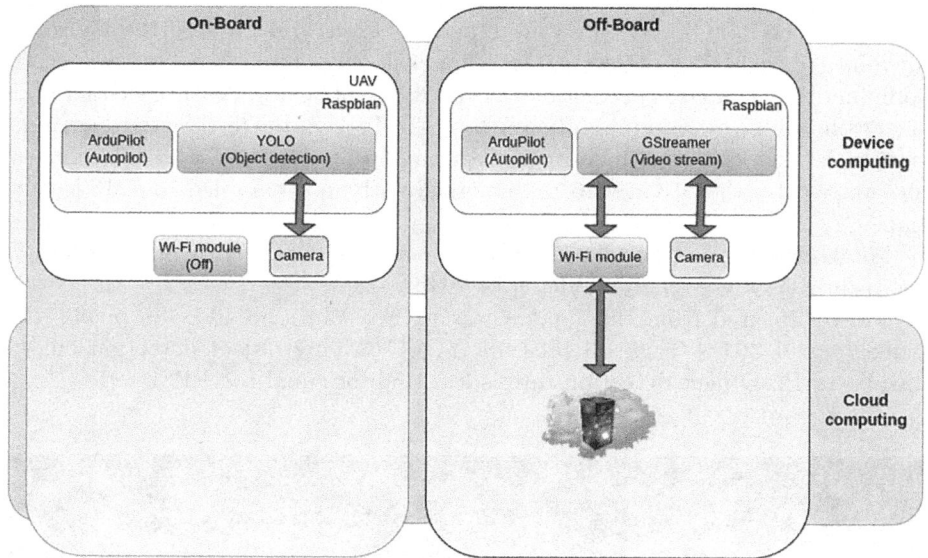

Fig. 2. On-Board and Off-Board scenario processing architecture

Some examples of practical use of On-Board and Off-Board processing are precision agriculture, traffic monitoring and redirecting system, wildlife surveillance, rescue missions etc.

Example of possibility to use a drone in precision agriculture is explored in [6]. In this application On-Board object detection on the drone is used for weed detection. The system was also tested for ArUco Marker detection and colour detection.

Drone can be used for traffic monitoring and redirecting [7]. In this practical case different architectures are used for collecting and processing traffic data.

Wildlife surveillance [8] using a drone is an other example of the possibility of using this devices in real life. In this case drones are used specifically for deer tracking, which are highly mobile in nature.

Drones can also be used in rescue missions [9]. In this case drone is able to do real-time object detection and once it detects signs of human presence autonomously navigates towards the suspicious location to get a better view in order to verify human presence. It processes images On-Board and sends selected frames to the operator for verification.

4 Image Processing and Object Detection

For this research YOLOv3-Tiny [4,10,11] model was deployed on Raspberry Pi 3B+ for object detection. Model architecture is presented in (Fig. 4).

YOLO v3-Tiny model is a YOLO v3 model variant with the decreased depth of the convolutional layer. Model running speed is increased but detection accuracy is reduced. YOLO v3-Tiny uses pooling layer and reduces the figure for convolution layer. It predicts a three-dimensional tensor that contains score, bounding box, and class predictions at two different scales. Feature extraction is in convolution layers and max-pooling layers utilized in the feed forward arrangement of YOLO v3-Tiny. Bounding boxes prediction occurs at two different feature map scales 13×13 and 26×26 merged with an upsampled 13×13 feature map.

Model was trained on COCO dataset using 80 classes with mAP of 33.1%. Pre-trained YOLOv3-Tiny model is 34 MB in size.

Video captured from the camera was with resolution 640×480 pixels with frame rate of 30 fps. Figure 3 presents YOLOv3-Tiny object detection on the Raspberry Pi. Object detection rate is less than or equal to 1FPS.

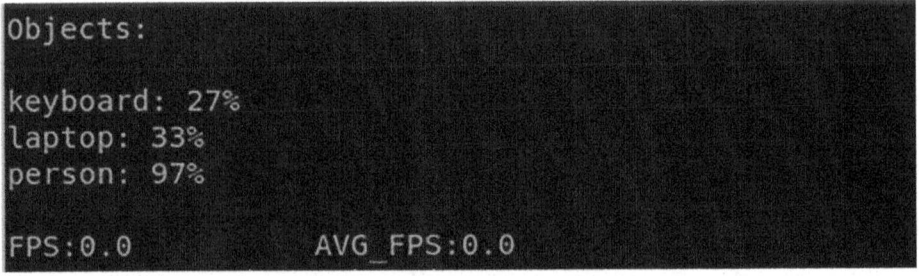

Fig. 3. YOLOv3-Tiny object detection on the Raspberry Pi

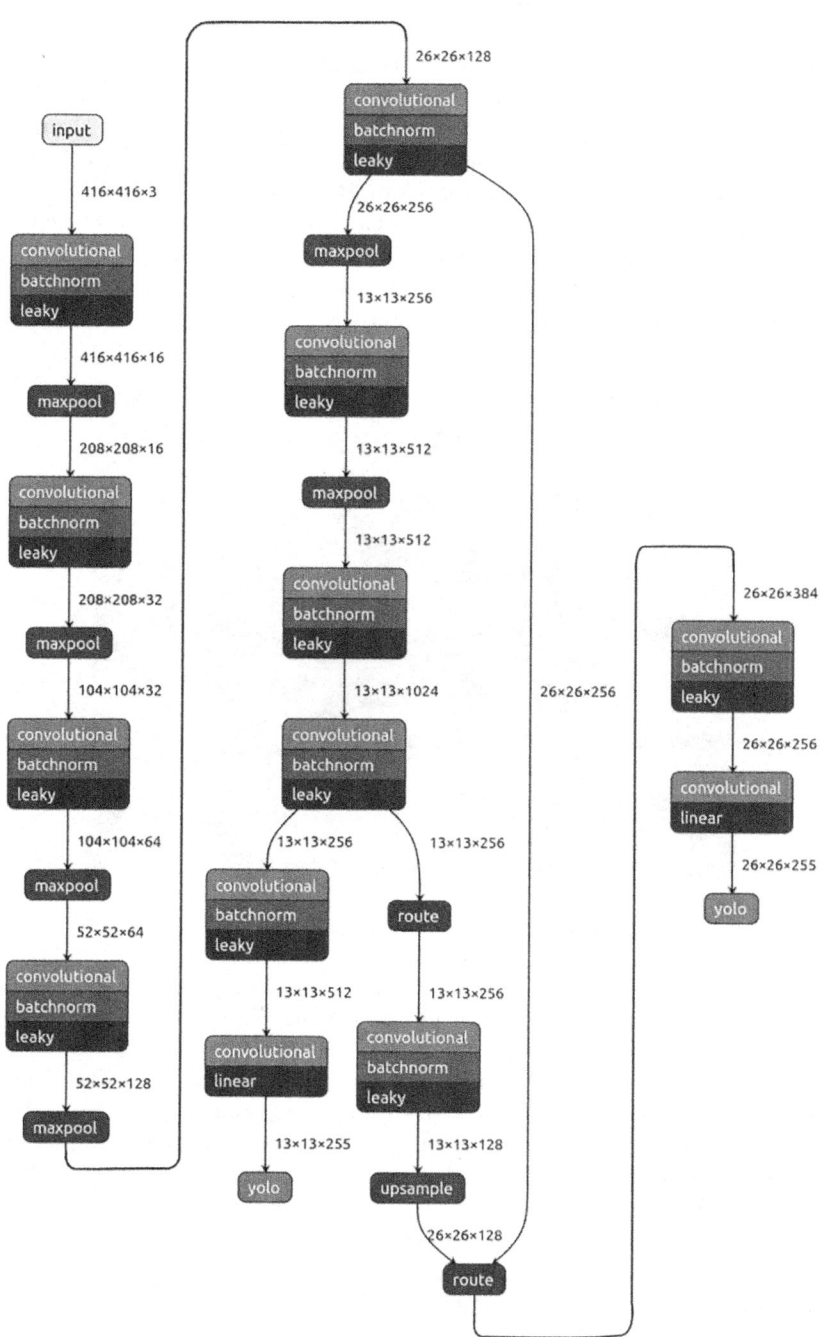

Fig. 4. YOLOv3-Tiny Architecture

5 Experimental Results

For this research custom hardware setup is used. UAV (drone) used (Fig. 1) and (Fig. 5), is custom build drone using Raspberry Pi 3B+ with 1 GB RAM [12] as a processing computational unit and Navio2 [13] as a flight controller. Drone frame is F330 4-Axis RC Quadcopter Frame [14], brushless motors BR2212 980KV 2-4S [15], electronic speed controller is Gool RC Simonk 20 AMP 20A SimonK Firmware w/3 A 5 V BEC [16], battery is Gens ace 3300 mAh 14.8 V 25C 4S1P [17], proppelers are Gemfan MR 8045 CW/CCW propellers [18], GPS/GNSS antenna is Emlid ANT102 [19] and radio reciver Flysky 2.4 G 6CH FS-iA6B [20]. Raspberry Pi Camera Module 3 camera is used for video capturing [21].

Fig. 5. Custom Drone build details

Raspberry Pi is using Raspbian OS based on Debian GNU/Linux Buster. Autopilot used is ArduPilot [22].

Two separate groups of measurements were performed for local video processing (object detection), and video streaming for external processing. For both groups of measurements five separate measurements were performed. For all measurements drone battery was charged to the MAX capacity (in our case ~16.8 V), and discharged to MIN power level ~14.2 V, at least one cell MIN power level 3.5 V recommended power to not damage the battery. Measurements results are presented in Table 1, Table 2 and (Fig. 6).

The first group of measurements is supposed to measure the power consumption of the device when there is no network transmission, video is not streamed. For this first group of measurements, Wi-Fi module was turned off and object detection was performed on the Raspberry Pi itself using Darknet YOLOv3-Tiny [3,10].

Table 1. On-Board YOLOv3-Tiny object detection measurements

Id	Date	Boot	Stop	Uptime	MAX charge(V) (Total)	MIN charge(V (Total)
1	05.04.2024	12:16:45	15:17:06	03:00:21	16.8	14.2
2	06.04.2024	15:18:16	18:17:07	02:58:51	16.8	14.2
3	08.04.2024	19:17:06	22:17:06	03:00:00	16.8	14.2
4	11.04.2024	09:17:07	12:17:07	03:00:00	16.8	14.2
5	12.04.2024	09:04:15	12:04:15	03:00:00	16.79	14.2

Table 2. Video streaming Off-board processing measurements

Id	Date	Boot	Stop	Uptime	MAX charge(V) (Total)	MIN charge(V (Total)
1	13.04.2024	14:04:00	17:06:46	03:02:46	16.78	14.2
2	15.04.2024	10:47:07	13:47:48	03:00:41	16.78	14.2
3	16.04.2024	09:14:29	12:17:06	03:02:37	16.78	14.2
4	17.04.2024	09:15:02	12:17:06	03:02:04	16.78	14.2
5	18.04.2024	09:18:10	12:17:12	02:59:02	16.78	14.2

The second group of measurements is supposed to measure the power consumption when there is no local image processing of the video captured from the camera, instead the video is streamed and the video processing is intended to be on external system.

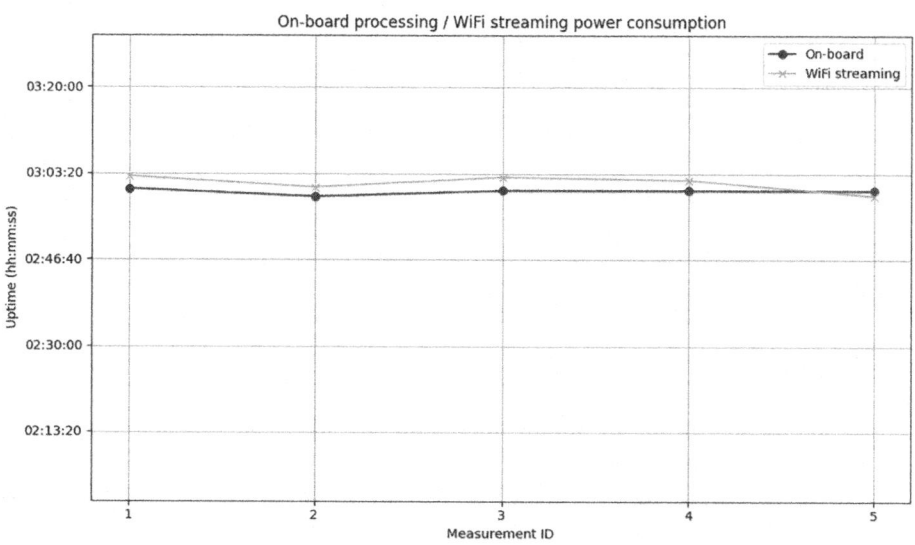

Fig. 6. On-board processing/Wi-Fi streaming power consumption

On both sides GStreamer is used to stream video from the camera mounted on the drone, and on a local computer to grab the video streamed from the drone.

In both scenarios the drone was not armed(motors are not spinning), only image processing operation was performed.

6 Discussion

In our study, we hypothesized that power consumption would be higher when video is streamed from the drone compared to when video is processed locally on the Raspberry Pi. This hypothesis was based on the assumption that the additional energy required for continuous wireless transmission would outweigh the energy demands of local processing. However, our experimental results revealed that the power consumption in both scenarios was very similar.

The similarity in power consumption between the two scenarios suggests that the energy cost of local video processing on the Raspberry Pi may be comparable to the energy cost of streaming video data wirelessly. This unexpected result prompts a deeper examination of the factors influencing power consumption in both cases.

For local video processing, the Raspberry Pi performs computationally intensive tasks, such as object detection. These tasks require substantial processing power, which in turn increases the overall power draw of the system. Despite being designed for low power consumption, the Raspberry Pi's CPU usage for real-time video processing can be considerable.

In the streaming scenario, while the Raspberry Pi offloads the computational workload to an external system, it incurs continuous wireless communication costs. Maintaining a stable and high-bandwidth connection for video streaming demands significant energy. The Wi-Fi module, operating at full capacity, consumes power at a rate that appears to rival the power used for local processing.

In both scenarios only processing unit was powered and drone was not armed (motors are not spinning), drone motors did not consume any power because our research explores only difference in processing power between on-board and off-board processing. In some future research computation power consumption share in comparison to the full power consumption when drone is armed can be measured.

7 Conclusion

Our research contributes to the understanding of power consumption dynamics in drone systems and underscores the need for a holistic approach to energy management. Future work should expand on this study by exploring a wider range of scenarios, including varying environmental conditions, different processing tasks, and diverse drone hardware. Additionally, investigating advanced power-saving techniques and algorithms could further enhance the energy efficiency of both local processing and streaming operations. An other future work could explore the power consumption in case when different transmission technology 4G, 5G etc. is used.

In summary, while our findings challenge the conventional expectation regarding the energy costs of video streaming versus local processing, they open new perspectives for optimizing drone performance and efficiency. By taking a comprehensive view of power consumption, we can better design drone systems that meet the demands of modern applications while conserving energy.

Acknowledgment. This work was partially financed by the Faculty of Computer Science and Engineering at the Ss. Cyril and Methodius University in Skopje.

References

1. Abeywickrama, H.V., Jayawickrama, B.A., He, Y., Dutkiewicz, E.: Comprehensive energy consumption model for unmanned aerial vehicles, based on empirical studies of battery performance. IEEE Access **6**, 58383–58394 (2018)
2. Kaup, F., Gottschling, P., Hausheer, D.: Powerpi: measuring and modeling the power consumption of the raspberry pi. In: 39th Annual IEEE Conference on Local Computer Networks, pp. 236–243. IEEE (2014)
3. Lorincz, J., Tahirovic, A., Risteska Stojkoska, B.: A novel real-time unmanned aerial vehicles-based disaster management framework (2021). https://doi.org/10.1109/TELFOR52709.2021.9653238
4. Adarsh, P., Rathi, P., Kumar, M.: Yolo v3-tiny: object detection and recognition using one stage improved model, pp. 687–694 (2020). https://doi.org/10.1109/ICACCS48705.2020.9074315
5. Zhai, X., Huang, Z., Li, T., Liu, H., Wang, S.: Yolo-drone: an optimized yolov8 network for tiny uav object detection. Electronics **12**(17) (2023). https://doi.org/10.3390/electronics12173664
6. Alsalam, B.H., Morton, K., Campbell, D., Gonzalez, L.: Autonomous uav with vision based on-board decision making for remote sensing and precision agriculture, pp. 1–12 (2017). https://doi.org/10.1109/AERO.2017.7943593
7. Kumar, A., Jain, S.: Drone-based monitoring and redirecting system. In: Development and Future of Internet of Drones (IoD): Insights, Trends and Road Ahead, pp. 163–183 (2021)
8. Chowdhury, S., Marufuzzaman, M., Tunc, H., Bian, L., Bullington, W.: A modified ant colony optimization algorithm to solve a dynamic traveling salesman problem: a case study with drones for wildlife surveillance. J. Comput. Des. Eng, **6**(3), 368–386 (2019)
9. Jayalath, K., Munasinghe, S.: Drone-based autonomous human identification for search and rescue missions in real-time. In: 2021 10th International Conference on Information and Automation for Sustainability (ICIAS), pp. 518–523. IEEE (2021)
10. Redmon, J., Farhadi, A.: YOLOv3: an incremental improvement (2018). https://doi.org/10.48550/arXiv.1804.02767
11. YOLOv3-Tiny. https://pjreddie.com/darknet/yolo/#google_vignette. Accessed 02 Apr 2024
12. Raspberry Pi 3 Model B+. https://datasheets.raspberrypi.com/rpi3/raspberry-pi-3-b-plus-productbrief.pdf. Accessed 25 Mar 2024

13. Autopilot HAT for Raspberry Pi Powered by ArduPilot and ROS. https://navio2.hipi.io/. Accessed 02 Mar 2024
14. DJI. https://www.dji.com/global. Accessed 13 Sept 2024
15. Racerstar. https://www.racerstar.com/. Accessed 13 Sept 2024
16. GoolRC. https://www.goolrc.com/. Accessed 13 Sept 2024
17. Gens ace. https://www.grepow.com/brands/gensace.html. Accessed 13 Sept 2024
18. Gemfanhobby. https://www.gemfanhobby.com/. Accessed 13 Sept 2024
19. Emlid. https://emlid.com/. Accessed 13 Sept 2024
20. Flysky. https://www.flysky-cn.com/. Accessed 13 Sept 2024
21. Raspberry Pi camera-module-3. https://www.raspberrypi.com/products/camera-module-3/. Accessed 13 Sept 2024
22. ArduPilot a trusted, versatile, and open source autopilot system. https://ardupilot.org/dev/docs/code-overviewobject-avoidance.html. Accessed 25 Mar 2024

Classification of Some Cosmological Images Using Deep Learning and Persistent Homology

Petar Sekuloski[✉] and Vesna Dimitrievska Ristovska

Faculty of Computer Science and Engineering, Ss. Cyril and Methodius University, Rugjer Boshkovikj, 16, 1000 Skopje, Macedonia
{petar.sekuloski,vesna.dimitrievska.ristovska}@finki.ukim.mk
http://www.finki.ukim.mk

Abstract. The study focuses on using models that integrate deep learning and Persistent Homology to classify images of stars and galaxies obtained from po-werful telescopes. The main result of this paper is that the persistent images captured the essential information from the dataset we used. We get slightly better evaluation results, if we use just persistent images obtained from the original images instead of the original images. Also, in this paper we evaluate models that incorporate Persistent Homology and Deep Learning that we have proposed earlier. From the evaluation, we can say that usage of topological characteristics of data improves the classification.

Keywords: Persistent Homology · Image Classification · Topological Data Analysis · Computational Topology · Deep Learning

1 Introduction

The universe is profoundly captivating and endlessly intriguing to the human mind, offering an endless realm of mystery and wonder. From the earliest days of civilization, humans have been fascinated by astronomical phenomena, seeking to understand the vast and complex cosmos. This enduring curiosity continues to drive scientific exploration and discovery, reflecting our deep desire to comprehend the mysteries of the universe.

Differentiating between stars and galaxies in telescope images is a key challenge in cosmology. As telescopes become more advanced and technology progresses, the need for accurate and efficient image classification has grown.

In astronomical images, stars and galaxies can appear similar but have different characteristics that are not always easy to distinguish by eye. Stars usually look like single points of light, whereas galaxies have complex, often intricate structures. Additionally, factors like movement, gravitational effects, and cosmic dust can alter the appearance of these objects, making classification even more challenging.

Using machine learning to classify stars and galaxies enhances the speed and accuracy of data analysis. It helps identify new types of galaxies, detect rare cosmic events, and deepens our understanding of the universe's structure and development. These advanced techniques are now integral to modern astrophysics, enabling researchers to analyze data from cutting-edge telescopes and space missions, and facilitating new discoveries in the cosmos.

The goal of this study is to investigate whether combining Convolutional Neural Networks (CNNs) and Persistent Homology (PH) can yield better results in classifying astronomical images, in the sense that the classification is about the nature of the objects.

The advantage of PH, is that it captures the geometric and topological features of objects in the image, such as their connectivity and shapes, which might not be easily discernible through conventional image processing techniques.

The goal of this paper is to investigate how tools of Topological Data Analysis, more concretely Persistent Homology, can improve classification of cosmological data.

We like to mention that the most of the models used in this paper are proposed in our previous work.

2 Related Work

Deep learning has transformed galaxy classification in cosmology, providing robust tools for analyzing astronomical data more accurately and efficiently. Techniques such as Convolutional Neural Networks (CNNs) have enabled automated extraction of complex features from galaxy images, enhancing our understanding of galaxy morpho-logy, spectral properties, and spatial distributions [1]. This advancement accelerates research into galaxy formation and evolution, contributing to broader cosmological studies [2]. Research and innovations in deep learning and cosmology, as evidenced by studies [3] and [4], promise continued improvements in classifying galaxies across va-rious cosmic scales, paving the way for deeper insights into the universe's structure and dynamics.

Neural network models are evaluated in [5], using one of the dataset used in our work.

In [6] transfer learning is used for classification of cosmological images and there are remarkable results. One of the models in our work incorporate transfer learning and Persistent Homology. The approaches to analyze cosmos data together with analysis are given in details in [7]. Convolutional Neural Network models showed a powerful results.

3 Persistent Homology

Persistent homology (PH) is a sophisticated mathematical tool in the field of Topological Data Analysis (TDA) that enhances our ability to understand and quantify the complex (for more details see [8, 16]), often subtle, topological features present in datasets. Unlike traditional methods that focus on static properties of data, persistent homology introduces a dynamic perspective by examining how these features persist and evolve across multiple scales or levels of detail. Shortly, computing PH of data can be observed in three phases:

Complex Construction: Persistent homology begins with constructing appropriate complexes (such as cubical or simplicial complexes) from the dataset under study. In the context of image analysis, for example, cubical complexes are often used where each pixel or voxel in an image corresponds to a cell in the complex. This structured representation captures the spatial relationships and intensities of the data.

Homology Calculation: Once the complex is constructed, homology groups are computed at each step of the filtration. Filtration [14, 15] refers to a series of nested subsets within a dataset. As this filtration process advances, new topological features appear, existing features may evolve, and some might vanish. The persistence of a feature is determined by the range of filtration values during which that feature remains present. Homology groups reveal the number and types of holes (or topological features) present in the data at different scales. For instance, $H0H\_0H0$ detects connected components, $H1H\_1H1$ identifies loops, and higher-dimensional homology groups capture more complex structures.

Persistent Image: Persistent images are stable vector representations derived from persistence diagrams. They encode essential information about the persistent homology groups of a specific dimension (usually 0 and 1 in image analysis). These representations provide a concise summary of the topological features extracted from the data, facilitating further analysis and interpretation. More on this topic can be found in [8, 9].

4 Methodology

As we mentioned, to investigate the impact of Persistent Homology in cosmological data, we will apply classification models, proposed in our previous work [10], based on deep learning and Persistent Homology in classification of cosmological images, with some modifications. In this section we shortly introduce the methodology and we will accent the modifications that we made for the purpose of this work.

4.1 Constructing Cell Complexes from a Digital Image and Computing Persistent Homology

The cornerstone of the methodology involves constructing cubical complexes and computing persistent homology. Initially, for each grayscale image in our dataset, we construct a cubic complex (see [19]) using a V-construction method (see [17, 18]). Subsequently, homology is computed on the resulting complex derived from the image. Once homology calculations are completed, we derive Persistent Images, which are stable vector representations of persistent homology. Persistent Images are obtained from persistence diagrams and serve as representations of the persistent homology groups of a specific dimension. Since the images used in this study are two-dimensional, the dimensions considered are 0 and 1. Therefore, for each image, we construct Persistent Images denoted as PI_0-which represents homological groups of dimension 0 and PI_1 – which represents homological groups of dimension 1.

4.2 Network Architectures

In this section we will give a description of the network architectures that are used in the experiments. The first model NA_0, which we used in these evaluations, has same architecture as NA_1, but as an input we used just the persistent images, without the original image and we compare with the evaluation if as an input we use the original

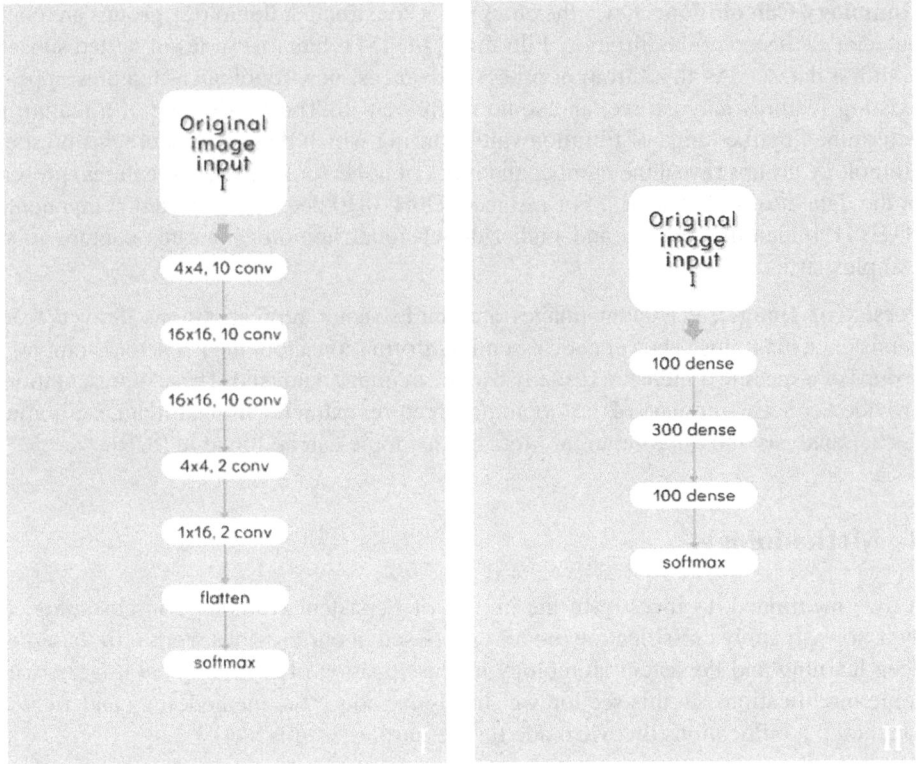

Fig. 1. Models without topological features, NA_1 and NA_2. These models are presented in [10]

images. Also, for the experiments we will use the network architectures, proposed in our previous work [10], that are shortly described in this section. We use two groups of neural networks: Fig. 1 depicts simpler models that directly process grayscale digital images. For the experiments for one of the datasets we will make a modification to this models.

Figure 2 presents more advanced models utilizing topological signatures from Persistent Homology. These models consist of three parts: Part 1 processes the original image, Part 2 uses the persistent images PI_0 for dimension 0, and Part 3 uses PI_1 for dimension 1. Outputs from these parts are concatenated into a single vector, refined with an activation layer, and fed into a classifier. This framework explores how integrating topological features can enhance neural network performance in image classification tasks.

NA_1 is a CNN with 4 convolutional layers and 2 dense layers, while NA_2 is a simple dense neural network, both used for classification without employing topological signatures from Persistent Homology. For more details see [10].

NA_3 is consisted of Part 1 - which is neural network NA_1 without two last layers dense and softmax and Part 2 and Part 3 as topological parts described above [10].

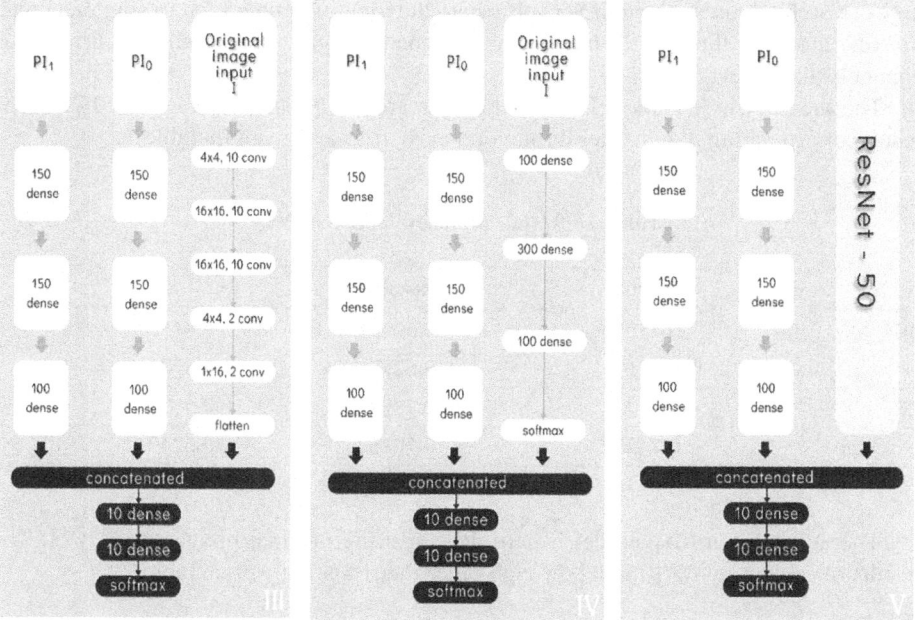

Fig. 2. Models with topological features, NA_3, NA_4 and NA_5 [10]

NA_4 is consisted of Part 1 - which is neural network NA_2 without two last layers dense and softmax and Part 2 and Part 3 as topological parts described above [10].

NA_5 is consisted of Part 1 - which is ResNet50, a transfer learning model, and Part 2 and Part 3 as topological parts described above [10].

5 Experiments and Results

In this section we will investigate how the models given in Sect. 4 work on cosmological dataset consisted of digital images of stars and galaxies. As a main focus of this work, we will evaluate the performance of models that utilize topological signatures from Persistent Homology against those that exclude them on cosmological data.

5.1 Galaxy, Star, Quasar Dataset

The dataset [11] used in this experiment 34,000 source RGB images, including 11543 galaxy sources – 0 class, 11967star sources – 1 class, and 10490 quasar sources - 2 class, originally obtained from SSSD 16 [12]. For our experiments, we rearrange the training set, validation set and test set in ratio 7:1:1. The verification set is used to fine-tune and revise the algorithm during training, while the test set evaluates the final model's generalization ability on unseen data. In the preprocessing phase we resize the pictures to 128×128. Then we computed persistent images PI_0 and PI_1 of the images in the dataset. The final step before training and evaluating our models was to modify the network architectures, given in Sect. 4, where the input was restricted to single-channel images, meaning the images have only one channel (grayscale) rather than multiple channels like RGB.

The accuracy in NA_0 model, where the input are just persistent images, is 0.754. The results of evaluating NA_0 model through other metrics are given in Table 1.

Table 1. Results of the evaluation of NA_0

Class	Precision	Recall	f1-score
0	0.7555	0.6998	0.6458
1	0.7211	0.6277	0.6787
2	0.7547	0.7858	0.5475
Average	**0.7438**	**0.7045**	**0.6240**

The accuracy in NA_1 model, where were just the original images, is 0.7621. The results of evaluating NA_1 model through other metrics are given in Table 2.

Table 2. Results of the evaluation of NA_1

Class	Precision	Recall	f1-score
0	0.7587	0.7871	0.7887
1	0.7322	0.7214	0.7125
2	0.7557	0.7618	0.7002
Average	**0.7486**	**0.7568**	**0.7338**

The accuracy in NA_2 model, where were just the original images, is 0.47. The results of evaluating NA_2 model through other metrics are given in Table 3.

Table 3. Results of the evaluation of NA_2

Class	Precision	Recall	f1-score
0	0.5211	0.5441	0.5147
1	0.4921	0.5412	0.3941
2	0.4781	0.4211	0.4315
Average	**0.4971**	**0.5021**	**0.4468**

The accuracy in NA_3 model, where were just the original images, is 0.851. The results of evaluating NA_3 model through other metrics are given in Table 4.

Table 4. Results of the evaluation of NA_3

Class	Precision	Recall	f1-score
0	0.8711	0.8547	0.9112
1	0.8879	0.9412	0.9012
2	0.8911	0.9015	0.8788
Average	**0.8834**	**0.8991**	**0.8971**

The accuracy in NA_4 model, where were just the original images, is 0.52. The results of evaluating NA_4 model through other metrics are given in Table 5.

Table 5. Results of the evaluation of NA_4

Class	Precision	Recall	f1-score
0	0.5546	0.4871	0.5781
1	0.5647	0.5141	0.5641
2	0.5777	0.4911	0.5471
Average	**0.5657**	**0.4974**	**0.561**

The accuracy in NA_5 model, where were just the original images, is 0.914. The average values for obtained results of evaluating NA_5 model through other metrics are given in Table 6.

Table 6. Average results of the evaluation of NA_5

Class	Precision	Recall	f1-score
Average	**0.9325**	**0.9169**	**0.9156**

The summary results of all used models are given in Table 7.

5.2 ARIES Star/Galaxy Dataset

The dataset [13] used in this study comprises 3986 images of stars and 3986 images of galaxies obtained from observations conducted at the ARIES observatory in Devasthal, Nainital, India. The images are divided into two classes: 942 of the images belong to the "Galaxy"-0 class, and 3044 of the images belong to the "Star"-1 class.

The summary results of the evaluation of all used models are given in Table 8. The values for precision, recall and f1-scores are average values of these metrics for each of the classifiers we used. For the classifiers that use persistent homology there is PH on the right (for example NA_0 (PH)).

Table 7. Summary results of the evaluation of the models

Classifier	Accuracy	Precision	Recall	f1-score
NA_0 (PH)	0.7540	0.7438	0.7045	0.6240
NA_1	0.7621	0.7486	0.7568	0.7338
NA_2	0.4700	0.4971	0.5021	0.4468
NA_3 (PH)	0.8510	0.8834	0.8991	0.8971
NA_4 (PH)	0.5200	0.5657	0.4974	0.5610
NA_5 (PH)	0.9140	0.9325	0.9169	0.9156

Table 8. Summary results of the evaluation of the models

Classifier	Accuracy	Precision	Recall	f1-score
NA_0 (PH)	0.6422	0.6253	0.6079	0.6154
NA_1	0.6160	0.6071	0.5918	0.5994
NA_2	0.4040	0.5049	0.3763	0.4322
NA_3 (PH)	0.8660	0.8036	0.8657	0.8335
NA_4 (PH)	0.7160	0.76257	0.6781	0.7184
NA_5 (PH)	0.9340	0.78587	0.9483	0.8595

6 Discussion

The results of evaluating NA_0 model, where the input are just persistent images, and NA_1 model that have the same neural network architecture, but differ in the input, the first showed slightly better results when in the experiments with the second dataset, and slightly worse results when in the experiments for the first dataset. The results are merely the same. From that, it can be concluded that the persistent images captured from the original images the essential information of the dataset.

From the experiments on both datasets, we can say that NA_3 model, which integrates Persistent Homology, outperforms the NA_1 model, which lacks it. This suggests that adding topological features boosts the network's performance. Conversely, the NA_4 model, which uses a dense neural network instead of a convolutional one, performed worse than NA_3. This likely occurs because convolutional networks are more effective at capturing spatial patterns, especially when combined with Persistent Homology. These results serve as an addition to [10] where the most of the models used in this work are evaluated in [10]. We can conclude that NA_3 showed significant improvement in relation to NA_1 model. It achieved at least 11% improvement on the first dataset and at least 32% improvement on the second dataset across all evaluation metrics.

Also, the transfer learning model NA_5 combining the pre-trained ResNet-50 with Persistent Homology achieved the best performance among all metrics, highlighting the effectiveness of integrating state-of-the-art networks with topological information for superior image classification results.

7 Conclusion and Further Work

Incorporating Persistent Homology and topological signatures notably improved image classification performance on this type of cosmological data.

Looking ahead, future research could explore integrating topological features into more complex classifiers and advanced models. Investigating how these features impact sophisticated algorithms could provide deeper insights and further enhance classification performance.

In our next plans is investigating transformer based methods, such as ViT, and how they might impact the performance.

Also, another way for further work is to apply some mathematical tools form algebraic geometry and topology to deep learning models, in order to interpret the deep the models.

Acknowledgement. The research presented in this paper is partly supported by the Faculty of Computer Science and Engineering, at the Ss. Cyril and Methodius University in Skopje.

References

1. Huertas-Company, M., et al.: Galaxy zoo: morphological classification of galaxy images from the sloan digital sky survey. Mon. Not. R. Astron. Soc. **443**(1), 778–792 (2015). https://doi.org/10.1093/mnras/stu1183
2. Dieleman, S., et al.: Rotation-invariant convolutional neural networks for galaxy morphology prediction. Mon. Not. R. Astron. Soc. **450**(2), 1441–1459 (2015). https://doi.org/10.1093/mnras/stv632
3. Barchi, P.H.P., et al.: Deep learning for galaxy morphology: a critical review. Publ. Astron. Soc. Pac. **132**(1010), 071101 (2020). https://doi.org/10.1088/1538-3873/ab8f01
4. Masters, D., et al.: Classifying the galaxy population using deep learning. Mon. Not. R. Astron. Soc. **487**(1), 819–838 (2019). https://doi.org/10.1093/mnras/stz1287
5. Leifer, E.: Image classification of stars and galaxies using different machine learning models (2023). https://doi.org/10.47611/harp.300
6. Praneeth, N., Kiran, K.C., Vijaya, K.: Stellar data analysis and deep space data analysis system. In: 2024 International Conference on Electronics, Computing, Communication and Control Technology (ICECCC) (2024)
7. Garg, P., Chandra, T., Ahlawat, R., Mittal, N., Ratnesh, R.K., Tripathi, S.K.: Star galaxy image classification via convolutional neural networks. In: 2022 3rd International Conference on Smart Electronics and Communication (ICOSEC), Trichy, India, pp. 1156–1161 (2022). https://doi.org/10.1109/ICOSEC54921.2022.9952065
8. Zomorodian, A., Carlsson, G.: Computing persistent homology. Disc. Comput. Geom. **33**, 249–274 (2005)
9. Adams, H., et al.: Persistence images: a stable vector representation of persistent homology. Found. Comput. Math. **18**, 1–35 (2018)

10. Sekuloski, P., Ristovska, V.D.: Image classification using deep neural networks and persistent homology. In: Mihova, M., Jovanov, M. (eds) ICT Innovations 2023. Learning: Humans, Theory, Machines, and Data. ICT Innovations 2023. Communications in Computer and Information Science, vol. 1991. Springer, Cham (2024). https://doi.org/10.1007/978-3-031-54321-0_1
11. Li, X.: Galaxy, star, quasar dataset [DS/OL]. V1. Science Data Bank, 2023 (2023). Accessed 26 July 2024. https://cstr.cn/31253.11.sciencedb.07177.CSTR:31253.11.sciencedb.07177
12. https://skyserver.sdss.org/dr16/en/home.aspx. Accessed 7 July 2024
13. https://www.kaggle.com/datasets/divyansh22/dummy-astronomy-data. Accessed 11 July 2024
14. Edelsbrunner, H., Harer, J.: Computational Topology: An Introduction. American Mathematical Society (2010)
15. Carlsson, G.: Topology and data. Bull. Am. Math. Soc. **46**(2), 255–308 (2009)
16. Munkres, J.R.: Elements of Algebraic Topology. Perseus Books, New York (2000)
17. Cohen-Steiner, D., Edelsbrunner, H., Harer, J.: Stability of persistence diagrams. Disc. Comput. Geom. **37**(1), 103–120 (2007)
18. Edelsbrunner, H., Letscher, D., Zomorodian, A.: Topological persistence and simplification. Disc. Comput. Geom. **28**(4), 511–533 (2002)
19. Kaczynski, T., Mischaikow, K., Mrozek, M.: Computational Homology. Springer, Heidelberg (2004)

Session 4

Mushroom Classification Using Machine Learning

Gulce Berfin Ercan, Melis Baran, Ecem Konca, Ilhan Mert Cetin, and Ilker Korkmaz(✉)

Izmir University of Economics, Izmir, Turkey
{gulce.ercan,melis.baran,ecem.konca,ilhan.mert}@std.ieu.edu.tr,
ilker.korkmaz@ieu.edu.tr

Abstract. This study aims to develop a robust system using image processing and machine learning to accurately differentiate poisonous and non-poisonous mushroom species, addressing the significant public health threat posed by poisonous mushroom consumption. Motivated by the urgent need for an efficient tool to aid mushroom enthusiasts, farmers, and healthcare professionals in real-time identification of harmful species, the research focuses on creating a mobile application capable of processing mushroom images, extracting pertinent features, and employing a well-trained machine learning model for precise toxic and non-toxic categorization. Through a diverse image dataset collection, preprocessing, feature extraction, and rigorous model evaluation, the study endeavors to enhance public safety and encourage the development of similar applications for species identification and environmental protection. Based on the experiments conducted, amongst many machine learning algorithms used to train a proper system to decide whether a mushroom is edible or poisonous, InceptionV3 deep learning model is chosen to be integrated into the mobile application implemented as the endpoint to the users. Additionally, a simple game is also embedded in the mobile app to make the users learn the poisonous mushrooms from their images.

Keywords: Mushroom classification · Machine learning · Image processing · Mobile application

1 Introduction

People have been fascinated by mushrooms for a long time because they come in all sorts of shapes and colors. Mushrooms are not just interesting to look at; they are also important for the environments they are in. On the other hand, some wild mushrooms may be harmful when eaten or even touched. Considering this issue, this study proposes a mobile application to make the users take pictures of mushrooms using their mobile phones and quickly understand whether they are edible or not.

The challenge which motivated to develop this project is the lack of accessible and reliable tools available to the general public for distinguishing between edible and poisonous mushrooms. For people who like picking wild mushrooms for food, it's often hard to know which mushrooms are safe. This uncertainty might lead to many cases of people getting poisoned by mushrooms. The fact that there is not a simple and reliable solution, like a user-friendly mobile app, to quickly tell if a mushroom is safe during outdoor foraging makes the problem worse. To present a solution to this problem, we propose a mobile smart application integrated with a machine learning model to classify the mushroom images as poisonous and non-poisonous. The goal is to provide the mushroom foragers with a helpful tool to stay safe and avoid dangerous situations.

The mobile application implemented is easy to use, just like any other app on a smartphone. The cool accessory is that the app comprises a simple game to teach people about the risks of eating the wrong mushrooms and help them learn which ones are safe. By this way, we aim to make mushroom picking safer and also take care of the environment by encouraging responsible picking and showing how amazing and diverse mushrooms are in nature.

The rest of this paper is structured in the following manner: Sect. 2 presents a detailed literature review, Sect. 3 explains the methodology employed, Sect. 4 shows the results of the experiments and describes the mobile application implemented, Sect. 5 concludes the paper.

2 Literature Review

Mushroom classification is a critical task with implications for human health and safety. In this section, various approaches for mushroom classification using machine learning and recognition are summarized.

Ahmed et al. [1] compared the following traditional models for mushroom classification: Decision Tree, Random Forest, K-Nearest Neighbors (KNN), Logistic Regression, Support Vector Machine (SVM), and Naive Bayes. The authors noted that Decision Tree, Random Forest, and KNN methods had achieved perfect accuracy amongst the models applied. To interpret the models, the authors used EXplainable Artificial Intelligence (XAI) models SHAP and LIME. SHAP provides global interpretability, highlighting the determined key features across all models. LIME offers local interpretability, explaining individual predictions and emphasizing the importance of odor, gill size, and gill color. The results stressed the importance of specific features in mushroom classification and highlighted the value of XAI in understanding model decision-making.

Zahan et al. [2] employed the deep learning approaches for mushroom classification, leveraging a dataset of diverse mushroom images. The authors evaluated the performance of the following Convolutional Neural Networks (CNNs) on both raw and contrast-enhanced images: VGG16, InceptionV3, and ResNet50. The authors showed that InceptionV3, enhanced by transfer learning and contrast-enhanced images, surpassed the other architectures with an accuracy of 88%. The effectiveness of using InceptionV3 model for mushroom classification when contrast-enhanced images are incorporated into the analysis was affirmed.

Chawathe [3] explored the classification of mushrooms as edible or poisonous considering observable properties like color, texture, and dimensions of mushroom sections. Data intensive methods using an augmented mushroom dataset were focused. The study quantified the dataset's merit for classification, evaluated the classification efficiency and explored the attribute selection methods. Results indicated the augmented dataset's richness, achieving high accuracy with efficient classifiers. The classifier algorithms examined includes Decision Tree (C4.5), Bayes Net, KStar, Random Forest and Naive Bayes. The paper concluded by emphasizing the need for a robust dataset in accurate mushroom classification.

Lee et al. [4] introduced a smartphone application, Purdue University Mushroom App (PUMA), employing CNNs for image classification to distinguish poisonous and edible mushroom species. PUMA incorporates three classification models and employs deep learning techniques, including ResNet with 152 layers. The training involved diverse image datasets, and then the trained models were integrated into the Android app for real-time classification. The results demonstrated high sensitivity and specificity, ranging from 89% to 100% for different models. The authors mentioned the challenges such as variations in image resolution and explored the implications of different smartphones and operating systems on classification accuracy.

Wang et al. [5] proposed an automated method for mushroom toxicity identification using the gcForest algorithm, highlighting the limitations in Logistic Regression and SVMs for accurate classification based on visual features. Emphasizing feature engineering and data preprocessing, including KNN for handling missing data, the authors evaluated the performances of gcForest, Logistic Regression, and SVM models regarding the accuracy, precision, recall, F1-score, and Receiver Operating Characteristic (ROC) curve. The gcForest model achieved approximately 98% accuracy. The study advocated for enhancements in the stability of the gcForest classifier to ensure reliable and consistent results.

Wibowo et al. [6] delved into the limited research on edible mushroom classification in Indonesia that boasts 13% of global mushroom species. The authors utilized machine learning and data mining approaches, and compared the following three classifier algorithms for identifying edible and poisonous mushrooms: C4.5, Naive Bayes, and SVM. Having conducted experiments using the Waikato Environment for Knowledge Analysis (WEKA) tool [7] with mushroom data from the Agaricus and Lepiota families, the authors measured 96% accuracy on the tests of Naive Bayes model, whereas they obtained almost a perfect accuracy on the tests of the C4.5 and SVM models.

Tutuncu et al. [8] conducted a study on edible and poisonous mushroom classification using the UCI mushroom dataset [9], focusing on four machine learning algorithms: Decision Tree, Naive Bayes, SVM, and AdaBoost. Based on their experimental findings, the authors mentioned that AdaBoost algorithm had achieved a perfect classification accuracy, outperforming the other models. The study highlighted the effectiveness of AdaBoost model in classifying mushrooms based on physical features, contributing to the development of reliable mushroom identification systems.

Kalyani and Manikanteswari [10] investigated the classification of poisonous and edible mushrooms using various learning approaches on a dataset with 22 features. The authors applied CNNs alongside traditional models like Decision Trees, Logistic Regression, Gaussian Naive Bayes, and Random Forest. The authors reported a perfect accuracy using Random Forest, whereas an accuracy of 99.98% using Decision Tree, and an accuracy of 95.25% using Gaussian Naive Bayes. The study demonstrated the effectiveness of combining deep learning models with traditional methods for reliable mushroom classification.

As a summary of the literature, Table 1 shows the methods used for mushroom classification in the aforementioned papers.

Table 1. Methods used for mushroom classification in the literature.

Methods	[1]	[2]	[3]	[4]	[5]	[6]	[8]	[10]
VGG16	✗	✓	✗	✗	✗	✗	✗	✗
InceptionV3	✗	✓	✗	✗	✗	✗	✗	✗
ResNet	✗	✓	✗	✓	✗	✗	✗	✗
Decision Tree	✓	✗	✓	✗	✗	✓	✓	✓
SVM	✓	✗	✗	✗	✓	✓	✓	✗
Random Forest	✓	✗	✓	✗	✗	✗	✗	✓
Logistic Regression	✓	✗	✗	✗	✓	✗	✗	✓
KNN	✓	✗	✗	✗	✓	✗	✗	✗
Naive Bayes	✓	✗	✓	✗	✗	✓	✓	✓
KStar	✗	✗	✓	✗	✗	✗	✗	✗
Bayes Net	✗	✗	✓	✗	✗	✗	✗	✗
AdaBoost	✗	✗	✗	✗	✗	✗	✓	✗

3 Methodology

This section explains in detail the methodology used to develop the machine learning model for classifying mushroom images as either poisonous or non-poisonous.

3.1 Data Collection and Image Processing

It is necessary to ensure that the dataset is meticulously categorized, covering a wide spectrum of mushroom species to enable the model to generalize effectively. The mushroom dataset [11] on Kaggle that encompasses a diverse range of images is utilized as the data source. This dataset serves as the cornerstone for training the machine learning model, ensuring that it learns from a wide array of mushroom types, both poisonous and non-poisonous.

The dataset used in our experiments was created by downloading original mushroom dataset [11] and manually cleaning up poor quality photos. Initially, any corrupted or mislabeled images were identified and removed from the dataset. This step ensured that the dataset only contained valid and correctly labeled images for training. The final dataset we used contains 6000 images (i.e., 3000 edible and 3000 poisonous). The entire data was split into training (80%) and test (20%) sets.

To prepare the dataset for model training a series of standardized preprocessing steps are adopted. These steps include resizing all images to a uniform size, normalizing the pixel values into a standardized range, promoting consistency in image dimensions and mitigating any bias towards certain image dimensions or color schemes.

The process of feature extraction involves the implementation of image processing techniques to identify and capture crucial features from the mushroom images. The attributes focused are color, texture, size, and shape, which are fundamental for distinguishing between different mushroom types.

3.2 Machine Learning Model

Convolutional networks have become dominant for recognition and detection tasks [12]. CNNs excel in classification because of their innate advantage on recognizing intricate patterns and features in images, a critical requirement for differentiating between diverse mushroom species. The architectural design of CNNs, which facilitates the automatic learning of spatial hierarchies of features, aligns seamlessly with our project's core objective of accurately classifying mushrooms, despite their wide-ranging visual characteristics.

3.3 Application Integration

The seamless integration of the trained machine learning model into a mobile application is pivotal. This integration would empower real-time classification of mushroom images uploaded by users, offering immediate feedback regarding the toxicity of the identified mushrooms.

The development of a user-friendly mobile application utilizing Flutter [13] is central to the project's achievement. Our application is designed to provide a convenient interface for users, enabling them to effortlessly upload or capture mushroom images and promptly receive classification results. Additionally, an educational game section is incorporated to educate people about mushroom safety.

4 Implementation and Results

In our project, we preferred to use ResNet50, VGG16, EfficientNetB0 and InceptionV3 pretrained CNN models and update them through transfer learning to create a proper model. InceptionV3 model, with its advanced architecture, has demonstrated exceptional effectiveness in image classification tasks, surpassing alternative models in accuracy and generalization when evaluated with the mushroom image dataset.

4.1 Training Methods

This section outlines the training methods and techniques we applied to develop robust deep learning models for mushroom recognition. Each model underwent the same preprocessing steps mentioned in Sect. 3.1, followed by training with methods designed to enhance performance and generalization.

Transfer learning was a pivotal process used for the pretrained ResNet50, VGG16, EfficientNetB0 and InceptionV3 models. These models, initially trained on the extensive ImageNet dataset [14], then were fine-tuned with the specific mushroom dataset. This allowed the models to benefit from prior knowledge and adapt to the new task with fewer training epochs.

Early stopping approach was utilized to prevent overfitting. By monitoring the validation loss, training was halted once the performance on the validation set ceased to improve.

Dynamic adjustment of the learning rate was implemented. Learning rate scheduling is a technique used to adjust the learning rate during training to enhance the convergence and performance of a model. In this respect, learning rate is a key hyperparameter in neural network training.

Dropout regularization was applied to the models during training. By randomly dropping neurons, this technique prevented the models from becoming overly reliant on specific pathways, thereby enhancing their generalization capabilities.

Each model was trained using these comprehensive techniques to ensure optimal performance and generalization. The combination of preprocessing, transfer learning, early stopping, learning rate scheduling, and dropout regularization collectively contributed to the development of robust mushroom recognition models. To provide a robust estimation of model performance, 5-fold cross-validation technique was used in the experiments.

4.2 Results of the Models

The initial experiments on transfer learning were conducted without cross validation. The epoch size to train the models was chosen 20 with a batch size of 32. Early stopping was used.

To assess the efficiency of the transfer learning, firstly a CNN architecture alone without a pretraining model (built and trained from scratch using TensorFlow and Keras) was employed on the dataset. The test accuracy of the base CNN model was measured as 75%, which was observed to be the worst compared to the transfer learning models of which the results are given in the rest of this section. So, the base CNN results without a transfer learning approach will not be mentioned within the rest of this paper.

Regarding the comparison of the transfer learning models based on the performance metrics, the experiment results were assessed using accuracy, recall, precision, F1 score and AUC. Additionally, the 2 × 2 confusion matrix was considered to comprehensively understand the performance of the classification. The test data were chosen to be 20% (600 images) of the image dataset.

Afterwards, extra experiments were conducted using 5-fold cross validation to fairly evaluate the performance of each model. It was considered to measure the mean values of the results to reach a more robust decision.

Table 2 shows the results of the initial experiments for the models. The rows are sorted with regards to the validation data accuracy results.

Table 2. Performance results of the corresponding models.

Algorithm	Accuracy	Recall	Precision	F1-score	AUC
InceptionV3	0.86	0.82	0.89	0.85	0.93
ResNet50	0.81	0.83	0.79	0.81	0.89
VGG16	0.77	0.69	0.82	0.75	0.86
EfficientNetB0	0.74	0.74	0.73	0.77	0.79

As seen in Table 2, InceptionV3 performed the best amongst the models with an accuracy of 86%. Furthermore, InceptionV3 outperformed the others in terms of precision, F1-score and AUC. Regarding recall value, InceptionV3 is not the best, however it is so close to ResNet50 that has the highest value. Figure 1 shows the confusion matrix results for all models.

As InceptionV3 model outperformed the other models with regards to the test accuracy ratio, it was used to be integrated into the mobile application. To evaluate InceptionV3 model in detail, its training and validation performances in terms of the accuracy and loss values are visualised at different epochs in Fig. 2.

In Fig. 2, training accuracy shows a consistently upward trend, indicating that the model is learning and improving its performance on the training dataset over epochs. It starts around 76% and reaches above 95%, which suggests a good fit on the training data. Validation accuracy line is more rougher, fluctuating as epochs increase. It begins to plateau around the 10th epoch, indicating that the model is not significantly improving on the validation dataset after this point. The gap between training and validation accuracy may be an indication of an overfitting.

In Fig. 2, as the training accuracy increases the corresponding training loss decreases, as expected. The validation loss is fluctuating between 35% and 40% before the 15th epoch. After the 15th epoch the gap between the traning and validation loss gets increased much.

Figure 3 shows the ROC curve for InceptionV3 model to illustrate its strong performance in classifying mushrooms as "Edible" or "Poisonous". The curve shows a high true positive rate across varying thresholds, and the AUC (Area

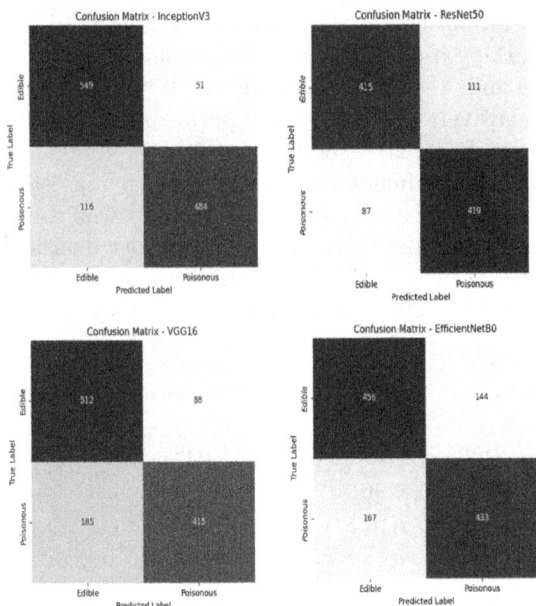

Fig. 1. Confusion matrix results for the corresponding models.

Under the Curve) value of 0.9345 indicates excellent discrimination ability. The curve's position well above the diagonal random classifier line confirms that the model is highly effective at distinguishing between the two classes.

Regarding the extra experiments conducted using 5-fold cross validation, the rest of this subsection gives the details of the experiments for each model.

ResNet50. The model was trained using the training set with a suitable optimizer (Adam) and a categorical cross-entropy loss function. The validation set was used for monitoring the model and preventing overfitting.

Table 3 lists the results obtained for each of 5-fold cross validation experiments on this model. The last row shows the mean value of the results within the corresponding column. To evaluate fairly, the mean values should be considered. As observed in Table 3, ResNet50 model achieved an accuracy of $\sim 81\%$ on average.

VGG16. In this model, VGG16 was used within the transfer learning approach to leverage ImageNet-pretrained weights for effective image recognition. Custom layers, including a Global Average Pooling Layer and Dense layers with 50% Dropout, were added into the architecture to reduce overfitting. The model was compiled using Adam optimizer and binary cross-entropy loss function, wihch is appropriate for binary classification task.

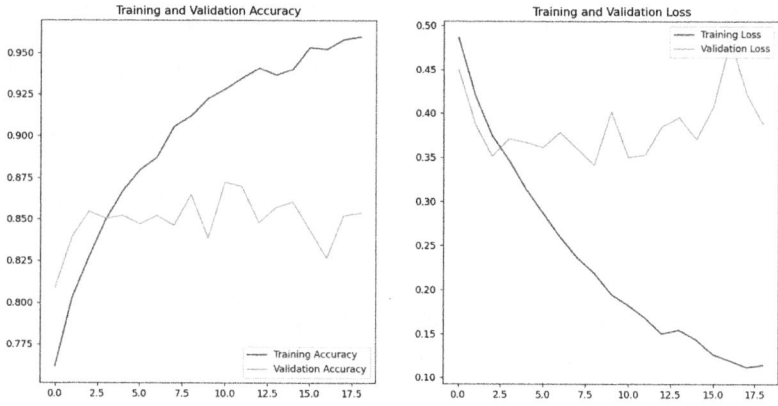

Fig. 2. Training and validation performances for InceptionV3 (a) Accuracy (b) Loss.

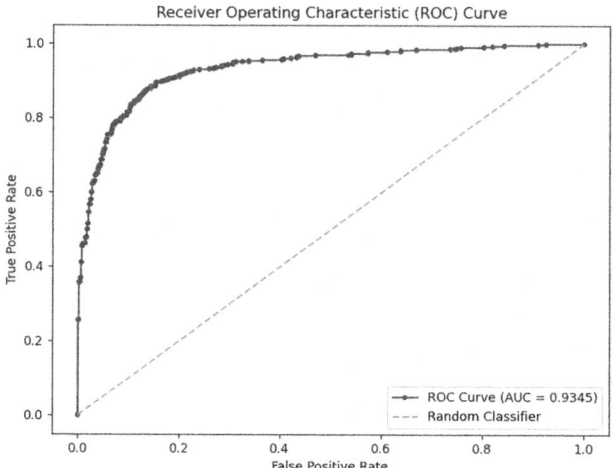

Fig. 3. ROC curve for InceptionV3.

Table 3. Cross-validation results for ResNet50 model.

Dataset	Accuracy	Recall	Precision	F1-score	AUC
Fold#1	0.8004	0.7490	0.8347	0.7896	0.8962
Fold#2	0.8175	0.8415	0.8051	0.8229	0.8825
Fold#3	0.8156	0.9312	0.7731	0.8448	0.8979
Fold#4	0.7975	0.8401	0.7703	0.8037	0.8880
Fold#5	0.8013	0.8368	0.8074	0.8218	0.8883
mean	**0.8065**	**0.8397**	**0.7981**	**0.8166**	**0.8906**

Training occured over 20 epochs with EarlyStopping to prevent overtraining, ModelCheckpoint to save the best model, and ReduceLROnPlateau to adjust the learning rate when validation loss stagnates. Fine-tuning was then performed by unfreezing the top 15 layers of VGG16, allowing the model to better adapt to the specific task of mushroom classification. The model was recompiled with a lower learning rate to refine its accuracy, ensuring both general feature learning and adaptation to the new dataset. As seen in Table 4, VGG16 model achieved a test accuracy of ∼65% on average.

Table 4. Cross-validation results for VGG16 model.

Dataset	Accuracy	Recall	Precision	F1-score	AUC
Fold#1	0.6469	0.7042	0.6318	0.6660	0.7089
Fold#2	0.6635	0.6604	0.6646	0.6625	0.7133
Fold#3	0.6240	0.6062	0.6285	0.6172	0.6706
Fold#4	0.6542	0.5354	0.7022	0.6076	0.7177
Fold#5	0.6604	0.6937	0.6504	0.6714	0.7250
mean	**0.6498**	**0.6400**	**0.6498**	**0.6449**	**0.7071**

EfficientNetB0. This model was based on the EfficientNetB0 architecture, and custom layers including GlobalAveragePooling2D, BatchNormalization, and Dense layers with Dropout were added. These modifications tailored the network to the corresponding binary classification task. The Dropout layers played a critical role in preventing overfitting by randomly omitting some features during training. The model was compiled using Adam optimizer and binary cross entropy loss function. As seen in Table 5, this model achieved a test accuracy of ∼67%.

Table 5. Cross-validation results for EfficientNetB0 model.

Dataset	Accuracy	Precision	Recall	F1-score	AUC
Fold#1	0.6718	0.6603	0.7093	0.6821	0.7512
Fold#2	0.6998	0.6887	0.7312	0.7097	0.7567
Fold#3	0.6513	0.6471	0.6585	0.6505	0.7105
Fold#4	0.6546	0.6402	0.6911	0.6679	0.7342
Fold#5	0.6698	0.6658	0.6792	0.6713	0.7379
Mean	**0.6695**	**0.6604**	**0.69386**	**0.6763**	**0.7381**

InceptionV3. This model utilized the InceptionV3 architecture, pre-trained on ImageNet, enhanced with additional layers including Global Average Pooling, Dense layers, and a 30% Dropout rate to prevent overfitting. The model was compiled using Adam optimizer and binary cross-entropy, and was trained for 20 epochs using callbacks like EarlyStopping, ModelCheckpoint, and ReduceLROnPlateau to optimize the performance.

Post-training, fine-tuning was performed by unfreezing layers after the 249th layer of InceptionV3 and reducing the learning rate to 0.00001, allowing the model to better adapt to the task.

Table 6 lists the results obtained for each of 5-fold cross validation experiments on this model. As observed in Table 6, InceptionV3 model achieved an accuracy of ~77% on average.

Table 6. Cross-validation results for InceptionV3 model.

Dataset	Accuracy	Recall	Precision	F1-score	AUC
Fold#1	0.7729	0.7812	0.7684	0.7748	0.8406
Fold#2	0.7854	0.8375	0.7585	0.7960	0.8670
Fold#3	0.7542	0.6813	0.7976	0.7348	0.8417
Fold#4	0.7510	0.9125	0.6898	0.7857	0.8682
Fold#5	0.8083	0.7875	0.8217	0.8043	0.8684
mean	**0.7744**	**0.8000**	**0.7672**	**0.7791**	**0.8572**

Although Resnet50 model outperformed InceptionV3 model with cross validation, InceptionV3 obtained the highest accuracy without cross validation as shown in Table 2. Since InceptionV3 performed the best test accuracy amongst the models, it was used to be integrated into the mobile application. The final model was exported in TensorFlow's SavedModel format and then it was converted to TensorFlow Lite for deployment on mobile or embedded devices.

4.3 Mobile Application

A mobile application that can identify poisonous and non-poisonous mushrooms using the InceptionV3 model designed with TensorFlow and trained on a comprehensive dataset has been implemented using Dart programming language on open source Flutter framework.

The application works as follows: After launching the app, users can select the camera to take a photo of any mushroom and find out if it is poisonous. A few seconds after taking the photo, the screen displays whether the mushroom is poisonous or non-poisonous.

Additionally, the app includes a simple game related to mushrooms. The game works in the following manner: Four mushroom images appear on the screen. The user tries to identify the edible mushroom. If the choice is correct, they will earn points and move on to the next set of four mushrooms. If they guess incorrectly, they can try again and continue until they identify the correct mushroom. Moreover, the application provides information about mushrooms periodically. Finally, the application provides a page that lists the names of mushrooms, showing them as red if they are poisonous, green if they are not poisonous.

Regarding the software system design, Fig. 4 illustrates the activity diagram of the mobile application.

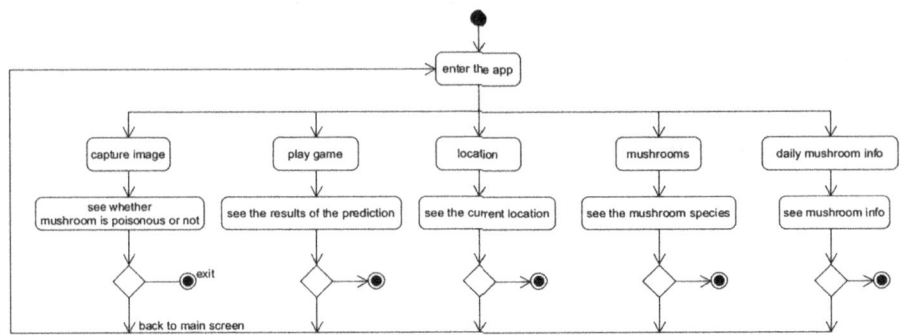

Fig. 4. Activity diagram of the mobile application.

The mobile application offers five main options to its users: Capture Image, Play Game, Location, Mushrooms, and Daily Mushroom Info. If user selects the Capture Image button, the camera will be enabled, allowing them to take a photo of a mushroom. The app then analyzes the taken photo (i.e., input test image) to determine whether the mushroom is edible or poisonous. For more information, user can access a detailed mushroom information page. Alternatively, selecting the Play Game button directs the user to a game where they identify edible mushrooms from a set of four images. Correct guesses earn points, and the user proceeds to the next set, with their score and the number of right and wrong answers displayed at the top. Furthermore the Location button opens a map where the user can mark the spot where they found a mushroom. The Mushrooms button leads to a list page, displaying the names and edibility of various mushrooms. Finally, the Daily Mushroom Info button provides daily updated information about mushrooms. After using any feature, the user can choose to exit the app or continue exploring its functionalities.

Regarding the system implementation and test demonstration, Fig. 5 shows a few number of sample screenshots from the mobile application.

4.4 Comparison with Other Mobile Apps

Table 7 presents a comparison of key features offered by different mushroom identification applications including our proposed app, "Mushroom Identify -

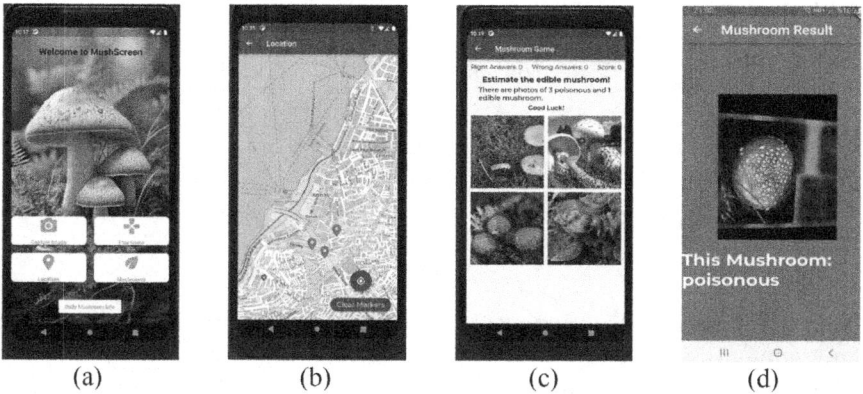

Fig. 5. Sample outputs from the mobile application (a) Home Page (b) Marked Location Map (c) Game Page (d) Decision Result Page.

Automatic" [15], "ShroomID" [16], and "Picture Mushroom - Mushroom ID" [17]. These compared applications are publicly available on Google Play with TRL9 level. The features to compare encompass recognition capabilities, gaming elements, integration with mushroom-related books, provision of location services, daily information updates, and the extent of their mushroom databases.

As seen in Table 7, our proposed app provides users with recognition functionality, integrating gaming elements, and offers location services and a comprehensive mushroom database. However, it does not include integration with mushroom-related books or daily information updates.

"Mushroom Identify - Automatic" encompasses recognition features, incorporates gaming elements, and seamlessly integrates with mushroom-related books. Additionally, it offers location services, yet it lacks daily information updates.

"ShroomID" includes integration with mushroom-related books, provides daily information updates, and offers users access to a substantial mushroom database. However, it lacks recognition capabilities and gaming elements.

"Picture Mushroom - Mushroom ID" incorporates recognition functionalities and integrates with mushroom-related books. It also provides access to a mushroom database. However, it does not include gaming elements or daily information updates.

Although our project does not offer a new algorithm to the machine learning area, it contributes to the literature by integrating the classification service of mushroom images with a game within an app so that the app would help the people be educated about the mushrooms and stay updated.

4.5 Possible Future Work

To improve test accuracy performance of the classification model and to provide the users with an improved application, the following may be the possible future works to consider:

Table 7. Comparison of the features provided by the other mobile applications.

Application	Recognition	Game	Book	Location	DailyInfo	Database
OurApp	Yes	Yes	No	Yes	Yes	No
MushroomIdentify [15]	No	Yes	Yes	Yes	Yes	Yes
ShroomID [16]	Yes	No	Yes	Yes	No	Yes
PictureMushroom [17]	Yes	No	Yes	Yes	No	Yes

- Enhanced Dataset: Continuously updating and expanding the dataset used for training the machine learning models is essential. This ensures the system's adaptability to new mushroom species.
- User Feedback Integration: Integrating user feedback and validation mechanisms can enhance the system's accuracy and reliability over time.
- Cross-Platform Compatibility: While the current focus is on Android, future work could explore expanding the application to iOS as well.
- Device and Resolution Adaptability: Possible techniques to enhance the application's adaptability to different smart devices and image resolutions.
- Integration of Image Storage (Database): The implementation of a feature that can store test images along with their corresponding identified labels (edible or poisonous) in a database would be considered.

5 Conclusion

In this study, it has been presented a strategy to develop a mobile application to effectively discern between toxic and non-toxic mushroom species using machine learning.

The data obtained from the Kaggle image dataset has been preprocessed to yield an efficient training. The performances of ResNet50, VGG16, EfficientNetB0 and InceptionV3 models used with transfer learning technique have been evaluated. The model trained with InceptionV3 has been integrated into the mobile app for mushroom classification.

The combination of TensorFlow and the InceptionV3 model produced a highly accurate system capable of providing real-time classification results.

Additionally, a simple educational game has been incorporated into the mobile application to enhance the users' awareness on mushroom safety in an engaging manner.

References

1. Ahmed, M.S., et al.: Comparative analysis of interpretable mushroom classification using several machine learning models. In: 2022 25th International Conference on Computer and Information Technology (ICCIT), Cox's Bazar, pp. 31–36 (2022). https://doi.org/10.1109/ICCIT57492.2022.10055555

2. Zahan, N., Hasan, M.Z., Malek, M.A., Reya, S.S.: A deep learning-based approach for edible, inedible and poisonous mushroom classification. In: 2021 International Conference on Information and Communication Technology for Sustainable Development (ICICT4SD), Dhaka, Bangladesh, pp. 440–444 (2021). https://doi.org/10.1109/ICICT4SD50815.2021.9396845
3. Chawathe, S.S.: Automated determination of mushroom edibility using an augmented dataset. In: 2022 IEEE World AI IoT Congress (AIIoT), Seattle, WA, USA, pp. 617–623 (2022). https://doi.org/10.1109/AIIoT54504.2022.9817321
4. Lee, J.J., Aime, M.C., Rajwa, B., Bae, E.: Machine learning-based classification of mushrooms using a smartphone application. Appl. Sci. **12**(22), 11685 (2022). https://doi.org/10.3390/app122211685
5. Wang, Y., Du, J., Zhang, H., Yang, X.: Mushroom toxicity recognition based on multigrained cascade forest. Sci. Program. **2020**(1), 8849011 (2020). https://doi.org/10.1155/2020/8849011
6. Wibowo, A., Rahayu, Y., Riyanto, A., Hidayatulloh, T.: Classification algorithm for edible mushroom identification. In: 2018 International Conference on Information and Communications Technology (ICOIACT), Yogyakarta, Indonesia, pp. 250–253 (2018). https://doi.org/10.1109/ICOIACT.2018.8350746
7. WEKA Software Workbench. https://ml.cms.waikato.ac.nz/weka/. Accessed 05 May 2024
8. Tutuncu, K., Cinar, I., Kursun R., Koklu, M.: Edible and poisonous mushrooms classification by machine learning algorithms. In: 2022 11th Mediterranean Conference on Embedded Computing (MECO), Budva, Montenegro, pp. 1–4 (2022). https://doi.org/10.1109/MECO55406.2022.9797212
9. UCI Machine Learning Repository: Mushroom. https://doi.org/10.24432/C5959T. Accessed 05 May 2024
10. Kalyani, C.S., Manikanteswari, D.S.L.: Classification of edible and poisonous mushrooms using machine learning algorithms. IJFMR **6**(2) (2024). https://doi.org/10.36948/ijfmr.2024.v06i02.18538
11. Predict Poison Mushroom By Photo as appeared on Kaggle. https://www.kaggle.com/datasets/stepandupliak/predict-poison-mushroom-by-photo. Accessed 05 May 2024
12. LeCun, Y., Bengio, Y., Hinton, G.: Deep learning. Nature **521**(7553), 436–444 (2015). https://doi.org/10.1038/nature14539
13. Flutter Homepage. https://flutter.dev/. Accessed 05 May 2024
14. ImageNet Homepage. https://www.image-net.org/. Accessed 05 May 2024
15. Annapurnapp Technologies: Mushroom Identify - Automatic as appeared on Google Play Store. https://play.google.com/store/apps/details?id=com.pingou.champignouf. Accessed 05 May 2024
16. AI App Experts: ShroomID - Identify Mushrooms! as appeared on Google Play Store. https://play.google.com/store/apps/details?id=com.shroomid. Accessed 05 May 2024
17. Next Vision Limited: Picture Mushroom - Mushroom ID as appeared on Google Play Store. https://play.google.com/store/apps/details?id=com.glority.picturemushroom. Accessed 05 May 2024

Towards a Framework for Promoting Student Engagement to Maximize Learning in Higher Education: A Case Study

Livinus Obiora Nweke[1,2](✉) [iD] and Rania El-Gazzar[2,3]

[1] Noroff University College, Tordenskjoldsgate 9, 4612 Kristiansand, Norway
livinus.nweke@noroff.no
[2] University of South-Eastern Norway (USN), Borre, Norway
[3] University of Agder, Kristiansand, Norway

Abstract. Student engagement is a well-established factor that impacts the achievement of learning outcomes in higher education and as such, it is being widely researched. This paper explores the efficacy of the Context Challenge Activity Feedback (CCAF) framework in an in-person classroom setting for third-year bachelor students in an Information Systems (IS) Security course. Adopting a quantitative research methodology, the study analyzes student surveys to evaluate the effect of the CCAF framework on aspects such as the applicability of course material to real-world scenarios, course structure and clarity, stimulation of intellectual curiosity, and overall student satisfaction. The findings reveal that a majority of students perceive the course material as highly relevant to real-world IS Security issues, indicating the framework's effectiveness in contextualizing theoretical concepts. The course's structure and content clarity were also well-received, with a majority of students reporting positive experiences, suggesting the framework's success in creating a clear and engaging learning environment. However, responses regarding the stimulation of intellectual curiosity showed variability, highlighting the necessity for adaptive teaching methods to cater to diverse learning preferences. Overall, the study reports high levels of student satisfaction, underscoring the potential of the CCAF framework in enhancing learning experiences in IS Security within traditional classroom settings. The study contributes to the existing studies by providing empirical support for the application and efficacy of the CCAF framework in in-person higher education settings, particularly in a technical and rapidly evolving field like IS Security. The findings have significant implications for pedagogical practices, suggesting that innovative, student-centered approaches like CCAF can effectively enhance student engagement and educational experiences in higher education.

Keywords: CCAF · Student-centered pedagogy · Student engagement · Active learning · Educational practice and policy · Information systems security

1 Introduction

The landscape of higher education is constantly evolving to incorporate new pedagogical strategies and learning environments. Among these, the Context Challenge Activity Feedback (CCAF) framework, as proposed by [2] has emerged as a noteworthy approach, primarily in online learning contexts. This framework, rooted in the theories of active learning and student-centered pedagogy, has gained widespread attention for its capacity to enhance student engagement and improve learning outcomes.

Research on student engagement, a complex construct encompassing emotional, behavioral, and cognitive dimensions, highlights its importance in the educational process [16,17]. Engaged students tend to excel academically, retain knowledge efficiently, and demonstrate heightened motivation and satisfaction [54,61]. The CCAF framework, with its emphasis on contextualizing content, presenting challenging activities, and providing immediate feedback, aligns well with the principles that promote engagement [56]. The study by [38] suggests that the CCAF framework can enhance learners' engagement to optimize learning in a synchronous online workshop. Similarly, the studies by [57] and [59] also support the idea that the framework aligns well with principles that foster student engagement.

Previous studies have shown that active learning environments, such as those promoted by the CCAF framework, contribute to greater student engagement and improved learning outcomes [10,48]. However, much of the existing literature focuses on online or hybrid learning contexts [38,57]. There is a noticeable gap in research regarding the implementation of the CCAF framework in purely physical classroom environments, especially among third-year bachelor students who are at a crucial stage of their academic journey.

Consequently, this current study seeks to explore the following research questions:

- How can the CCAF framework be applied to maximize student engagement and learning outcomes in an in-person classroom setting?
- How does the CCAF framework affect student engagement, perceived relevance of the course material, and overall satisfaction?

Accordingly, this study provides empirical validation for the efficacy of the CCAF framework in fostering student engagement to maximize learning in an in-person classroom setting for third-year bachelor students in an Information Systems (IS) Security course. This exploration is pivotal, as it extends the application of the CCAF framework beyond the realm of online learning and adds to the wider conversation on enhancing educational practices in physical classroom environments.

The results of this study are anticipated to offer valuable perspectives for educators and curriculum developers, emphasizing the importance of engagement-focused teaching strategies in higher education. The rest of this article is structured as follows. Section 2 presents the literature review, where the concept of student engagement, the importance of innovative teaching methods in higher education, the CCAF framework, and the related work are explored. Section 3 outlines the method employed in this study. Section 4 discusses the findings of this study, while Sect. 5 presents the discussion and implications of the results. Lastly, Sect. 6 concludes the paper.

2 Literature Review

This section presents an analysis of student engagement, the significance of novel teaching methods in higher education, and the CCAF framework, offering the necessary background for understanding the research undertaken in this study. Additionally, an overview of works related to the current study is presented.

2.1 Student Engagement

The notion of student engagement has been thoroughly explored in educational research, especially within the realm of higher education. This body of literature presents a multifaceted view of engagement, encompassing behavioral, emotional, and cognitive dimensions, each contributing to students' overall educational experience and outcomes.

Behavioral engagement refers to students' involvement in academic tasks, both inside and outside the classroom. This includes attendance, contribution to discussions, and involvement in extracurricular pursuits. Astin's Theory of Student Involvement [3] is a foundational work in this area, emphasizing the significance of proactive participation in fostering learning and development. Studies by [29,32,63], and [28] further elaborate on this, highlighting the link between engagement in academic and campus pursuits and higher academic achievement.

Emotional engagement covers students' affective responses to their learning environment, including their feelings of interest, enthusiasm, and sense of belonging. The work by [60] on student retention underscores the importance of emotional engagement, particularly the sense of belonging, in student persistence and success in higher education. Research by [26,43,55], and [53] also examine the influence of emotions in engagement, noting how positive emotional experiences enhance motivation and commitment to learning.

Cognitive engagement involves the depth of cognitive and metacognitive approaches employed by students in their learning. The model of self-regulated learning proposed by [65] highlights how cognitive engagement is linked to the application of sophisticated learning methods and critical thinking skills. Similarly, [10,44,62], and [46] discuss how cognitive engagement is vital for deep learning and understanding, as it involves students actively processing and evaluating information.

The existing studies identify several factors influencing student engagement. Teaching strategies, such as the use of active learning methods, have been shown to significantly impact engagement [18,49]. The institutional environment, including support services and campus culture, also plays a crucial role

[51,61]. Moreover, individual student factors, like motivation, prior educational experiences, and learning styles, influence levels of engagement [30,34].

The impact of student engagement is far-reaching. Research by [15,31], and [37] demonstrate that engagement is strongly correlated with positive academic outcomes, including higher grades and retention rates. Furthermore, studies by [33,42], and [41] show that engagement is linked to overall student well-being and satisfaction.

The existing literature on student engagement provides an overarching understanding of its multidimensional nature and its critical role in student learning and development in higher education. It highlights the necessity for educational institutions to foster environments and teaching methodologies that enhance all aspects of student engagement to improve educational outcomes.

2.2 Importance of Innovative Teaching Approaches in Higher Education

In recent years, the adoption of innovative teaching methods in higher education has become increasingly important. Traditional lecture-based teaching approaches are often criticized for their limited ability to engage students actively and meaningfully in the process of learning [18,19]. In contrast, innovative teaching methods, which include a variety of pedagogical approaches such as collaborative learning, problem-based learning, and the application of technology in learning, have been shown to enhance student engagement, and improve learning outcomes [20,39,48].

The importance of these methods stems from their alignment with contemporary educational theories, such as constructivism, which emphasize the active role of learners in constructing their own knowledge [8]. By engaging students in hands-on, collaborative, and problem-solving activities, these methods foster a better understanding of the subject matter and help them develop critical thinking skills [14]. Additionally, they are more attuned to the needs of today's heterogeneous student population, which includes digital natives who are accustomed to interactive and technology-driven environments [47].

Innovative teaching methods are also important for preparing students for the challenges of the 21st-century workplace. Employers increasingly value skills such as critical thinking, problem-solving, teamwork, and adaptability, that can be fostered through these modern pedagogical approaches [5,40]. Furthermore, as higher education institutions face growing pressure to demonstrate their relevance and value, adopting innovative teaching methods can enhance student satisfaction and success, thereby improving institutional reputation and competitiveness [12].

The shift towards innovative teaching methods in higher education comes with its challenges. For example, it requires educators to adopt new roles, from being knowledge transmitters to facilitators of learning [4]. It also demands substantial institutional support in terms of training, resources, and culture change [27]. Despite these challenges, the move towards innovative pedagogies is essential for creating engaging, relevant, and effective learning outcomes that meet the needs of today's students and prepare them for career success.

2.3 The CCAF Framework

The CCAF framework is a pedagogical model that emphasizes creating a learning environment where students are actively engaged in the learning undertaking through contextualized content, challenging activities, and timely feedback [2]. This framework, emerging from contemporary educational theories, aligns with the principles of constructivist learning, which advocate for student-centered, active learning environments [25].

The first component, 'Context', refers to the integration of learning content within a relevant and meaningful framework. It aims to connect theoretical concepts with real-world applications, thereby enhancing the student's ability to relate to and comprehend the material [22]. Through situating learning in a context that mirrors authentic scenarios, students are more likely to find the material engaging and relevant.

The 'Challenge' aspect of the CCAF framework involves presenting students with tasks that demand them to apply their knowledge and skills in problem-solving scenarios. These challenges are crafted to be thought-provoking and to stimulate advanced thinking skills, such as analysis, synthesis, and judgment [6]. The idea is to move beyond rote memorization to a more profound understanding and application of concepts.

The 'Activity' component highlights the significance of active engagement in the learning process. This involves hands-on activities, collaborative projects, and interactive discussions, all aimed at promoting active participation and engagement in learning [7]. These activities are designed to be diverse and inclusive, catering to different learning styles and preferences.

Finally, 'Feedback' is a crucial element of the framework. Timely and constructive feedback is provided to students to guide their learning process [21]. Effective feedback enables students to understand their performance, identify areas for improvement, and feel supported in their learning journey. This component is essential for fostering a continuous learning loop where students can iteratively improve their understanding and skills.

The CCAF framework is a holistic approach to teaching and learning, emphasizing an active, contextualized, and student-centered educational experience. It is particularly effective in higher education settings, where cultivating critical thinking, problem-solving abilities, and practical applicability of knowledge are paramount.

2.4 Related Work

The CCAF framework has been a subject of interest in educational research, especially within the context of online and blended learning environments. Existing research highlights the significance of the CCAF framework's components in fostering active learning, and improving learning outcomes. For example, the work by [22] and [23] emphasize the role of authentic contexts in motivating students and improving their ability to apply knowledge in real-world scenarios. Studies have shown that when students perceive the learning material as relevant to their lives or future careers, their engagement and retention rates improve significantly [9,35].

Previous studies by [25] and [24] have demonstrated that challenging tasks encourage students to apply, analyze, and evaluate information, leading to a more profound understanding of the learning content. This approach aligns with Bloom's taxonomy of cognitive objectives, fostering critical thinking and problem-solving abilities [6,58]. Also, studies by [7,48], and [13] support the notion that active learning strategies, as hands-on activities and collaborative projects, improve student engagement and knowledge retention. Research by [21] and [36] found that effective feedback, both from instructors and peers, significantly enhances learning outcomes. The CCAF framework's emphasis on timely and constructive feedback supports the development of self-regulation skills and continuous improvement in learners.

The CCAF framework has been used in various educational settings to enhance student engagement and learning outcomes. Case studies indicate that integrating CCAF can lead to significant improvements in how students interact with and absorb educational content. For example, a study by [57] applied the CCAF framework to explore the link between learners' cognitive styles and their attitudes toward e-learning. The results showed that the framework helped promote positive attitudes, suggesting a greater level of engagement with online learning. Another study by [38] investigated the efficacy of the CCAF framework in promoting learners' engagement during a synchronous online workshop on the fundamentals of cybersecurity for cloud computing. The results indicated that the CCAF framework successfully engaged students, fostered collaboration, and improved learning outcomes. Similarly, a study by [59] used the CCAF framework to shape problem-based learning activities in a college course, revealing that it encouraged collaboration among students, boosted critical thinking, and facilitated deeper knowledge application. Overall, these studies indicate that the CCAF framework can be an essential instrument for designing educational experiences that actively engage students and support effective learning outcomes.

Whilst there is existing research on both the CCAF framework, student engagement, and the teaching of IS Security in higher education, specific gaps in the literature justify the need for this research. Most studies on the CCAF framework have focused on general online or blended learning environments, with limited exploration in the specific context of IS Security education. This gap signifies a need to investigate how the CCAF framework can be adapted and applied effectively in the context of teaching IS Security, which has distinct challenges and requirements. The CCAF framework has been predominantly explored in online settings, however, its application and effectiveness in traditional, in-person classroom settings, especially for technical subjects like IS Security, are less documented. This study hopes to address this gap by evaluating the framework's impact on student engagement and learning outcomes in a physical classroom setting. Also, the rapidly evolving nature of cybersecurity threats necessitates a curriculum that is not only theoretically sound but also contextually relevant and updated. Research on the integration of real-world scenarios and the latest industry trends within the IS Security curriculum is limited, presenting an opportunity to explore the effectiveness of the CCAF framework in enhancing this aspect.

Furthermore, active learning strategies are critical in technical fields like IS Security. However, there are limited studies that quantify the impact of such pedagogies on student engagement and learning experiences in IS Security courses. This study seeks to address this by providing empirical evidence on the effectiveness of the CCAF framework's active learning components. It addresses the critical need to explore the use of the CCAF framework in the specific context of IS Security courses in higher education, particularly in in-person settings. By filling these gaps, the study aims to contribute to the existing literature and provide practical insights for teachers and curriculum developers in the area of IS Security.

3 Methodology

This section presents the research background, the implementation of the CCAF, the questionnaire design and data collection methods, and the questionnaire validation and data analysis methods.

3.1 Research Background

This study employed a case study approach and a quantitative research design to evaluate the effectiveness of the CCAF framework in an in-person classroom setting for a third-year bachelor's course in IS Security. This design was chosen for its ability to provide objective, numerical data that could be statistically analyzed to draw conclusions about the framework's impact on student learning experiences. The study was conducted during the Autumn 2023 semester (August 14 - December 21, 2023) at University of South-Eastern Norway (USN)'s Campus Vestfold in Borre, Norway. The campus features modern classrooms and laboratory facilities designed to support interactive and collaborative learning. The classroom setting was equipped with technology for presentations, online collaboration tools, and hands-on exercises. This setting facilitated the implementation of the CCAF framework during class sessions and group activities.

The participants in this study were third-year bachelor students enrolled in the IS Security course at USN. They comprised of 42 students, with a diverse mix of genders, nationalities, and educational backgrounds. Due to university policies and ethical considerations, demographic data such as age and gender were not collected. The course spanned an academic semester, typically lasting 19 weeks. During this period, the curriculum integrated the CCAF framework into various components, including lectures, group discussions, and practical exercises. The course structure allowed for continuous assessment and iterative feedback to enhance student learning.

3.2 CCAF Framework Implementation

The CCAF framework was carefully implemented in the third-year bachelor's IS Security course to maximize student engagement and learning outcomes. The course was structured over a semester, and each element of the CCAF framework was integrated into various aspects of the curriculum and teaching methodology.

The course began each module with current, real-world scenarios related to IS Security. These scenarios were designed to provide context to the theoretical concepts being taught, making them more relatable and understandable. Guest speakers from the IS Security industry were invited to discuss how course topics are applied in their professional roles, further enhancing the real-world context. Students were engaged in problem-based learning tasks where they had to apply theoretical knowledge to solve complex, real-world IS Security problems. Collaborative group projects were assigned, requiring students to work together to address challenging IS Security scenarios, thereby fostering teamwork and practical application of knowledge.

Instructors provided immediate and ongoing feedback on assignments and class activities, helping students understand their progress and areas for improvement. A system of peer review was implemented for certain class activities, enabling students to receive diverse perspectives and feedback from their classmates. The course included hands-on labs where students could practice IS Security techniques in a controlled environment, enhancing their practical skills. Regular class discussions and debates on contemporary IS Security issues were conducted, encouraging active participation and critical thinking.

The CCAF components were not implemented in isolation but were integrated to complement and reinforce each other throughout the course. For instance, real-world scenarios (Context) were used as the basis for problem-based learning activities (Challenge), which were then followed by immediate instructor and peer feedback (Feedback), culminating in a hands-on lab or group project (Activity). This integrated approach ensured that students remained engaged and could see the practical relevance of their learning.

Assessments were designed to reflect the CCAF approach, focusing on students' ability to apply knowledge in real-world scenarios, their engagement in problem-solving, and the quality of their participation in active learning activities. The implementation of the CCAF framework was aimed at creating an engaging, interactive, and practical learning environment, aligning with the specific needs and challenges of teaching IS Security in higher education.

For educators aiming to implement similar strategies, it is crucial to make sure that the content is contextually relevant and that activities are both challenging and aligned with learning objectives. Additionally, feedback should be timely and constructive, enabling students to participate in a continuous cycle of learning and improvement. This holistic approach can enhance the efficacy of teaching strategies in higher education.

3.3 Questionnaire Design and Data Collection

The questionnaire used for this research was designed as a mix of Likert-scale questions and open-ended items, with questions structured to draw out the answers needed to address the research questions. English was employed in the design of the questionnaire. The investigators used it for data collection. Since all the students had access to the Internet, a questionnaire tool with automatic data aggregation and analysis called Nettskjema (developed by the University of Oslo) was used for data collection. The link to the questionnaire was forwarded to all the students through the learning platform used for the course (Canvas). The students were given one week to fill out and submit the questionnaire forms, with periodic reminders sent to them.

3.4 Questionnaire Validation and Data Analysis Methods

To ensure the accuracy of the questionnaire and confirm that it was free of errors and solicited the required type of answers/responses from the students, the questionnaire was shared with an expert who is well-versed in the concept the questionnaire was intended to evaluate. The feedback received indicated that the questionnaire items were effectively capturing the concept meant to be examined and that the items were comprehensive enough to cover the area of interest for the research.

The population for the study consisted of 42 students enrolled in the IS security course. Because of the small size of the population, the study included all participants, removing the need for sampling. Thus, the sample size was effectively 100% of the population. Out of the 42 students who took part in the course and received the questionnaire, 28 responded, representing a 66.7% response rate. This rate was considered sufficient for the study.

Lastly, the data analysis involved descriptive statistics to outline survey responses and inferential statistics to detect significant patterns and correlations. The study adhered to strict ethical standards to ensure the confidentiality and protection of participant data. All student responses were anonymized, and data was securely stored in compliance with institutional data protection policies. Informed consent was obtained from all participants, who were informed of the study's purpose and their right to opt-out at any time without consequence. These measures ensured that the research was conducted in an ethically responsible manner, respecting the privacy and rights of all participants.

4 Results

The quantitative analysis of the CCAF framework's application in an in-person class setting for third-year bachelor students yielded significant insights. Results from Fig. 1 indicate that the majority of students (64.3%) acknowledged the high relevance of the course material to real-world information security issues, underscoring the practical applicability of the curriculum. This finding aligns

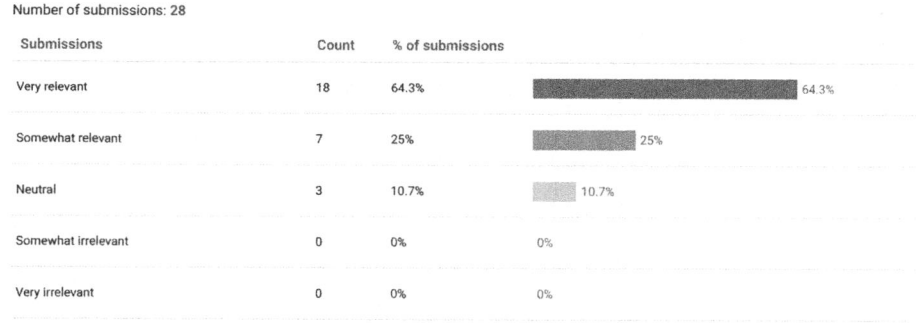

Submissions	Count	% of submissions	
Very relevant	18	64.3%	64.3%
Somewhat relevant	7	25%	25%
Neutral	3	10.7%	10.7%
Somewhat irrelevant	0	0%	0%
Very irrelevant	0	0%	0%

Number of submissions: 28

Fig. 1. Relevance of course material to real-world applications

with the framework's goal of contextual relevance, a critical factor in enhancing student engagement.

Figure 2 shows that the course's structure and content clarity were highly rated, with 64.3% of students agreeing and 21.4% strongly agreeing on its effectiveness. This overwhelmingly positive response suggests a successful implementation of the CCAF framework, emphasizing its utility in providing a clear and structured learning environment.

Number of submissions: 28

Submissions	Count	% of submissions	
Strongly Agree	6	21.4%	21.4%
Agree	18	64.3%	64.3%
Neutral	4	14.3%	14.3%
Disagree	0	0%	0%
Strongly Disagree	0	0%	0%

Fig. 2. Course structure and clarity

The stimulation of intellectual curiosity as depicted in Fig. 3 presents a more varied picture. While over half of the students (55.6%) agreed that the course stimulated their intellectual curiosity, a notable proportion (25.9%) remained neutral. This variability points to opportunities for further refining the framework to enhance engagement for all students.

The results as seen in Fig. 4 were predominantly positive, with 39.3% of students reporting being "Very satisfied" and 42.9% "Somewhat satisfied". This high level of satisfaction indicates the framework's effectiveness in fostering a conducive learning environment in physical classrooms.

Number of submissions: 27

Submissions	Count	% of submissions	
Strongly Agree	5	18.5%	18.5%
Agree	15	55.6%	55.6%
Neutral	7	25.9%	25.9%
Disagree	0	0%	0%
Strongly Disagree	0	0%	0%

Fig. 3. Simulation of intellectual curiosity

Number of submissions: 28

Submissions	Count	% of submissions	
Very satisfied	11	39.3%	39.3%
Somewhat satisfied	12	42.9%	42.9%
Neutral	5	17.9%	17.9%
Somewhat unsatisfied	0	0%	0%
Very unsatisfied	0	0%	0%

Fig. 4. Overall student satisfaction

Furthermore, the descriptive statistics indicated that students generally found the course material highly relevant to real-world IS Security issues, with a mean relevance rating of 4.1 out of 5. Further, an independent t-test revealed a statistically significant difference in engagement levels between students who rated the course relevance highly and those who did not ($t(26) = 2.13, p = 0.043$). An ANOVA test comparing overall satisfaction across different course sections also indicated a significant difference ($F(2, 25) = 3.58, p = 0.041$), suggesting variability in student experiences. Pearson correlation analysis showed a strong, positive relationship between the clarity of course structure and overall satisfaction ($r = 0.65, p = 0.002$). Finally, regression analysis demonstrated that course relevance, structure, and intellectual curiosity collectively explained 52% of the variance in overall satisfaction ($R^2 = 0.52, p < 0.01$).

These results collectively demonstrate the CCAF framework's positive impact on student learning experiences, suggesting its potential as a versatile and effective teaching strategy in higher education.

5 Discussion

This section presents a discussion of the results obtained, the comparison of the results with previous research on the CCAF framework in an online setting, and the implications of the findings for educational practice and policy.

5.1 Interpretation of Results in the Context of Existing Literature

The study's results, indicating a positive impact of the CCAF framework in a physical classroom setting, correspond with and build upon existing literature in educational pedagogy. The strong correlation of course material with real-world applications is a key finding, resonating with the work of [31,34], and [50], which emphasize the significance of practical relevance in enhancing student engagement and learning outcomes. This relevance is vital for third-year bachelor students, who are transitioning from theoretical learning to practical application.

The course structure and clarity findings support [11] work on theories of good practice in undergraduate education and studies by [1] and [64], which underscore the importance of clear organization and well-structured courses in effective teaching. The variation in intellectual curiosity stimulation among students reflects the observation by [56] and [45] that engagement strategies might need to be tailored to individual learning preferences, highlighting the complexity of student engagement in diverse learning environments. Furthermore, providing varied forms of feedback, including peer and self-assessment opportunities, could enhance the engagement of students who might not respond as expected to traditional feedback methods. These insights suggest potential adjustments to the CCAF framework that could make it even more effective across a broader range of students.

The overall satisfaction levels, aligning with the research by [33,52,61], suggest that students perceive the quality and effectiveness of their learning experience positively when engaged through structured and relevant frameworks like the CCAF. However, the neutral responses in some areas indicate room for further enhancement, possibly through more personalized engagement strategies or addressing specific student needs.

These results add to the broader discussion on student engagement in higher education, especially in the context of adapting online teaching strategies to physical classrooms. They underscore the capacity of frameworks like the CCAF to improve the learning experience and outcomes in traditional classroom settings, a crucial consideration for educators and policymakers aiming to improve educational practices in higher education.

Although this research was conducted within the context of an in-person Information Systems Security course, the CCAF framework's design principles suggest that its effectiveness is not limited to this particular setting. The framework's focus on contextualized learning, challenging activities, and timely feedback is applicable across various disciplines and delivery formats. As such, the findings of this research may be generalizable to other higher education contexts, including online as well as hybrid learning environments. This broader applicability underscores the CCAF framework's capacity to improve student engagement and learning outcomes across diverse educational settings.

5.2 Comparison with Previous Findings on the CCAF Framework in Online Settings

The comparison of this research's findings with previous study on the CCAF framework in online settings reveals notable insights. Earlier studies have highlighted the effectiveness of the CCAF framework in online education, especially in terms of fostering student engagement and enhancing learning outcomes [38,57]. These studies emphasized the framework's ability to create an interactive and immersive online learning experience, which was instrumental in engaging students.

In contrast, the current study extends this understanding to a physical classroom setting, showing that the framework's principles are equally effective in a traditional learning environment. This suggests a versatile applicability of the CCAF framework, transcending the boundaries of online and in-person education. The current study's findings are particularly significant given the different dynamics and interactions inherent in physical classroom settings, underscoring the framework's adaptability and effectiveness in various educational contexts.

5.3 Implications for Educational Practice and Policy

The implications of this study for educational practice and policy are multifaceted. Firstly, the successful application of the CCAF framework in a physical classroom setting underscores the potential for cross-modal pedagogical approaches. This adaptability is pivotal for educators, suggesting that methods developed for online learning can be effectively integrated into traditional classroom environments. It encourages educators to focus on creating engaging, contextually relevant activities, fostering a more dynamic and interactive learning experience.

For policy makers, these findings highlight the importance of supporting flexible and innovative curriculum designs that can adapt across different learning environments. This approach is particularly relevant in the current educational landscape, where hybrid models of learning are becoming more prevalent. Policies should therefore encourage and facilitate the integration of diverse teaching strategies, bridging the gap between online and in-person learning experiences.

Furthermore, the study's insights into student engagement and satisfaction provide valuable guidance for continuous improvement in teaching methodologies. Educational institutions should consider these factors in their strategic planning and resource allocation, emphasizing the development of engaging and effective learning environments. Overall, this study contributes to the discourse on enhancing higher education practices, offering practical insights for educators and policymakers in their efforts to improve educational outcomes and student experiences.

6 Conclusion

This study provides valuable insights into the effectiveness of the CCAF framework in a physical classroom setting for third-year bachelor students. Key findings indicate the framework's success in enhancing course relevance to real-world scenarios, structuring course content effectively, and achieving high levels of student satisfaction. However, variations in intellectual curiosity stimulation suggest room for further refinement.

While this study provides valuable insights into the effectiveness of the CCAF framework in a third-year IS Security course, its findings are limited by the focus on a single course in a specific discipline. This limitation restricts the generalizability of the results to other educational contexts. Future research should consider conducting longitudinal studies to assess the long-term impact of the CCAF framework on student learning outcomes, knowledge retention, and overall academic success. Such studies would provide a more robust understanding of how the framework influences students over time and across different courses and disciplines. The study acknowledges potential biases that could influence the findings. One such bias is self-reporting bias, where students may have provided responses that they perceived as favorable rather than reflecting their true experiences. Additionally, researcher bias may have influenced the interpretation of data, particularly in the feedback sessions where instructors were also part of the research team. To mitigate these biases, future studies should consider incorporating third-party evaluations and objective measures of student engagement and learning outcomes. This research lays the foundation for future explorations into effective pedagogical strategies in higher education, emphasizing the importance of adaptable and engaging teaching approaches.

Data Availibility Statement. The data that support the findings of this study are available from the corresponding author upon reasonable request.

Disclosure of Interests. The authors have no competing interests to declare that are relevant to the content of this article.

References

1. Ahuja, V.: Faculty Development for Dynamic Curriculum Design in Online Higher Education, pp. 155–175. IGI Global (2023). https://doi.org/10.4018/978-1-6684-8646-7.ch009
2. Allen, M.W.: Michael Allen's Guide to e-Learning: Building Interactive, Fun, and Effective Learning Programs for Any Company. Wiley, Hoboken (2016)
3. Astin, A.W.: Student involvement: a developmental theory for higher education. J. Coll. Student Personnel (1984)
4. Barr, R.B., Tagg, J.: From teaching to learning—a new paradigm for undergraduate education. Change Mag. High. Learn. **27**(6), 12–26 (1995). https://doi.org/10.1080/00091383.1995.10544672
5. Binkley, M., et al.: Defining Twenty-First Century Skills, pp. 17–66. Springer (2011). https://doi.org/10.1007/978-94-007-2324-5_2

6. Bloom, B.S., Engelhart, M.D., Furst, E.J., Hill, W.H., Krathwohl, D.R.: Taxonomy of educational objectives: the classification of educational goals. McKay (1956)
7. Bonwell, C.C., Eison, J.A.: Active learning: creating excitement in the classroom. 1991 Ashe-Eric higher education reports. Technical report, ERIC Clearinghouse on Higher Education, The George Washington University (1991)
8. Bruner, J.: The Culture of Education. Harvard University Press (1997). https://doi.org/10.2307/j.ctv136c601
9. Canning, E.A., Harackiewicz, J.M., Priniski, S.J., Hecht, C.A., Tibbetts, Y., Hyde, J.S.: Improving performance and retention in introductory biology with a utility-value intervention. J. Educ. Psychol. **110**(6), 834–849 (2018). https://doi.org/10.1037/edu0000244
10. Chi, M., Wylie, R.: The ICAP framework: linking cognitive engagement to active learning outcomes. Educ. Psychol. **49**(4), 219–243 (2014). https://doi.org/10.1080/00461520.2014.965823
11. Chickering, A.W., Gamson, Z.F.: Appendix a: seven principles for good practice in undergraduate education. New Dir. Teach. Learn. **1991**(47), 63–69 (1991). https://doi.org/10.1002/tl.37219914708
12. Christensen, C.M., Eyring, H.J.: The Innovative University: Changing the DNA of Higher Education from the Inside Out. Jossey-Bass (2011)
13. David Agwu, U., Nmadu, J.: Students' interactive engagement, academic achievement and self concept in chemistry: an evaluation of cooperative learning pedagogy. Chem. Educ. Res. Pract. **24**(2), 688–705 (2023). https://doi.org/10.1039/d2rp00148a
14. Fink, L.D.: Creating Significant Learning Experiences: An Integrated Approach to Designing College Courses. Jossey-Bass (2013)
15. Finn, J.D., Zimmer, K.S.: Student Engagement: What Is It? Why Does It Matter?, pp. 97–131. Springer, Cham (2012). https://doi.org/10.1007/978-1-4614-2018-7_5
16. Fredricks, J.A., Blumenfeld, P.C., Paris, A.H.: School engagement: potential of the concept, state of the evidence. Rev. Educ. Res. **74**(1), 59–109 (2004). https://doi.org/10.3102/00346543074001059
17. Fredricks, J.A., Filsecker, M., Lawson, M.A.: Student engagement, context, and adjustment: addressing definitional, measurement, and methodological issues. Learn. Instr. **43**, 1–4 (2016). https://doi.org/10.1016/j.learninstruc.2016.02.002
18. Freeman, S., et al.: Active learning increases student performance in science, engineering, and mathematics. Proc. Natl. Acad. Sci. **111**(23), 8410–8415 (2014). https://doi.org/10.1073/pnas.1319030111
19. French, S., Kennedy, G.: Reassessing the value of university lectures. Teach. High. Educ. **22**(6), 639–654 (2016). https://doi.org/10.1080/13562517.2016.1273213
20. Gumartifa, A., Syahri, I., Siroj, R.A., Nurrahmi, M., Yusof, N.: Perception of teachers regarding problem-based learning and traditional method in the classroom learning innovation process. Indonesian J. Learn. Adv. Educ. (IJOLAE) **5**(2), 151–166 (2023). https://doi.org/10.23917/ijolae.v5i2.20714
21. Hattie, J., Timperley, H.: The power of feedback. Rev. Educ. Res. **77**(1), 81–112 (2007). https://doi.org/10.3102/003465430298487
22. Herrington, J., Reeves, T.C., Oliver, R.: Authentic Learning Environments, pp. 401–412. Springer, New York (2013). https://doi.org/10.1007/978-1-4614-3185-5_32
23. Hwang, W.Y., Hariyanti, U., Chen, N.S., Purba, S.: Developing and validating an authentic contextual learning framework: promoting healthy learning through learning by applying. Interact. Learn. Environ. **31**(4), 2206–2218 (2021). https://doi.org/10.1080/10494820.2021.1876737

24. Jensen, J.L., McDaniel, M.A., Woodard, S.M., Kummer, T.A.: Teaching to the test...or testing to teach: exams requiring higher order thinking skills encourage greater conceptual understanding. Educ. Psychol. Rev. **26**(2), 307–329 (2014). https://doi.org/10.1007/s10648-013-9248-9
25. Jonassen, D.H., Rohrer-Murphy, L.: Activity theory as a framework for designing constructivist learning environments. Educ. Tech. Res. Dev. **47**(1), 61–79 (1999). https://doi.org/10.1007/bf02299477
26. Kahu, E., Stephens, C., Leach, L., Zepke, N.: Linking academic emotions and student engagement: mature-aged distance students' transition to university. J. Furth. High. Educ. **39**(4), 481–497 (2014). https://doi.org/10.1080/0309877x.2014.895305
27. Kezar, A., Sam, C.: Understanding the New Majority: Contingent Faculty in Higher Education, vol. I. Jossey Bass (2010)
28. Khan, H., Gul, R., Zeb, M.: The effect of students' cognitive and emotional engagement on students' academic success and academic productivity. J. Soc. Sci. Rev. **3**(1), 322–334 (2023). https://doi.org/10.54183/jssr.v3i1.141
29. Kimbark, K., Peters, M.L., Richardson, T.: Effectiveness of the student success course on persistence, retention, academic achievement, and student engagement. Commun. Coll. J. Res. Pract. **41**(2), 124–138 (2016). https://doi.org/10.1080/10668926.2016.1166352
30. Krause, K., Coates, H.: Students' engagement in first-year university. Assess. Eval. High. Educ. **33**(5), 493–505 (2008). https://doi.org/10.1080/02602930701698892
31. Kuh, G.D., Cruce, T.M., Shoup, R., Kinzie, J., Gonyea, R.M.: Unmasking the effects of student engagement on first-year college grades and persistence. J. High. Educ. **79**(5), 540–563 (2008). https://doi.org/10.1353/jhe.0.0019
32. Kuh, G.D., Kinzie, J., Schuh, J.H., Whitt, E.J.: Assessing Conditions to Enhance Educational Effectiveness: The Inventory for Student Engagement and Success. Jossey-Bass (2005)
33. Lewis, A.D., Huebner, E.S., Malone, P.S., Valois, R.F.: Life satisfaction and student engagement in adolescents. J. Youth Adolesc. **40**(3), 249–262 (2010). https://doi.org/10.1007/s10964-010-9517-6
34. Li, J., Xue, E.: Dynamic interaction between student learning behaviour and learning environment: meta-analysis of student engagement and its influencing factors. Behav. Sci. **13**(1), 59 (2023). https://doi.org/10.3390/bs13010059
35. Lombardi, M.M., Oblinger, D.G.: Authentic learning for the 21st century: an overview. Educause Learn. Initiative **1**(2007), 1–12 (2007)
36. Mandouit, L., Hattie, J.: Revisiting "the power of feedback" from the perspective of the learner. Learn. Instr. **84**, 101718 (2023). https://doi.org/10.1016/j.learninstruc.2022.101718
37. Meng, Q., Zhang, Q.: The influence of academic self-efficacy on university students' academic performance: the mediating effect of academic engagement. Sustainability **15**(7), 5767 (2023). https://doi.org/10.3390/su15075767
38. Nweke, L.O.: Promoting Learners' Engagement to Maximize Learning in a Synchronous Online Workshop: A Case Study Analysis from Different Perspectives, pp. 407–426. Springer, Cham (2023). https://doi.org/10.1007/978-3-031-47451-4_29
39. Nweke, L.O., Bokolo, A.J., Mba, G., Nwigwe, E.: Investigating the effectiveness of a hyflex cyber security training in a developing country: a case study. Educ. Inf. Technol. **27**(7), 10107–10133 (2022). https://doi.org/10.1007/s10639-022-11038-z
40. Ornellas, A., Falkner, K., Edman Stålbrandt, E.: Enhancing graduates' employability skills through authentic learning approaches. High. Educ. Skills Work-Based Learn. **9**(1), 107–120 (2019). https://doi.org/10.1108/heswbl-04-2018-0049

41. Pandita, A., Kiran, R.: The technology interface and student engagement are significant stimuli in sustainable student satisfaction. Sustainability **15**(10), 7923 (2023). https://doi.org/10.3390/su15107923
42. Pascarella, E.T., Terenzini, P.T.: How College Affects Students: A Third Decade of Research, vol. 2. Jossey-Bass, An Imprint of Wiley (2005)
43. Pekrun, R., Linnenbrink-Garcia, L.: Academic Emotions and Student Engagement, pp. 259–282. Springer, Cham (2012). https://doi.org/10.1007/978-1-4614-2018-7_12
44. Pintrich, P.R.: Motivation and classroom learning (2003). https://doi.org/10.1002/0471264385.wei0706
45. Plak, S., van Klaveren, C., Cornelisz, I.: Raising student engagement using digital nudges tailored to students' motivation and perceived ability levels. Br. J. Edu. Technol. **54**(2), 554–580 (2022). https://doi.org/10.1111/bjet.13261
46. Poort, I., Jansen, E., Hofman, A.: Promoting university students' engagement in intercultural group work: the importance of expectancy, value, and cost. Res. High. Educ. **64**(2), 331–348 (2022). https://doi.org/10.1007/s11162-022-09705-8
47. Prensky, M.: Digital natives, digital immigrants part 1. Horizon **9**(5), 1–6 (2001). https://doi.org/10.1108/10748120110424816
48. Prince, M.: Does active learning work? A review of the research. J. Eng. Educ. **93**(3), 223–231 (2004). https://doi.org/10.1002/j.2168-9830.2004.tb00809.x
49. Qureshi, M.A., Khaskheli, A., Qureshi, J.A., Raza, S.A., Yousufi, S.Q.: Factors affecting students' learning performance through collaborative learning and engagement. Interact. Learn. Environ. **31**(4), 2371–2391 (2021). https://doi.org/10.1080/10494820.2021.1884886
50. Rafiq, A.A., Triyono, M.B., Djatmiko, I.W.: The integration of inquiry and problem-based learning and its impact on increasing the vocational student involvement. Int. J. Instr. **16**(1), 659–684 (2023). https://doi.org/10.29333/iji.2023.16137a
51. Rowe, A.D., Jackson, D., Fleming, J.: Exploring university student engagement and sense of belonging during work-integrated learning. J. Vocational Educ. Train. **75**(3), 564–585 (2021). https://doi.org/10.1080/13636820.2021.1914134
52. Sadoughi, M., Hejazi, S.Y.: The effect of teacher support on academic engagement: the serial mediation of learning experience and motivated learning behavior. Curr. Psychol. **42**(22), 18858–18869 (2022). https://doi.org/10.1007/s12144-022-03045-7
53. Santos, A.C., et al.: Social and emotional competencies as predictors of student engagement in youth: a cross-cultural multilevel study. Stud. High. Educ. **48**(1), 1–19 (2022). https://doi.org/10.1080/03075079.2022.2099370
54. Schunk, D.H., Mullen, C.A.: Self-Efficacy as an Engaged Learner, pp. 219–235. Springer, Cham (2012). https://doi.org/10.1007/978-1-4614-2018-7_10
55. Skinner, E.A., Pitzer, J.R.: Developmental Dynamics of Student Engagement, Coping, and Everyday Resilience, pp. 21–44. Springer, Cham (2012). https://doi.org/10.1007/978-1-4614-2018-7_2
56. Smith, P.L., Ragan, T.J.: Instructional Design. Wiley, Hoboken (2004)
57. Sozcu, O.F.: The relationships between cognitive style of field dependence and learner variables in e-learning instruction. Turkish Online J. Distance Educ. **15**(2) (2014). https://doi.org/10.17718/tojde.11039
58. Stevani, M., Tarigan, K.E.: Evaluating English textbooks by using bloom's taxonomy to analyze reading comprehension question. SALEE: Study Appl. Linguist. Engl. Educ. **4**(1), 1–18 (2022). https://doi.org/10.35961/salee.v0i0.526

59. Tawfik, A.A., Lilly, C.: Using a flipped classroom approach to support problem-based learning. Technol. Knowl. Learn. **20**(3), 299–315 (2015). https://doi.org/10.1007/s10758-015-9262-8
60. Tinto, V.: Classrooms as communities: exploring the educational character of student persistence. J. High. Educ. (1997)
61. Trowler, V.: Student Engagement Literature Review. The Higher Education Academy (2010)
62. Venn, E., Park, J., Andersen, L.P., Hejmadi, M.: How do learning technologies impact on undergraduates' emotional and cognitive engagement with their learning? Teach. High. Educ. **28**(4), 822–839 (2020). https://doi.org/10.1080/13562517.2020.1863349
63. Xerri, M.J., Radford, K., Shacklock, K.: Student engagement in academic activities: a social support perspective. High. Educ. **75**(4), 589–605 (2017). https://doi.org/10.1007/s10734-017-0162-9
64. Zhu, M., Berri, S., Koda, R., Wu, Y.: Exploring students' self-directed learning strategies and satisfaction in online learning. Educ. Inf. Technol. (2023). https://doi.org/10.1007/s10639-023-11914-2
65. Zimmerman, B.J.: Self-regulated learning and academic achievement: an overview. Educ. Psychol. **25**(1), 3–17 (1990). https://doi.org/10.1207/s15326985ep2501_2

Session 5

Blood Oxygen Saturation Estimation Using PPG Signals from the MIMIC-III Database

Nenad Petrovikj[✉], Bojana Mishkovska, Bojana Koteska, and Ana Madevska Bogdanova

Faculty of Computer Science and Engineering, University Ss. Cyril and Methodius, Rugjer Boskovikj 16, Skopje 1000, North Macedonia
{nenad.petrovikj,bojana.mishkovska}@students.finki.ukim.mk,
{bojana.koteska,ana.madevska.bogdanova}@finki.ukim.mk

Abstract. Photoplethysmogram (PPG) signals are pivotal in cardiovascular monitoring, offering real-time insights into heart rate and oxygen saturation (SpO2). This study explores the creation of deep learning and machine learning models - specifically Convolutional Neural Networks (CNNs), Bidirectional Long Short-Term Memory (BiLSTM) networks, Recurrent Neural Networks (RNNs), and Random Forest Regressors (RFRs)-to estimate SpO2 levels from single-channel PPG data. Another point is developing algorithms for using the data sourced from the PhysioNet MIMIC-III database. The patients used for training and testing are distinct, ensuring no overlap between the datasets and enabling rigorous model evaluation. A comprehensive analyses reveal that LSTM-based model achieve significant accuracy in SpO2 estimation, with R-squared value reaching up to 0.59. Specifically, the LSTM model demonstrated an MAE of 1.26, MSE of 3.11 and RMSE of 1.76. These results demonstrate the potential of machine learning techniques in advancing clinical monitoring and decision-making processes within critical care environments, thereby enhancing patient care outcomes.

Keywords: photoplethysmogram · blood oxygen saturation · machine learning · estimation

1 Introduction

A photoplethysmogram (PPG) signal is generated using infrared light to detect changes in blood volume in tissue microvasculature. This technique captures the heart's pumping action by sensing blood flow rate [1]. Each peak in the PPG graph corresponds to a heartbeat, allowing for direct measurement of heart rate-60 peaks indicating 60 beats per minute. PPG signals are vital in diagnosing, monitoring, and screening cardiovascular conditions like heart attack, stroke, and heart failure [2].

Oxygen saturation (SpO2) measurement employs a pulse oximeter to measure the percentage of oxygen-saturated hemoglobin in blood. Normal SpO2 levels range from 97% to 100%, with levels below 95% necessitating immediate medical attention. SpO2 signals are pivotal in detecting various lung diseases by assessing blood oxygen levels [3,4]. During anesthesia and surgery, SpO2 monitoring ensures patients receive adequate oxygen supply.

Calculating SpO2 involves assessing the ratio of AC to DC components in the PPG signal, reflecting heart rate during systole and diastole, and respiratory rate over typically 60 s. This method estimates SpO2 by establishing correlations over specific durations, crucial in medical settings for monitoring oxygen levels. Calibration of PPG signals for different pulse oximeter sensors involves empirical approaches, including in vitro measurement of SpO2 through co-oximetry [5].

Artificial Neural Networks (ANNs) and Machine Learning models enhance SpO2 prediction from single-channel PPG signals, surpassing traditional R-value based methods in signal processing [6]. These advancements enable rapid and accurate SpO2 estimation in clinical applications.

In this paper, we explore the application of machine learning models to estimate SpO2 levels from PPG signals. We leverage a dataset sourced from the PhysioNet MIMIC-III clinical database, focusing on PPG segments and corresponding SpO2 values extracted from patient records.

The paper is structured as follows: Sect. 2 reviews related work in the field of SpO2 estimation using PPG signals and discusses existing methodologies and advancements. In Sect. 3, we describe our methodology and experimental setup, including data preprocessing steps and feature extraction techniques applied to our dataset. Section 4 presents the results of our experiments, where we evaluate and compare the performance of various machine learning models trained on different dataset sizes. Finally, Sect. 5 concludes with a discussion of our findings, their implications for healthcare applications, and avenues for future research.

2 Related Work

The field of blood oxygen saturation estimation using PPG signals has gained significant attention due to its potential for non-invasive and continuous monitoring. Various approaches have been explored to enhance the accuracy and reliability of SpO2 estimation from PPG signals, employing machine learning algorithms, signal processing techniques, and wearable devices.

A novel approach using machine learning models has demonstrated effective SpO2 estimation by incorporating meaningful PPG signal features, leading to a Gaussian process regression model with an RMSE of 0.98 and an MAE of 0.57. This work highlights the use of machine learning for respiratory rate (RR) and SpO2 estimation by utilizing feature selection algorithms to reduce computational complexity and trained 19 models for both RR and SpO2 estimation, with the Gaussian process regression model outperforming all the other methods [7].

Wearable devices, particularly wrist-worn devices, have also shown promise. A low-power wrist wearable device with a PPG array sensor and Multiple Linear Regression model effectively estimated SpO2 parameters by reducing motion artifacts [8]. Another study developed a wearable pulse oximeter (PO) that separated red and green PPG signals to estimate SpO2, achieving an MAE of 2.08 without motion and 3.66 with motion [9]. A recent study proposed a smart patch-like device with an embedded PPG sensor for real-time SpO2 monitoring, using a deep neural network regressor trained on data from 52 subjects with SpO2 levels varying from 83% to 100%. The model achieved a mean absolute percentage error (MAPE) of 2.00% for SpO2 levels between 83% and 95%, with 7.21% of errors classified as significant (absolute percentage errors equal to or greater than 5%) [10]. Additionally, in emergency situations, a SmartPatch device with an embedded PPG sensor was proposed for continuous SpO2 monitoring, using Python toolkits and machine learning regressors. The Random Forest regressor achieved the best results with an MAE of 1.45 [11].

An innovative approach utilized PPG signals from mobile video to determine SpO2 levels. This method involved extracting PPG signals from smartphone camera frames and preprocessing them for enhanced quality. The Logistic Regression Model exhibited the lowest MAE of 0.845, demonstrating the potential of smartphones as a cost-effective and portable alternative for SpO2 monitoring [12].

Advances in blood oxygen saturation (SpO2) estimation using PPG signals reflect significant advances in non-invasive, continuous health monitoring. Various innovative approaches have emerged, ranging from machine learning models to improve SpO2 accuracy to wearable devices that improve real-time monitoring even in demanding environments. These developments are particularly promising in clinical settings, including in low-resource settings, where they have the potential to improve patient outcomes and safety. Additionally, the use of smartphones and wearable technology demonstrates the accessibility and versatility of modern SpO2 monitoring methods.

3 Materials and Methods

3.1 Patient Data Source

The patient data utilized in this study was sourced from the PhysioNet MIMIC-III clinical database. The MIMIC-III database is a comprehensive repository containing detailed medical records of patients admitted to the critical care units of the Beth Israel Deaconess Medical Center from 2001 to 2012. This extensive dataset includes 67,830 record sets from approximately 30,000 patients, featuring physiological waveforms and time-series data of vital signs captured from bedside patient monitors. The database, which is a collaborative effort between the Massachusetts Institute of Technology, Beth Israel Deaconess Medical Center, and Philips Healthcare, was published in April 2020 and has an uncompressed size of about 6.7 TB [13–15].

PPG signals were sampled at 125 Hz, while SpO2 values were recorded either once per second or once per minute. The extracted PPG segments were normalized and subjected to several filtering steps to ensure data quality. These steps included checks for NaN values, flat lines, and flat peaks, with a maximum threshold of 20% for flat lines and flat peaks. The filtered data, amounting to approximately 340 GB and covering around 2100 patients, was stored on a private ownCloud server [16].

For this research, the focus was specifically on PPG signals and SpO2 values. We used files from the database with one-minute interval SpO2 entries. Each file contains a one minute PPG signal and a SpO2 value measured at the end of the interval. The database used for this study is available at https://sp4life.finki.ukim.mk/ and is available upon request.

3.2 Construction of Patient Train/Test Dataset

A Python script [17] was developed for downloading and preprocessing the data from the OwnCloud Server to create a patient train/test dataset with applicable entries for a machine learning model. The procedure involved multiple steps, as outlined below.

3.2.1 Interpolation and Deletion of Entries

Before developing the preprocessing script, we conducted an initial statistical analysis to better understand the data. We found that file sizes ranged from 16 MB to 2000 MB. Files could contain entries from multiple patients spread across different files. Some PPG signals contained a small percentage of NaN values, which we opted not to delete in the initial filtering.

Based on these observations, we decided to include patient entries with no more than 5% NaN values in the PPG signal. This decision aimed to retain as much data as possible, given that only a small fraction of entries exceeded this threshold. We used linear interpolation as an imputation technique to fill in missing data points, preserving the overall data trend in segments where interpolation occurred. This approach helped maintain valuable entries, especially where SpO2 values might have been below the 95 threshold.

An alternative approach would have been to interpolate every entry with missing PPG values. However, with more than 5% NaN values (or more than 375 out of 7500), a significant portion of the signal would have been compromised, potentially leading to misleading measurements.

3.2.2 Selecting Patient Entries

An algorithm was developed to create a balanced dataset of patient entries based on their SpO2 levels. Initially, the algorithm processed the patient's data by filtering out entries where the SpO2 level was below 95%. These filtered entries were then gathered into a separate dataset.

To ensure balance, an equal number of entries with SpO2 levels of 95% or higher were randomly selected from the remaining data. If there were fewer entries with SpO2 $\geq$ 95% than those with SpO2 $<$ 95%, all available entries with SpO2 $\geq$ 95% were included. Subsequently, these two subsets were merged to form a dataset with roughly equal numbers of entries representing low and high SpO2 levels.

This balanced approach was pivotal for subsequent analysis or model training to prevent any bias towards specific SpO2 ranges, thereby enhancing the overall accuracy and reliability of the models developed.

3.2.3 Feature Extraction

Respiratory-related features extracted from the PPG signals, as detailed in [10], encompass a range of characteristics:

Amplitude characteristics involve parameters such as the **mean, maximum, minimum,** and **standard deviation** of the respiratory signal's amplitude. These metrics quantify the typical values and variability of the respiratory signal amplitudes.

Time-domain features include descriptors like **breath duration**, which measures the duration of each respiratory cycle, and **respiratory rate**, representing the number of breaths per minute.

Variability features encompass the mean and standard deviation of **respiratory rate variability**, providing insights into how respiratory rate varies over time.

Shape-related features describe aspects such as **rise time** and **fall time** of the PPG signal, indicating the duration from the start to the peak and from the peak to the end of each respiratory cycle, respectively.

3.3 Machine Learning Models
3.3.1 Train/Test Split

Once the patient dataset was constructed, the final preparatory step before commencing training was to split the data appropriately.

The dataset was divided to ensure that each patient's entries appeared exclusively in either the training or testing set, preventing any overlap. To ensure a sufficiently large training set, approximately 80% of the patients were assigned to it, while the remaining 20% were assigned to the testing set. The assignment of patients to these sets was done randomly. However, because each patient might contribute a different number of entries, the overall ratio of entries between the training and testing sets could vary. Thus, while the patients were split 80/20, the distribution of entries might differ accordingly.

3.3.2 ML Models

To estimate SpO2 levels from PPG signals, several machine learning models were utilized, leveraging both deep learning and traditional techniques suited to the data:

Convolutional Neural Networks (CNNs), Bidirectional Long Short-Term Memory (BiLSTM) networks, Recurrent Neural Networks (RNNs), and Random Forest Regressors (RFRs) were employed.

To ensure consistent results for every model trained, the dataset was split into identical training and testing sets for each individual model.

4 Results

4.1 Results for the Dataset of Files with Sizes Below 250 MB

As described in Sect. 3, the preprocessing stage involved creating an initial train/test dataset, thereby preparing the raw measured PPG signals for application in ML models. The dataset used to obtain the initial results comprised 156 different patients and involved processing 57 files, each below 250 MB in size. It contained a total of 64,719 entries, with approximately 56% (36,116 entries) having SpO2 levels below 95%, and 44% (28,603 entries) having SpO2 levels of 95% or higher. The dataset included 15 columns, with 10 columns representing respiratory features, 4 columns containing metadata (which were excluded during model training), and 1 target column indicating the SpO2 value.

In Table 1, the results for the metrics are presented for all models. The LSTM model achieves an R-squared value of ≈0.56, indicating a moderate level of predictive accuracy in capturing underlying data patterns. The R-squared values of the other models show closer alignment to the LSTM's performance, highlighting their comparable effectiveness in explaining data variability. However, the CNN model demonstrates the lowest R-squared value among the models, specifically ≈0.32, implying its partial success in capturing the inherent data patterns and trends.

Table 1. Performance metrics for all models trained and tested on the database constructed with respiratory feature extraction from files smaller than 250 MB

Model	MAE	MSE	RMSE	R2
LSTM	1.26072	3.10660	1.76256	0.55614
BiLSTM	1.44269	3.59930	1.89718	0.48575
CNN	1.62274	4.76043	2.18184	0.31985
RFR	1.41697	3.56261	1.88749	0.49099

Our next step involved increasing the dataset size to provide the models with more training data to assess the impact on model performance.

4.2 Results for the Dataset of Files with Sizes Below 600 MB

The final dataset consisted of 682 different patients and involved processing 181 files, with only those below 600 MB in size included due to RAM constraints during testing. The dataset comprised a total of 456,719 entries, with approximately 58% (263,047 entries) having SpO2 levels below 95%, and the remaining 42% (193,672 entries) having SpO2 levels of 95% or higher. The dataset included 15 columns, with 10 columns representing respiratory features, 4 columns containing metadata (which were excluded during model training), and 1 target column indicating the SpO2 value.

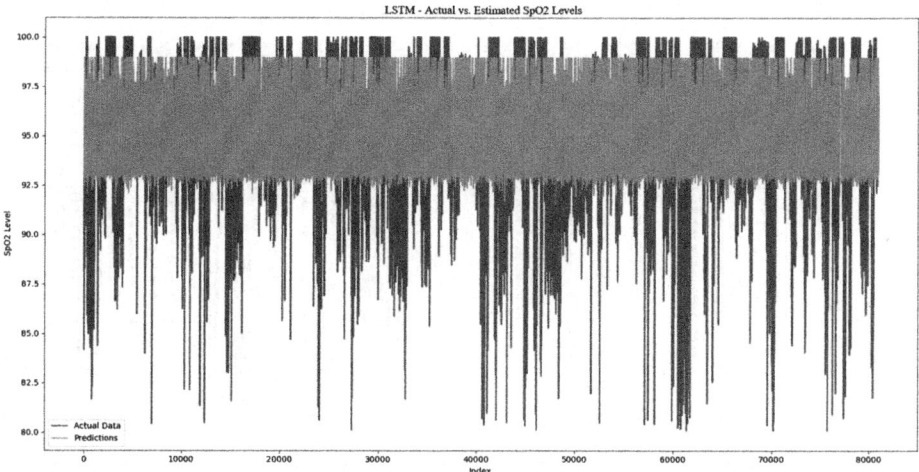

Fig. 1. Plot of the actual versus estimated SpO2 levels for the LSTM model (Color figure online)

In Table 2, all metrics are displayed as before, and the results are fairly similar. The R-squared metric shows consistent results across all models, with the highest value observed for the LSTM model at approximately 0.59.

Table 2. Performance metrics for all models trained and tested on the database constructed with respiratory feature extraction from files smaller than 600 MB

Model	MAE	MSE	RMSE	R2
LSTM	1.32152	3.09243	1.75853	0.58759
BiLSTM	1.33044	3.17016	1.78049	0.57723
CNN	1.37391	3.19322	1.78696	0.57415
RFR	1.43917	3.23543	1.79873	0.56852

In Figs. 1, 2, 3, and 4, for each model the estimated values (red) are plotted against the actual values (blue) for all entries in the testing set.

Based on the generated results, the findings indicate that increasing the dataset size resulted in modest improvements without yielding a significant impact.

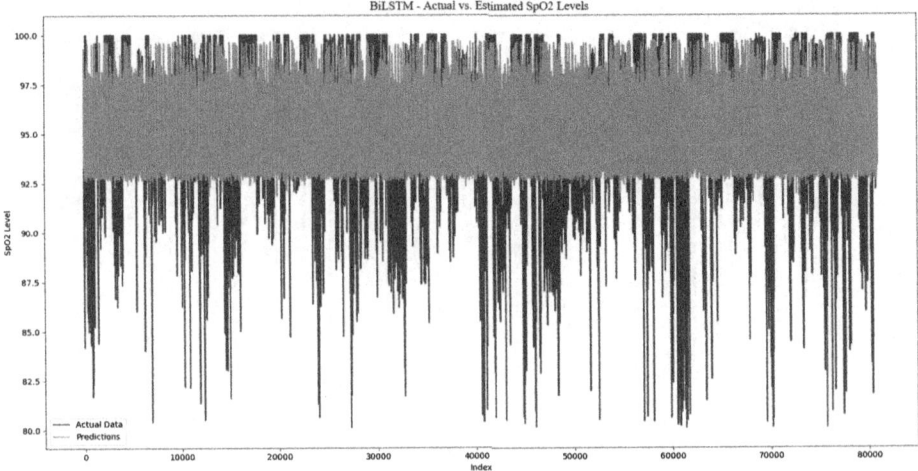

Fig. 2. Plot of the actual versus estimated SpO2 levels for the BiLSTM model (Color figure online)

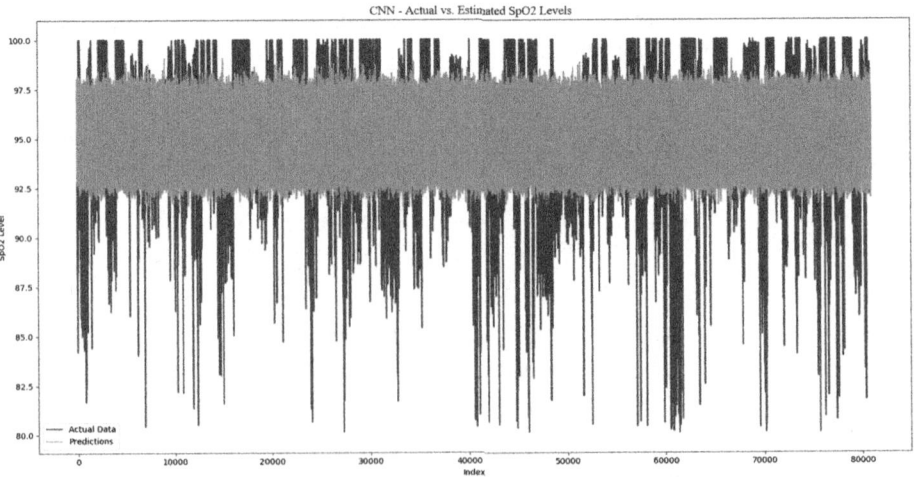

Fig. 3. Plot of the actual versus estimated SpO2 levels for the CNN model (Color figure online)

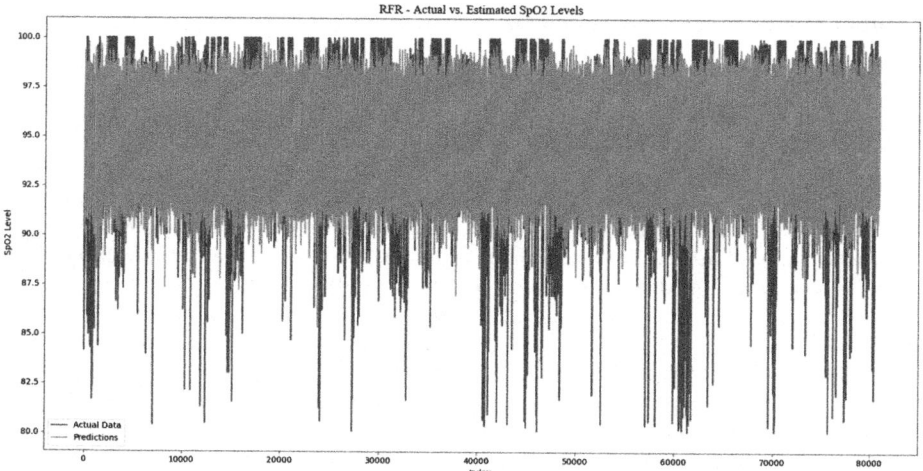

Fig. 4. Plot of the actual versus estimated SpO2 levels for the RFR model (Color figure online)

5 Conclusion

This study explored the effectiveness of DNN and ML models in estimating oxygen saturation - SpO2 levels directly from PPG signals using data from the PhysioNet MIMIC-III database. Our findings demonstrate that models such as Convolutional Neural Networks (CNNs), Bidirectional Long Short-Term Memory (BiLSTM) networks, Recurrent Neural Networks (RNNs), and Random Forest Regressors (RFRs) can accurately estimate SpO2 levels. Specifically, LSTM-based models achieved R-squared values up to 0.59, indicating potential in the predictive performance.

Importantly, the study ensured that patients in the training and test sets were entirely distinct, improving the generalizability of the models. These results highlight the potential of leveraging PPG signals and machine learning for real-time monitoring of SpO2. Such predictive models offer valuable insights into respiratory and cardiovascular health, facilitating early intervention and personalized patient care.

Future research could include expanding the dataset to validate and refine model performance, exploring additional features derived from PPG signals, and developing new algorithms to extract deeper insights from the MIMIC-III database. These steps would contribute to advancing the accuracy and applicability of SpO2 estimation models in clinical and on-field settings.

Acknowledgments. This work was supported in part by the NATO Science for Peace and Security Program under project SP4LIFE, number G5825. This work was supported in part by the Faculty of Computer Science and Engineering in Skopje, North Macedonia under project BIOX.

Conflicts of Interest. The authors declare no conflict of interest regarding the publication of this manuscript.

References

1. Cheriyedath, S.: Photoplethysmography (PPG). News-Medical. net (2019)
2. Castaneda, D., Esparza, A., Ghamari, M., Soltanpur, C., Nazeran, H.: A review on wearable photoplethysmography sensors and their potential future applications in health care. Int. J. Biosens. Bioelectron. **4**(4), 195 (2018)
3. Šoštarić, D., Mester, G., Dorner, S.: Mobile ECG and SPO2 chest pain subjective indicators of patient with GPS location in smart cities. Interdisc. Description Complex Syst.: INDECS **17**(3-B), 629–639 (2019)
4. Haque, C.A., Hossain, S., Kwon, T.-H., Kim, K.-D.: Comparison of different methods to estimate blood oxygen saturation using PPG. In: 2021 International Conference on Information and Communication Technology Convergence (ICTC), pp. 792–794. IEEE (2021)
5. Nitzan, M., Romem, A., Koppel, R.: Pulse oximetry: fundamentals and technology update. Med. Dev. Evidence Res. 231–239 (2014)
6. Venkat, S., et al.: Machine learning based SPO 2 computation using reflectance pulse oximetry. In: 2019 41st Annual International Conference of the IEEE Engineering in Medicine and Biology Society (EMBC), pp. 482–485. IEEE (2019)
7. Shuzan, M.N.I., et al.: Machine learning based respiration rate and blood oxygen saturation estimation using photoplethysmogram signals. Bioengineering **10**(2) (2023). https://doi.org/10.3390/bioengineering10020167
8. Krizea, M., Gialelis, J., Kladas, A., Theodorou, G., Protopsaltis, G., Koubias, S.: Accurate detection of heart rate and blood oxygen saturation in reflective photoplethysmography. In: 2020 IEEE International Symposium on Signal Processing and Information Technology (ISSPIT), pp. 1–4 (2020). https://doi.org/10.1109/ISSPIT51521.2020.9408845
9. Banik, P.P., Hossain, S., Kwon, T.-H., Kim, H., Kim, K.-D.: Development of a wearable reflection-type pulse oximeter system to acquire clean PPG signals and measure pulse rate and SPO2 with and without finger motion. Electronics **9**(11) (2020). https://doi.org/10.3390/electronics9111905
10. Koteska, B., Bodanova, A.M., Mitrova, H., Sidorenko, M., Lehocki, F.: A deep learning approach to estimate SPO2 from PPG signals. In: Proceedings of the 9th International Conference on Bioinformatics Research and Applications, pp. 142–148 (2022)
11. Koteska, B., Mitrova, H., Bogdanova, A.M., Lehocki, F.: Machine learning based SPO2 prediction from PPG signal's characteristics features. In: 2022 IEEE International Symposium on Medical Measurements and Applications (MeMeA), pp. 1–6 (2022). https://doi.org/10.1109/MeMeA54994.2022.9856498
12. Tonmoy, A.S., Ahmed, M.S., Chowdhury, A., Chowdhury, M.H.: Estimation of oxygen saturation from PPG signal using smartphone recording. In: 2024 International Conference on Advances in Computing, Communication, Electrical, and Smart Systems (iCACCESS), pp. 1–6 (2024). https://doi.org/10.1109/iCACCESS61735.2024.10499498
13. Goldberger, A.L., et al.: Physiobank, physiotoolkit, and physionet: components of a new research resource for complex physiologic signals. Circulation **101**(23), 215–220 (2000)

14. Johnson, A., et al.: MIMIC-III, a freely accessible critical care database sci. Data **3**(160035), 10-1038 (2016)
15. Moody, B., Moody, G., Villarroel, M., Clifford, G., Silva, I., III.: MIMIC-III waveform database (version 1.0). PhysioNet **3** (2020)
16. Meglenovski, V., et al.: Extraction and preprocessing of PPG data from the MIMIC III database (2023)
17. Petrovikj, N.: Predicting-SpO2-from-PPG. https://github.com/nenadpetrovikj/Predicting-SpO2-from-PPG. Accessed 27 June 2024

Novel Methodology for Gaining New Insights Into the Pharmacological Mechanisms of *Cannabis sativa* and *Alzheimer's Disease* Through Signaling Pathway Analysis Using Bioinformatics Tools

Filip Donev(✉), Nevena Ackovska, and Ana Madevska Bogdanova

Faculty of Computer Science and Engineering, "Ss. Cyril and Methodius" University,
Rugjer Boskovikj 16, 1000 Skopje, North Macedonia
donev.filip@students.finki.ukim.mk,
{nevena.ackovska,ana.madevska.bogdanova}@finki.ukim.mk

Abstract. Signaling pathway analysis is a crucial approach in understanding the pharmacological mechanisms of various cell functions. This paper presents a comprehensive framework for analyzing the signaling pathways of bioactive compounds from *C. sativa* and their potential interactions with Alzheimer's disease-related proteins. The study offers a new methodology utilizing a series of bioinformatics tools and databases to screen for active compounds, predict target genes, perform enrichment analysis, network analysis and molecular docking. The results reveal a significant intersection of genes between the bioactive compounds of *C. sativa* and Alzheimer's disease, indicating potential therapeutic implications. Moreover, network analysis identifies key proteins involved in the interactions, highlighting the potential targets for further investigation.

Keywords: Cannabis sativa · Alzheimer's disease · Signaling pathway · Enrichment analysis · Molecular docking

1 Introduction

Bioinformatics has revolutionized the field of pharmacology by providing researchers powerful tools to explore the complex interactions between the bioactive molecules and target compounds such as proteins, enzymes etc. One area of significant interest for research is the study of signaling pathways, which are essential for regulating cellular responses to external signals and are frequently dysregulated in various diseases. *C. sativa* has been used for centuries for its medicinal properties. Recent studies have shown that the bioactive compounds in *C. sativa*, such as cannabinoids and terpenes, have diverse anti-inflammatory, neuroprotective, and antidiabetic effects [1].

Alzheimer's disease (AD), on the other hand, is a neurodegenerative condition, typically associated with progressive decline in cognitive function as well as memory impairment [2]. Recent research has suggested that dysregulation of certain signaling pathways, such as the insulin signaling pathway, may be crucial for the development and progression of the disorder [3]. Given the potential therapeutic effects of *C. sativa*, there is growing interest to understand the underlying pharmacological mechanisms of *C. sativa* and its relation to AD.

In this study, bioinformatics tools and databases were utilized to analyze the signaling pathways of bioactive compounds from *C. sativa* and their potential interactions with AD-related proteins. By doing so, a novel methodology for better understanding the pharmacological mechanisms of *C. sativa* and it's relation with AD was developed.

2 Related Work

Previous studies have extensively explored the pharmacological effects of *C. sativa* and its bioactive compounds, particularly cannabinoids and terpenes, due to their potential therapeutic benefits. Research has shown these compounds exhibit anti-inflammatory, neuroprotective, and antidiabetic properties, primarily through the modulation of various signaling pathways [4,5]. In the context of AD, dysregulation of certain signaling pathways (e.g. the insulin signaling pathway) has been associated with the development of neurodegenerative processes [6]. Various bioinformatics tools and databases including the Traditional Chinese Medicine Systems Pharmacology (TCMSP), SwissTarget Prediction and STRING, have been utilized to predict target genes and construct protein-protein interaction (PPI) networks to better understand the molecular mechanisms causing these effects [7–9]. Enrichment analysis tools such as ShinyGO 0.80 have also been employed to identify significant biological processes (BP), molecular functions (MF) and cellular components (CC).

Molecular docking (MD) studies further validate these interactions by simulating protein-ligand interactions to predict binding affinities and potential therapeutic targets. The use of MD tools and software in these studies has provided insights into the binding mechanisms and efficacy of *C. sativa* compounds at a molecular level [11]. Moreover, integration with databases like PubChem, UniProt, and DisGeNet has enriched the understanding of disease-gene and compound-target interactions, facilitating a more comprehensive mapping of pharmacological pathways [12–14]. Recent studies have highlighted the importance of multi-omics approaches to provide a holistic view of the therapeutic potential of natural compounds [15].

Advanced computational techniques and machine learning algorithms have also been increasingly applied to predict the pharmacokinetics and pharmacodynamics of bioactive compounds from *C. sativa* [16]. These approaches enhance the predictive accuracy of ADME (Absorption, Distribution, Metabolism, and Excretion) properties, which are critical for drug development [17].

While some of the steps have been utilized in previous studies [18,19], this research offers a more concise and comprehensive methodology by systematically integrating advanced bioinformatics tools into a unified workflow. By leveraging the abovementoined tools and methodologies, this study expands the knowledge on the pharmacological mechanisms of natural compounds. offering potential methodology for the drug development process for complex diseases such as AD.

The methodology presented in this study not only includes the best practices from previous work, but also introduces a coherent framework that facilitates the exploration of Cannabis sativa's pharmacological mechanisms in the context of AD. By offering a concise yet robust approach, this study provides a valuable blueprint for future research in this field.

3 Materials and Methods

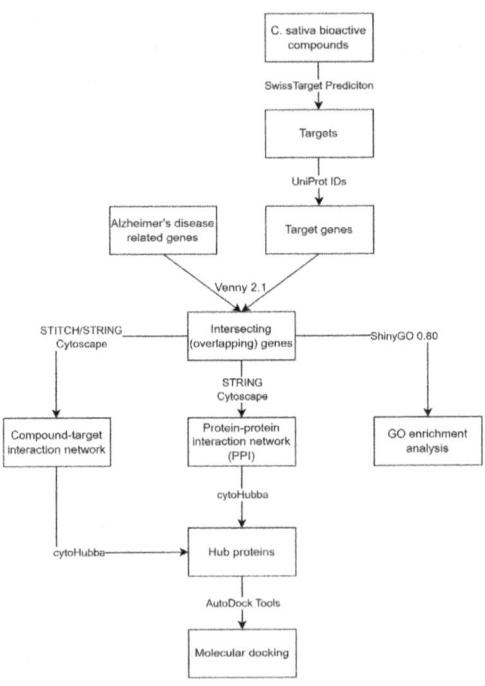

Fig. 1. Basic workflow of the new methodology

The workflow of the novel methodology, as shown in Fig. 1, comprises several steps. The first step includes screening for active compounds in *C. sativa*, followed by predicting target genes for these compounds, hence identifying the overlap with AD-related genes. The enrichment analysis helps us understand the biological functions and importance of the intersecting genes. Moreover, constructing and visualizing PPI and compound-target interaction networks plays a crucial role in identifying key proteins involved in the interaction network. Running MD with the hub proteins and their corresponding compounds to validate the results is the final step of this workflow. In the following subsections, the workflow of the developed methodology is explained in details.

3.1 Screening for Active Compounds, Target Genes and Finding Overlapping Genes

TCMSP was utilized to identify bioactive compounds in *C. sativa*. ADME evaluation was performed to avoid unfavorable pharmacokinetic properties, which can lead to costly late-stage failures in drug development. The active compounds from *C. sativa* were selected based on the ADME criteria.

The molecular structures were obtained directly from TCMSP in a *mol2* format. SwissTargetPrediction does not support this molecule format, so *OpenBabel* was utilized to convert the structures into SMILES molecule format. OpenBabel is a collaborative, open-source project designed to facilitate the conversion of chemical data from one format to another, including chemical file formats as mol2 to SMILES [20].

Bioactive molecules in SMILES format were submitted to SwissTargetPrediction. SwissTargetPrediction's "engine" operates by calculating the similarity between the input compounds and ones that have been experimentally confirmed to exhibit a certain biological activity against a specific target.

3.2 Enrichment Analysis

Gene Ontology (GO) analysis was conducted on the intersecting genes. *Gene Ontology* (GO) is a widely used bioinformatics resource that aims to unify the representation of gene and gene product attributes [21]. *GO enrichment analysis* is a method used to determine whether a particular set of genes or proteins has a statistically significant overrepresentation of GO terms compared to a background set. The result of GO enrichment analysis is a list of significantly enriched GO terms, along with statistical measures like *p-values* or *false discovery rates* (*FDRs*), which indicate the likelihood if the enrichment is due to random chance.

BP Enrichment Analysis focuses on identifying overrepresented GO terms related to biological processes. It helps in understanding the specific biological activities or events that the genes or proteins in a given dataset are involved in.

MF Enrichment Analysis identifies overrepresented GO terms related to the molecular functions of genes or proteins. It helps in explaining the specific biochemical activities that the genes or proteins perform, such as enzyme catalysis or protein binding.

CC Enrichment Analysis identifies overrepresented GO terms related to the cellular components where genes or proteins are active. It helps in understanding the specific locations within the cell where the gene products are localized or active, such as the nucleus, cytoplasm, or cell membrane.

GO enrichment analysis was conducted using the Shiny GO 0.8 tool, an intuitive graphical tool for enrichment analysis.

3.3 Protein-Protein and Target-Compound Interaction Network

The *PPI* network of the plant-disease overlapping genes was constructed from the *STRING* database, providing the UniProt IDs of the intersecting genes. *Cytoscape* was utilized to visualize the network [22], which was then analyzed with the *CytoHubba* plugin to identify significant nodes based on the node degree [23]. Furthermore, a compound-disease network from *C. sativa* bioactive compounds and the intersecting genes was constructed.

3.4 Molecular Docking

Protein structures were obtained from the *RCSB Protein Data Bank* in *pdb* format by entering the protein UniProt IDs [24]. *AutoDock Tools* is selected as a suitable software for conducting the docking analysis. The MD protocol includes several steps. Firstly, the macromolecule (protein) is imported in the GUI, followed by deleting small molecules (ligands) that are docked to the protein by default. Ligands are then imported one by one outputting each molecule in the same format as the protein (pdbqt). The grid box is then set for each protein, separately, creating grid box files. These files are then used as input to the *Autogrid4* in order to generate the three-dimensional grid map of each protein.

The actual MD calculations were computed by the *Autodock4* component which evaluates the free energy of different ligand conformations [25]. *Lamarckian Genetic Algorithm* (*LGA*) is selected as a scoring function to evaluate the affinity of different ligand conformations when interacting with the protein. LGA is designed to estimate the free binding energy of the protein-ligand interaction, taking into account various factors [26].

4 Results

TCMSP was utilized for screening and filtering bioactive compounds. The compounds were filtered based on ADME criteria by defining 2 parameters: *oral bioavailability(OB)* ≥ 20 and *drug likeliness(DL)* ≥ 0.10, thus there were 13 compounds remaining. After converting the structures to SMILES format using OpenBabel, these compounds were then submitted to SwissTargetPrediction to predict potential targets. Only targets specific to *Homo sapiens* with a probability threshold of 0.1 were considered, and the total number was 990. Duplicate targets were eliminated, their UniProt IDs were compared, and 352 unique targets were taken into consideration.

Genes related to AD were obtained from *DisGeNet*. 3077 AD-related genes were found, and after extracting the unique UniProt IDs, 3103 genes were obtained.

The intersection was visualized using a Venn diagram generated with *Venny 2.1* [27]. This was done before processing both datasets (Fig. 2A), after processing target genes dataset (Fig. 2B) and with processed data of both target and AD-related genes dataset (Fig. 2C).

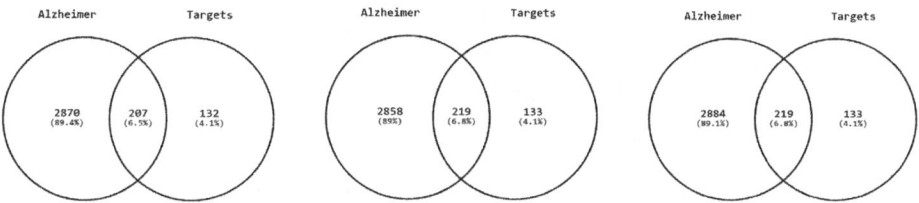

Fig. 2. A: Venn diagram before processing the datasets; B: Venn diagram with curated target genes dataset; C: Venn diagram after processing both datasets

One can conclude that the Venn diagram intersection has changed significantly by processing target genes data. On the other hand, processing AD-related genes dataset didn't affect the intersection of the diagram. Therefore, the Venn diagram of the processed targets dataset and original AD-related genes dataset should be prioritized. The total number of genes from the intersection was 219.

The GO enrichment analysis using ShinnyGO 0.80 was conducted by selecting Homo sapiens as target species from genome assembly $GRCh38.p13$ ($ENSEMBL$ database). The statistical significance is defined by FDR < 0.05, and the results were sorted by FDR, simultaneously being on the x-axis. The color corresponds to the *Fold Enrichment* and the size corresponds to the number of genes. The top 15 results were displayed on each analysis:

GO Biological Process. According to the results of the BP analysis using ShinnyGO 0.80, shown in Fig. 3, the bioactive compounds were involved in the response to oxygen-, organic (cyclic)- and nitrogen-containing compounds as well as in the response to lipids, and chemical and endogenous stimulus.

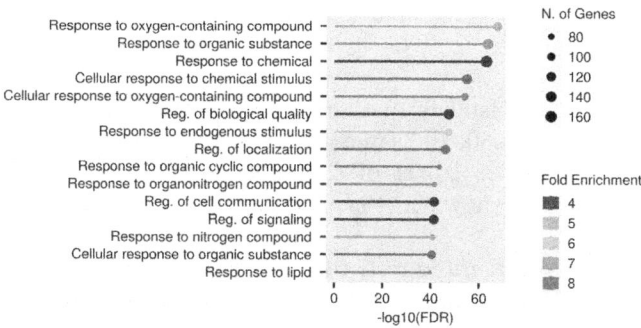

Fig. 3. GO Biological Process analysis results

GO Cellular Component. Other than BP, CC analysis was also conducted (Fig. 4), and it showed that most of the intersecting genes coded proteins in the plasma membrane, presynaptic membrane and neuron parts.

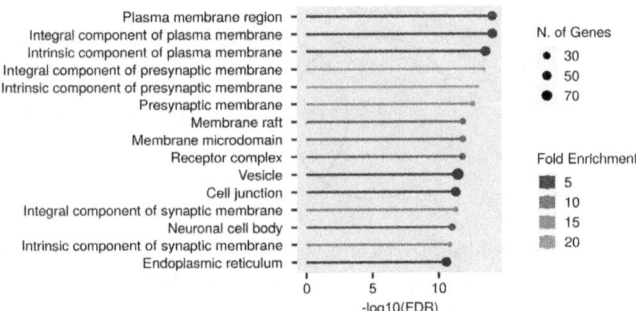

Fig. 4. GO Cellular Component analysis results

GO Molecular Function. MF analysis (Fig. 5) implicates that the intersecting genes are involved in binding with small molecules, nuclear receptors, transcription factors and kinases.

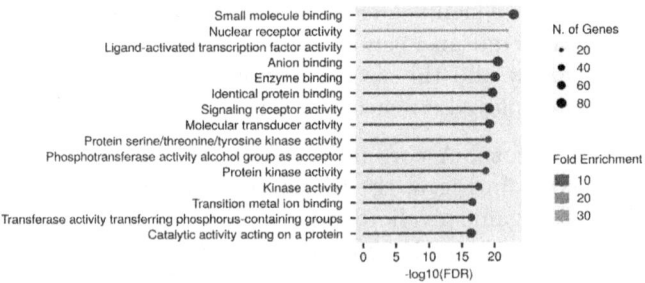

Fig. 5. GO Molecular Function analysis results

To evaluate the consistency and reliability of the GO enrichment analysis results using ShinyGO 0.80, the same analyses were conducted again using the *clusterProfiler* package as a part of the Bioconductor project in R [28]. The corresponding plots are shown in Fig. 6, Fig. 7 and Fig. 8 for BP, CC and MF, respectively.

Scatter plots were generated in R to compare the FDR values obtained from each tool for each GO category (Fig. 9). The Spearman's correlation coefficients between the results from ShinyGO and clusterProfiler were calculated to assess the level of agreement.

The BP scatter plot (Fig. 9A) demonstrates a strong positive correlation between the FDR values from ShinyGO and clusterProfiler, with a correlation coefficient of *0.833509*. This indicates a high level of consistency between the two tools in identifying significant biological processes.

The MF scatter plot (Fig. 9B) also shows a positive correlation between the FDR values from ShinyGO and clusterProfiler, with a correlation coefficient

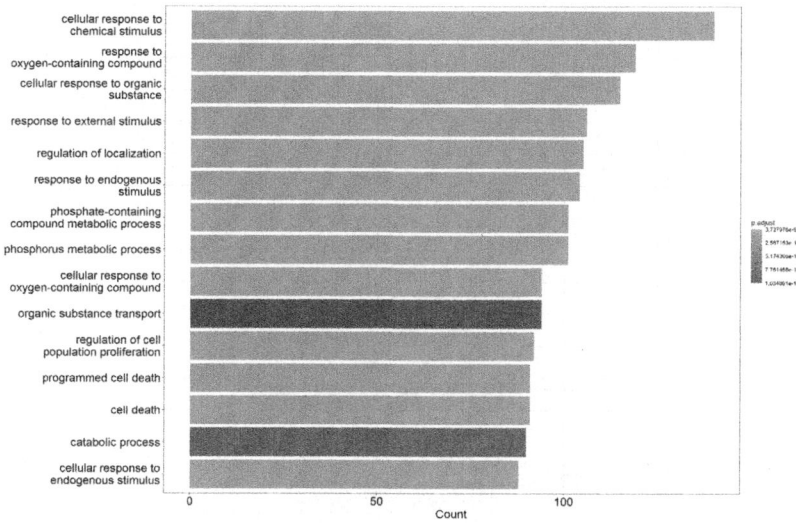

Fig. 6. GO Biological Process analysis using clusterProfiler

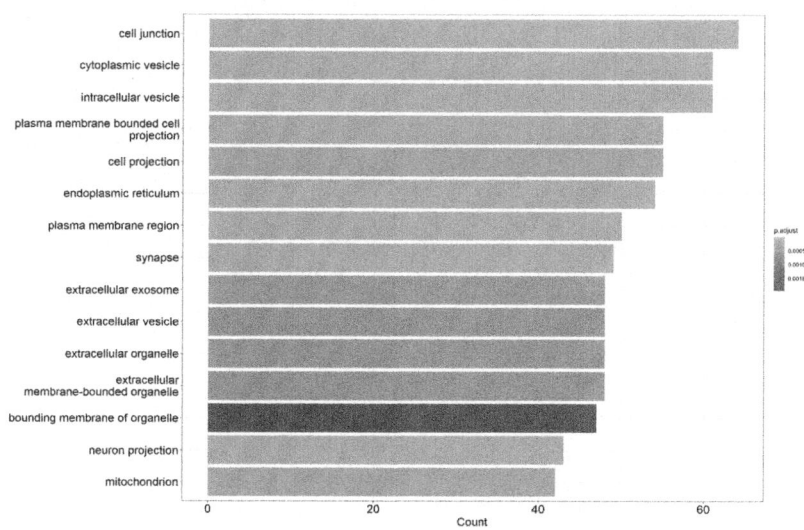

Fig. 7. GO Cellular Component analysis using clusterProfiler

of *0.797528*. Although slightly lower than the BP, this coefficient still reflects a substantial agreement between the tools in identifying significant molecular functions.

The CC scatter plot (Fig. 9C) reveals the highest correlation among the three categories (correlation coefficient of *0.8999844*). This strong positive correlation

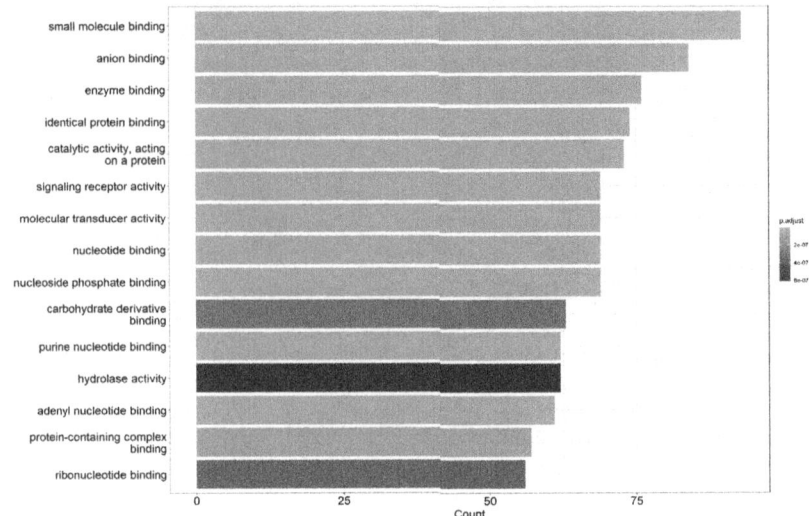

Fig. 8. GO Molecular Function analysis using clusterProfiler

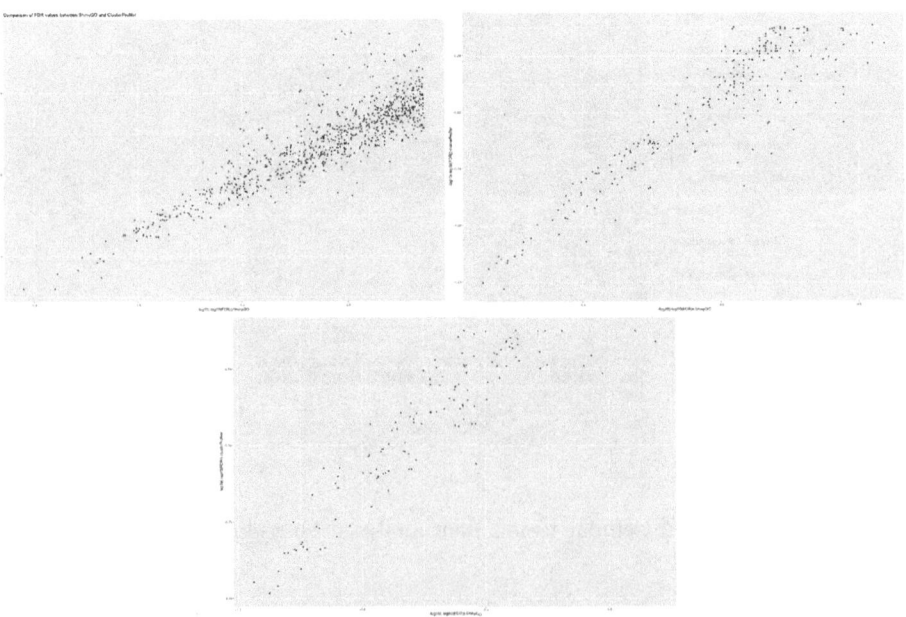

Fig. 9. A: Scatter plot comparing FDR values for Biological Process terms between ShinyGO and clusterProfiler (correlation = 0.833509); B: Scatter plot comparing FDR values for Molecular Function terms between ShinyGO and clusterProfiler (correlation = 0.797528); C: Scatter plot comparing FDR values for Cellular Component terms between ShinyGO and clusterProfiler (correlation = 0.8999844)

suggests a high level of correspondence between ShinyGO and clusterProfiler in identifying significant cellular components.

The high correlation coefficients across all three GO categories indicate that ShinyGO and clusterProfiler provide consistent results in terms of identifying significant GO terms. This consistency supports the reliability of using either tool for GO enrichment analysis, with minor variations that can be attributed to differences in underlying algorithms and databases used by each tool.

In the next step, PPI network was constructed using STRING from the intersecting genes limited to Homo sapiens, with a medium FDR stringency (5%) and interactions with highest confidence score i.e. exceeding 0.9. In this network (Fig. 10), the nodes are representing target proteins and the edges represent the interactions between them.

Nodes with highest degree i.e. proteins with most interactions were: TP53, HSP90AA1, HSP90AB1, ESR1, PIK3R1, AKT1, MAPK1, MAPK3, EGFR and PTGS2.

Akin to the PPI network, compound-target interaction network was analyzed with the CytoHubba plugin to identify significant nodes. Proteins with most interactions were: AKT1, IL6, TP53, BCL2, PPARG, EGFR, PTGS2, ESR1, MAPK3 and HSP90AA1. The overlap of hub proteins between the PPI and compound-target interaction network is considered highly significant. It is reasonable to assume that some of these proteins have major role in the development of AD.

This hypothesis should be confirmed through MD analysis. For the sake of the simplicity in this paper, only one target from the overlap (alpha serine/threonine protein kinase - AKT1) has been selected to simulate a protein-ligand interaction. AKT1 is found to be a target for two of the *C. sativa* compounds (luteolin and apigenin). Molecular structures of the corresponding ligands are obtained directly from TCMSP in a mol2 format.

Before conducting the MD to support the hypothesis, i.e. before running Autodock4, the parameters for the LGA are defined: number of runs was set to 10, population size of 250, number of evaluations was set to 2,500,000, number of generations 27,000, mutation rate 0.02 and crossover rate 0.8. There are numerous types of analysis one can conduct on the output file (*.dlg*) which contains the docking results. Our target is the information incorporated in the RMSD table of each file. The top result (docked conformation) with lowest energy is displayed in Table 1 for AKT1-luteolin docking.

As shown in the docking results, the free energy of the interaction between AKT1 and luteolin is -5.20 kcal/mol. ΔG indicates the thermodynamic favorability of a process. When $\Delta G < 0$, the process is thermodynamically favored. In other words, because of the negative value of ΔG, AKT1 binds spontaneously to luteolin. It is important to note that the MD only considers interaction between two selected molecules, when in reality, many molecules compete for the protein's binding site. For that reason, experimental validation of the result is an essential step of this type of research.

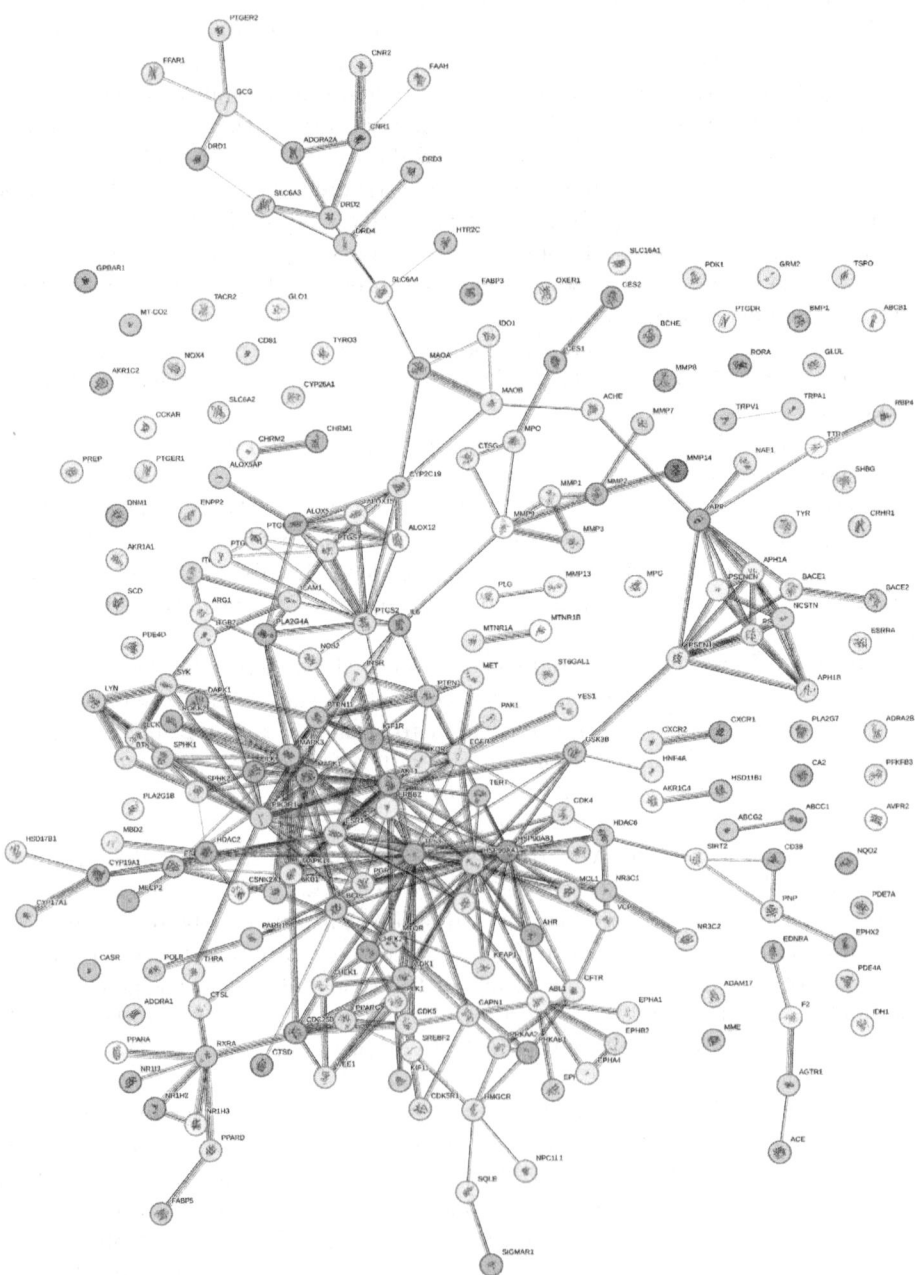

Fig. 10. PPI network of the intersecting protein genes

Table 1. Docked conformation with the lowest energy in the highest ranked cluster from AKT1 and luteolin docking

Parameter	Value
Cluster Rank	1
Number of conformations in this cluster	12
RMSD from reference structure	11.379 Å
Estimated Free Energy of Binding	−5.20 kcal/mol
Estimated Inhibition Constant (K_i)	154.48 μM ($T = 298.15$ K)
Final Intermolecular Energy	−9.67 kcal/mol
van der Waals + Hydrogen bond + desolvation Energy	−9.37 kcal/mol
Electrostatic Energy	−0.30 kcal/mol
Final Total Internal Energy	−1.38 kcal/mol
Torsional Free Energy	4.47 kcal/mol
Unbound System's Energy	−1.38 kcal/mol

5 Discussion

This study offers a novel methodology for getting insights into the pharmacological mechanisms of *C. sativa* and its potential therapeutic applications, particularly in the context of AD. The integration of various bioinformatics tools allows one to identify and validate key bioactive compounds, their target genes, and the associated signaling pathways these are involved in.

AKT1 is a key downstream effector of the PI3K/AKT pathway which is essential for cell survival, growth and metabolism, and the dysregulation of the PI3K/AKT pathway has been implicated in AD.

Activating AKT1 by bioactive compounds in *C. sativa* is a potential key to a *C. sativa*-AD relation, where PI3K/AKT, HIF-1 and cancer related signaling pathways play a crucial role. Therefore, the discovered essential protein may be therapeutic target of *C. sativa* in treating AD.

The correlation between the enrichment analysis results obtained using the ClusterProfiler and the ShinyGO web tool underscores the reliability and robustness of the analysis. This consistency indicates that both tools can effectively identify significant BP, MF, and CC associated with the bioactive compounds studied. This finding is crucial, as it validates the approach and suggests that combining different computational tools can enhance the accuracy and reliability of bioinformatics analyses.

The study also highlights the importance of the MD in validating the interactions between bioactive compounds and their target proteins. The high binding affinities observed in the docking simulations suggest potential therapeutic targets and support the hypothesis that the abovementoined compounds can modulate key proteins involved in the AD pathogenesis. This adds a layer of confidence to the predictive analyses and provides a basis for future experimental validation.

The use of comprehensive databases such as TCMSP, UniProt, and DisGeNet facilitated a detailed understanding of the disease-related proteins and bioactive compounds interactions. This integrative approach not only enhances the understanding of the molecular mechanisms at play but also provides a holistic view of the potential therapeutic effects of natural compounds.

However, this study is not without limitations. One significant constraint is the reliance on the TCMSP, since it primarily includes molecules documented in Chinese literature. This may result in a limited scope of bioactive compounds considered, potentially overlooking relevant compounds identified in other regions or studies. Additionally, the predictive nature of computational tools necessitates further experimental validation to confirm the biological activity and therapeutic efficacy of the identified compounds.

Another limitation is the inherent variability in the quality and completeness of the data across different databases. While databases like PubChem and UniProt are well-curated, there can still be discrepancies and gaps that might affect the accuracy of the predictions. Moreover, the complexity of multi-omics data integration poses significant challenges, requiring sophisticated computational techniques and extensive validation to ensure the reliability of the findings.

6 Conclusion

This study provides a new methodology and a comprehensive analysis of the pharmacological mechanisms underlying the *C. sativa*-AD relation. By employing a systematic workflow that includes active compound screening, target prediction, disease-related gene screening, network analysis, enrichment analysis and MD, compelling evidence of the plant's bioactive compounds interacting with key genes associated with AD pathology has been revealed.

The novelty of the methodology lies in its ability to seamlessly integrate data from diverse sources, providing a holistic view of the pharmacological potential of natural compounds. This approach enhances the predictive accuracy and reliability of bioinformatics analyses, making it a valuable tool for future pharmacological research.

Despite its limitations, this study provides a robust framework for future research, bridging traditional medicinal knowledge with modern computational biology. The findings lay the groundwork for more comprehensive studies and encourage the integration of multi-omics data and approach to further elucidate the complex interactions between bioactive compounds and their targets.

Future research should focus on experimental validation of the predicted interactions and therapeutic effects, addressing any limitations related to database scope and data variability. Additionally, applying this integrative approach to other medicinal plants could further advance the field of natural compound pharmacology. Overall, this study paves the way for conducting similar studies on other plants and developing novel treatments for complex diseases, leveraging the power of advanced bioinformatics tools and methodologies.

References

1. Bonini, S.A., et al.: Cannabis sativa: a comprehensive ethnopharmacological review of a medicinal plant with a long history. J. Ethnopharmacol. **5**(227), 300–315 (2018). https://doi.org/10.1016/j.jep.2018.09.004
2. Knopman, D.S., Amieva, H., Petersen, R.C., et al.: Alzheimer disease. Nat. Rev. Dis. Primers. **7**, 33 (2021). https://doi.org/10.1038/s41572-021-00269-y
3. Cedernaes, J., Osorio, R.S., Varga, A.W., Kam, K., Schiöth, H.B., Benedict, C.: Candidate mechanisms underlying the association between sleep-wake disruptions and Alzheimer's disease. Sleep Med. Rev. **31**, 102–111 (2017). https://doi.org/10.1016/j.smrv.2016.02.002
4. Mechoulam, R., Parker, L.A.: The endocannabinoid system and the brain. Annu. Rev. Psychol. **64**, 21–47 (2013). https://doi.org/10.1146/annurev-psych-113011-143739
5. Hampson, A.J., Grimaldi, M., Axelrod, J., Wink, D.: Cannabidiol and (-)Delta9-tetrahydrocannabinol are neuroprotective antioxidants. Proc. Natl. Acad. Sci. U.S.A. **95**(14), 8268–8273 (1998). https://doi.org/10.1073/pnas.95.14.8268
6. Talbot, K., et al.: Demonstrated brain insulin resistance in Alzheimer's disease patients is associated with IGF-1 resistance, IRS-1 dysregulation, and cognitive decline. J. Clin. Invest. **122**(4), 1316–1338 (2012). https://doi.org/10.1172/JCI59903
7. Ru, J., et al.: TCMSP: a database of systems pharmacology for drug discovery from herbal medicines. J. Cheminform. **6**(1), 1–6 (2014). https://doi.org/10.1186/1758-2946-6-13
8. Daina, A., Michielin, O., Zoete, V.: SwissTargetPrediction: updated data and new features for efficient prediction of protein targets of small molecules. Nucleic Acids Res. **47**(W1), W357–W364 (2019). https://doi.org/10.1093/nar/gkz382
9. Szklarczyk, D., et al.: The STRING database in 2023: protein-protein association networks and functional enrichment analyses for any sequenced genome of interest. Nucleic Acids Res. **51**(D1), D638–D646 (2023). https://doi.org/10.1093/nar/gkac1000
10. Ge, S.X., Jung, D., Yao, R.: ShinyGO: a graphical gene-set enrichment tool for animals and plants. Bioinformatics **36**(8), 2628–2629 (2020). https://doi.org/10.1093/bioinformatics/btz931
11. Hourfane, S., Mechqoq, H., Bekkali, A.Y., Rocha, J.M., El Aouad, N.: A comprehensive review on cannabis sativa ethnobotany, phytochemistry, molecular docking and biological activities. Plants (Basel) **12**(6), 1245 (2023). https://doi.org/10.3390/plants12061245
12. Kim, S., et al.: PubChem substance and compound databases. Nucleic Acids Res. **44**(D1), D1202–D1213 (2016). https://doi.org/10.1093/nar/gkv951
13. The UniProt Consortium: UniProt: the universal protein knowledgebase in 2023. Nucleic Acids Res. **51**(D1), D523–D531 (2023). https://doi.org/10.1093/nar/gkac1052
14. Piñero, J., et al.: The DisGeNET knowledge platform for disease genomics: 2019 update. Nucl. Acids Res. (2019). https://doi.org/10.1093/nar/gkz1021
15. Vijay, A., et al.: Emergence of phytochemical genomics: integration of multi-omics approaches for understanding genomic basis of phytochemicals. In: Swamy, M.K., Kumar, A. (eds.) Phytochemical Genomics. Springer, Singapore (2022). https://doi.org/10.1007/978-981-19-5779-6_9

16. Ma'ayan, A., Rouillard, A.D., Clark, N.R., Wang, Z., Duan, Q., Kou, Y.: Lean Big Data integration in systems biology and systems pharmacology. Trends Pharmacol. Sci. **35**(9), 450–460 (2014). https://doi.org/10.1016/j.tips.2014.07.001
17. Yang, H., et al.: AdmetSAR 2.0: web-service for prediction and optimization of chemical ADMET properties. Bioinformatics **35**(6), 1067–1069 (2019). https://doi.org/10.1093/bioinformatics/bty707
18. Ma, H., Xu, F., Liu, C., Seeram, N.P.: A network pharmacology approach to identify potential molecular targets for cannabidiol's anti-inflammatory activity. Cannabis Cannabinoid Res. **6**, 288–299 (2020). https://doi.org/10.1089/can.2020.0025
19. Guzmán-Flores, J.M., Pérez-Vázquez, V., Martínez-Esquivias, F., Isiordia-Espinoza, M.A., Viveros-Paredes, J.M.: Molecular docking integrated with network pharmacology explores the therapeutic mechanism of cannabis sativa against type 2 diabetes. Curr. Issues Mol. Biol. **45**, 7228–7241 (2023). https://doi.org/10.3390/cimb45090457
20. O'Boyle, N.M., Banck, M., James, C.A., Morley, C., Vandermeersch, T., Hutchison, G.R.: Open babel: an open chemical toolbox. J. Cheminform. **3**, 33 (2011). https://doi.org/10.1186/1758-2946-3-33
21. Gene Ontology Consortium. The gene ontology project in 2008. Nucleic Acids Res. **36**(Database issue), D440-4 (2008). https://doi.org/10.1093/nar/gkm883
22. Shannon, P., et al.: Cytoscape: a software environment for integrated models of biomolecular interaction networks. Genome Res. **13**(11), 2498–2504 (2003). https://doi.org/10.1101/gr.1239303
23. Chin, C.H., Chen, S.H., Wu, H.H., Ho, C.W., Ko, M.T., Lin, C.Y.: CytoHubba: identifying hub objects and sub-networks from complex interactome. BMC Syst. Biol. **8**(Suppl. 4), S11 (2014). https://doi.org/10.1186/1752-0509-8-S4-S11
24. Berman, H.M., et al.: The protein data bank. Nucleic Acids Res. **28**, 235–242 (2000). https://doi.org/10.1093/nar/28.1.235
25. Morris, G.M., et al.: Autodock4 and AutoDockTools4: automated docking with selective receptor flexiblity. J. Comput. Chem. **16**, 2785–2791 (2009). https://doi.org/10.1002/jcc.21256
26. Morris, G.M., et al.: Automated docking using a lamarckian genetic algorithm and and empirical binding free energy function. J. Comput. Chem. **19**, 1639–1662 (1998). https://doi.org/10.1002/(SICI)1096-987X(19981115)19:14<1639::AID-JCC10>3.0.CO;2-B
27. Oliveros, J.C.: Venny, an interactive tool for comparing lists with Venn's diagrams (2007–2015). https://bioinfogp.cnb.csic.es/tools/venny/index.html. Accessed 10 Mar 2024
28. Yu, G., Wang, L., Han, Y., He, Q.: ClusterProfiler: an R package for comparing biological themes among gene clusters. OMICS: J. Integrative Biol. **16**(5), 284–287 (2012). https://doi.org/10.1089/omi.2011.0118

Session 6, 7

AI Cardiologist: Arrhythmia Detection by Transformer-Based Language Model

Marjan Gusev

Faculty of Computer Science and Engineering, Sts Cyril and Methodius University in Skopje, Skopje, North Macedonia
`marjan.gushev@finki.ukim.mk`

Abstract. Wearable electrocardiogram sensors have emerged on the market but lack the technology to interpret the results with high performance. This paper explores the potential for innovation in developing a heart-based language and applying natural language processing to detect and classify dangerous arrhythmia. In the AI Cardiologist project, we propose real-time heart monitoring and alerting dangerous arrhythmia using AI-based technology using RoBERTa transformers in large language models. The focus is on analyzing the domains of medical applications and introducing new business models to exploit the technology's results. We confirmed that foundation models for ECG records created by deep learning methods could greatly enhance precision and effectiveness of heart health monitoring and detection of atrial fibrillation, enabling better healthcare and well-being. No similar solution is available on the market outside of hospital environments, giving the project results a high potential for commercialization.

Keywords: Transformers · Deep Learning · Self-supervised Learning · ECG · AFIB

1 Introduction

Small electrocardiogram (ECG) sensors have become affordable wearable technology, providing long-term, real-time measurement of heart activity while patients perform everyday activities at their homes and offices outside hospitals. Most cloud-based or standalone computer solutions are based on processing the complete measurement, and only a few refer to remote real-time monitoring [16] as a doctor tool. So far, no customer-related solutions are on the market.

World Health Organization statistics on mortality rate show that approximately 1/3 of deaths are due to cardiovascular diseases. In contrast, premature heart disease and stroke can be avoided in 80% of cases [37]. In addition, The American Heart Association [2] estimates up to 90% of cardiovascular diseases may be preventable. The proposed AI Cardiologist solution will not prevent a heart attack but reduce the risk of severe damage and consequences. More than hundreds of millions are diagnosed with cardiovascular disease in Europe and the

USA. Early detection with the help of continuous heart monitoring will reduce the death rate and increase life expectancy.

Supporting systems in cardiology are primarily based on medical knowledge implemented either directly with rule-based signal processing algorithms or by Artificial Intelligence (AI) with Machine Learning (ML) techniques, including Neural Networks (NNs) based on mathematical models and deep learning (DL) architectures with multiple neuron layers. The AI Cardiologist project [15] project is based on state-of-the-art AI technology with Large Language Models (LLMs) as AI programs that recognize and generate text and transformers that use NN architecture to process sequences of tokens or other structured data through layers with self-attention mechanisms that capture dependencies and relationships within input sequences. Transformer-based LLMs follow a two-phase training process: first, they undergo pretraining using a self-supervised learning method on vast quantities of unannotated datasets and the fine-tuning stage on a smaller annotated dataset. The achieved popularity of Generative Pre-trained Transformer (GPT) technology [27] influences researchers to develop new models. Our approach builds a new heart-based language by developing tokens from ECG samples and heartbeat annotations.

In our earlier paper [12], we have discussed the challenges, benefits, and disadvantages of an AI-based self-monitoring home healthcare application. The rise of edge and dew computing technology makes such an application serve as an AI-based cardiologist for the customer. This paper presents the innovation potential of a dew computing solution based on signal processing algorithms and transformer-based AI technology for a heart-based language capable of autonomous performance (B2C). This solution transfers all measured ECG data to the cloud and provides doctors with a sophisticated tool to set a proper diagnosis and recommend treatment. Realizing such a solution with an alerting center for detecting and classifying dangerous arrhythmia simulates intensive monitoring of outpatients.

The AI Cardiologist project [15] outcome brings an AI doctor closer to the end-user, alerting dangerous arrhythmia and preventing severe heart damage, with results understandable by patients. The system consists of a wireless and mobile ECG biosensor, a data center, a smartphone, and a web application that provides a remote 24/7 healthcare monitoring service. Our approach aims to build an autonomous arrhythmia detection system that patients can use directly to alert them when dangerous arrhythmia is detected, which is an entirely new concept. In contrast to other solutions that aim to establish a diagnosis after the measurement is finished as a support system for a doctor, this concept works in real-time, as the patient is constantly monitored in the intensive care unit.

Our state-of-the-art analysis shows that Large Language Models (LLM) mostly report a concept without building a prototype or a solution. In contrast, in this paper, we report a use-case solution [34] to detect Atrial Fibrillation as one of the most dangerous arrhythmias. A transformer-based language model technology is applied for ECG-based rhythm episode detection and classification. In addition, self-attention generates contextualized representations and relative context between ECG samples and heartbeat annotations. Our goal is to build a very efficient tool for the real-time monitoring of thousands of incoming ECG streams using sophisticated, scalable, serverless technology.

The available heart monitor products on the market are event (on-demand) monitors (smart watches or handheld devices), which can capture smaller ECG measurements (up to 30 s), uncomfortable Holter devices that measure 24 h or 48 h ECG, or wearable patches, which do not provide real-time monitoring and alerting for patients outside a hospital. An alternative is implantable loop recorders (ILR) as an invasive medical device.

The advancement of such a solution beyond the state-of-the-art includes wireless light wearable ECG monitors for personal use in everyday life and a remote real-time monitoring system that alerts in case of dangerous arrhythmia with features usual for intensive care units. The novelty lies in that no such solution is available outside of hospital environments, and the AI Cardiologist solution addresses the mass population at affordable prices. This paper presents the underlying concepts of the AI Cardiologist project, which was approved by the European Network of AI Excellence Centres (ELISE).

Section 2 presents related work on articles integrating AI algorithms in cardiology, especially language models in detecting and classifying ECGs. Implementation details in Sect. 3 refer to the architectural approach. Section 3 discusses the Innovation potential, including the functional analysis, business potential, and limitations in performance evaluation. Finally, Sect. 5 presents the conclusions and gives directions for future work.

2 Related Work

2.1 AI in Cardiology

Many research papers review the application of AI in cardiology [7,17,18, 20,21,23,24,28]. The analyzed cardiology domains include echocardiography, cardiac/coronary computer tomography, cardiac magnetic resonance imaging, nuclear cardiac imaging, heart failure, detection of arrhythmias from ECG, and other clinical data from electronic health records. In this paper, we focus on the application of AI in detecting arrhythmias, which is one of the most promising areas. All analyzed applications serve as a support system for doctors to establish a diagnosis of the related heart problem more efficiently, faster, and more accurately. In contrast, our approach is intended for customers, similar to medical devices that measure temperature, and patients can understand the measured information.

Many essential needs of healthcare professionals remain unaddressed [9], such as the absence of explanations in clinically relevant language, managing unknown medical conditions, and ensuring transparency about the system's limitations, both in statistical performance and in identifying scenarios where the system's predictions do not apply. Performance is critical for evaluating the AI-based medical device, especially analyzing the significance, uncertainty estimation, relevance of the generalization approach, and handling unknown medical conditions. Identifying a heterogeneous mix of known and unknown arrhythmias from

ECG is also noted in our research analyzing the data quality, especially when dealing with real-time detection in contrast to detection after the measurement is finished. Reaching a robust system is a significant concern in ECG interpretation, considering that there are no two identical out of 8B living persons or identical human ECG forms due to the heart's unique anatomy and physiology. The heart is a dynamic organ that continuously adjusts to the body's needs for oxygen and nutrients, ensuring a steady flow of blood through its valves, chambers, and vessels. In addition, the real-time monitoring addressed in our research means that the analyzed ECG comes from a physically active person. Muscle noise corrupts the ECG with many artifacts and noise, which can mislead proper detection and classification.

The analysis of AI applications in cardiology shows several potentials and limitations. Although AI potential is enormous, data quality still has limitations considering the explainability and data bias [28]. To avoid these problems, in our research, we follow international standards to evaluate the performance of the realized AI-based system and comply with medical device requirements.

The Apple Watch measures an ECG and photoplethysmography (PPG), feeding a convolutional NN to classify the rhythm as sinus or AFIB [36]. A clinical study on the Apple Watch AFIB detection performance shows that it is not precise for a population with known AF, and the algorithm appears accurate for AF screening as currently FDA-cleared, but increased sensitivity and wear times are necessary [35]. Although a combination of ECG beat detection and PPG rate measuring gives information on an irregular rhythm, in addition to 24/7 monitoring, our approach considers the beat class, which offers an in-depth analysis of the underlying rhythm.

Predictive analytics based on AI and ML represent a significant advancement that can optimize resource allocation and enhance clinical decision-making, benefiting those affected by heart conditions [1], mainly focusing on aortic aneurysm growth and rupture, carotid stenosis and extremity arterial disease. In the proposed solution, we rely on real-time monitoring, alerting on detecting dangerous arrhythmia, and predictive analytics will be an added value.

Besides being an integral part of clinical decision support systems, the impact of AI in current clinical practice is still limited [20], but promising. The application of emerging wearable technology generates a vast volume of medical data [21], including unlabeled ECGs and a much smaller volume of annotated ECGs. Therefore, this new transformer-based approach is a preferred technology for detecting patterns in the vast volume of data while developing foundation models. A smaller dataset will fine-tune a model to detect a specific arrhythmia.

The day when AI replaces a cardiologist is not yet in sight, nor may it ever be [7]. However, AI will continue to assist cardiologists worldwide daily to make faster and better diagnoses and image interpretations. In contrast, our innovation brings AI closer to the patient than doctors.

The rise of big data and open science [23], impacted AI with vast amounts of data powering advanced algorithms to create statistical models that often match or surpass human capabilities. However, AI can still make substantial errors in interpretation, validity, and generalizability, leading to safety and ethical issues that must be proactively addressed.

Despite some potential pitfalls, it is becoming evident that the best way to make decisions based on data is by applying techniques drawn from AI [17]. AI will drive improved patient care because physicians can interpret more data in greater depth than ever.

2.2 Generative Pre-trained Large Language Models for Arrhythmia Detection

Generative AI has the potential to diagnose heart disease and recommend management options suitable for the patient by assisting physicians, allowing them to spend more time on patient care [10] lacking the most up-to-date with the latest research and presenting outdated information, which may lead to an adverse event.

Generative AI can help diagnose heart disease and suggest appropriate patient management options by supporting physicians. This allows doctors to focus more on patient care [10]. However, AI could lead to adverse events if it uses outdated information and needs the latest research. Examples on use of revolutionizing cardiology with LLMs include automated generation of clinical notes and discharge letters, medical term coding for billing, medical chatbots both for patients and clinicians, data enrichment in the identification of disease symptoms or diagnosis, cohort selection for clinical trials, and auditing purposes [5]. Other examples address personalized assistance in learning, managing electronic medical records, and clinical decision-making in cardiology [4]. ChatGPT is demonstrated in correctly answering undergraduate medical exams, including the exams in cardiology, such as the European Exam in core cardiology [33].

Generative LLMs [19] offer numerous benefits, such as enhancing patient satisfaction and outcomes by freeing up doctors' time by automatically generating clinical notes, emails, and patient letters. They also help patients comprehend their medical records, improve the accuracy and consistency of medical documentation, and provide reminders about guidelines, next steps, and differential diagnoses based on specific symptoms. Advanced applications involve organizing and examining unstructured data, including medical notes and test results, which could transform data management and analysis in cardiovascular research [25].

Limitations include the generation of incorrect or plagiarized content, facing challenges in handling tasks without detailed prompts, and lack of originality, and therefore, need human oversight [5]. Besides the technical considerations, human interface, and regulations, the limitations include potential biases, reasoning opacity, and the need for rigorous validation in medical contexts [25]. Quality of data used to train is crucial, and users need to be aware of the potential data biases, inaccuracies, or misconceptions [31].

Generative LLMs are transforming the medical field by creating a dynamic environment with immense potential to revolutionize patient care, interactions, education, and research. By processing vast amounts of data and producing contextually relevant responses, they are paving the way for a new era in healthcare [25].

Their future depends on effectively balancing their strengths and limitations, ensuring they become reliable and effective tools for healthcare professionals and patients. This balance is crucial to prevent harm and protect privacy [25].

Even though many obstacles remain, generative AI in cardiology can completely alter how medical care is provided, improving patient outcomes and enhancing physician productivity [10]. Although LLMs showed promising results in interpreting and applying complex clinical guidelines when answering vignette-based clinical queries, with a potential for enhancing patient outcomes through personalized advice, they should be utilized with a grain of salt as supplementary tools in clinical cardiology [26]. Generative AI LLMs must be fine-tuned explicitly on medical data to produce accurate or relevant information in clinical settings. Consequently, the general pre-trained models cannot substitute the judgment and expertise of human healthcare professionals [19].

A challenge with LLMs is their ability to confidently regurgitate "facts" regardless of whether they are accurate [30], such that even minor inaccuracies in clinical information may lead to critical consequences. Using statistical learning, these AI models are trained to identify the most likely response to a query, but not necessarily the correct ones, appropriate for established guideline recommendations and may reinforce existing biases.

One challenge with LLMs is their tendency to confidently present "facts" without verifying their accuracy. This can lead to critical consequences if even minor inaccuracies are present in clinical information [30]. These AI models use statistical learning to predict the most likely response to a query, but this doesn't guarantee the correctness or alignment with established guidelines, potentially reinforcing existing biases.

ChatGPT and other large language models are double-edged swords [31] to provide some guidance, as they were not designed to offer direct medical advice with integrated safety rules. Moreover, users must formulate questions or prompts, including detailed information about the clinical scenario and any potential contraindications, to prevent receiving inaccurate or incomplete responses.

Our approach does not use a generative pre-trained transformer large language model but uses the technology. We create a heart-based language to feed the new model and serve as an AI Cardiologist for the patient.

2.3 Transfomer-Based Technology for Arrhythmia Detection

Most researchers on transformers detecting arrhythmia [38] use a training methodology with fine-tuning randomly initialized models, mainly based on ECG samples as input. We found that this research needs to be more balanced due to the low data volume or the use of only one benchmark for fine-tuning (training) and testing. Most researchers do not specify train/test data split and use lightweight transformers [38], N-cross validation instead of patient-wise evaluation of multiple classes from 12-lead ECGs without unsupervised training followed by fine-tuning [3], use of 10-cross validation mixed data split instead patient-wise on a single lead ECG [14], detection of multiple classes with 9/1 mixed train/test data split from 12 lead ECGs [6].

Although a similar method with unsupervised pre-training and labeled fine-tuning methodology is used on inputs created by ECG samples from 12-lead ECGs with 10-fold cross-validation to classify multiple classes [39], our approach uses self-supervised pre-training and single-lead ECG with patient-wise data split on a binary classification task.

3 Implementation Concepts

3.1 Architectural Approach

Our earlier paper presented details on the underlying architecture with a wearable ECG sensor and a smart device to communicate with the sensor, user, and cloud [12]. This paper introduces a dew computing solution (Fig. 1) such that the ECG sensor transmits ECG samples to the smart device to process them. The processing uses signal processing algorithms to detect heartbeats and their classes and feeds the AI-based module for arrhythmia detection. The innovation potential in this architectural approach is in providing a sophisticated technology that can be implemented in a smart device and correspond with a human.

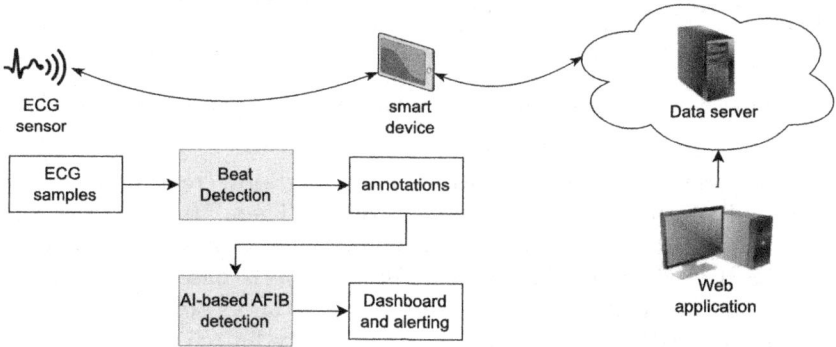

Fig. 1. The dew computing approach to develop an AI Cardiologist solution

The AI-based module is based on an approach to building a heart-based language and using transformer-based LLM technology to detect and classify arrhythmia. The AI Cardiologist's solution [34] leverages Google's BERT model for natural language processing, transforming ECG features into tokens to create a heart-specific language. Transformers examine sequences of time-series ECG data, enabling the detection and classification of arrhythmias.

The AI Cardilogist's solution [34] uses Google's BERT as a deep learning model for natural language processing, developing tokens from ECG features to build a heart-based language. This approach uses transformers to analyze sequences of time-series ECG data to detect and classify dangerous arrhythmia. A BERT model was trained using a self-supervised method with vast tokenized ECG recordings without the need for manual annotations or labeled data, followed by fine-tuning to classify AFIB.

The approach (Fig. 2) consists of two development phases, the first to develop a foundation model and the second to fine-tune a specific model for AFIB detection. Details of the used approach with the bidirectional RoBERTa model [22] are presented in [34] to detect AFIB in the AI Cardiologist solution. The achieved foundation model creates patterns for complex data and connections from high-volume unlabelled data, so training a smaller volume of labeled data in a supervised manner can successfully detect AFIB. The foundation model is trained on a dataset comprising 11 ECG benchmark databases from the PhysioBank archive [11], including AFDB, AHADB, CUDB, EDB, INCARTDB, LTAFDB, LTDB, LTSTDB, MITDB, NSRDB, and SVDB. MITDB trains the fine-tuned model, and all other databases, excluding MITDB, are used for testing.

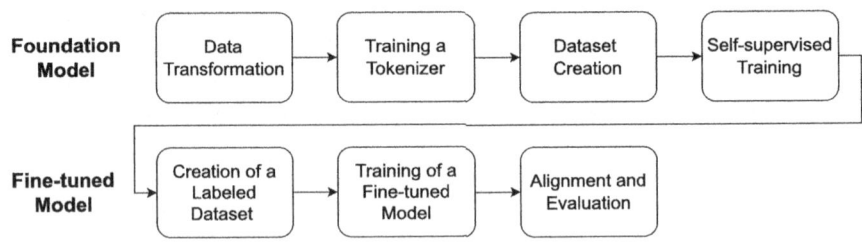

Fig. 2. Approach to develop a foundation and fine-tuned model of heart-based language

The development of the foundation model starts by creating a heart-based alphabet and transforming it into a text corpus. There are two approaches to creating the alphabet: analyzing the ECG samples or heartbeat annotations. The approach used in the AI Cardiologist solution is based on providing differences from two consecutive inter-heartbeat distances to reflect the rhythm change, as the AFIB is detected as an irregular rhythm. To realize this feature engineering task, we created a mapping that converts a range of values into a specific character. Once the alphabet is made, a tokenizer is trained on a text corpus of 128 characters to input the foundation model and create the extensive training dataset. After all these preparations, the foundation model is trained by a self-supervised approach without labels.

Developing the fine-tuned model starts by creating the labeled training dataset (MITDB) and testing datasets (all other datasets), followed by training. The created model outputs a label for each input, and the final development step aligns these labels to form an uninterrupted series of labels satisfying several signal-processing rules to annotate the start and end of a rhythm episode.

4 Discussion

This section discusses the benefits and limitations of the new proposed approach.

4.1 Functional Analysis

At present, individuals diagnosed with heart conditions, those at risk of high blood pressure, and all elderly adults are unable to monitor their heart health unless they are admitted to a hospital. The AI Cardiologist solution provides an independent measuring device, so the patient can measure their heart function as simply as measuring temperature. In addition, the system alerts when dangerous arrhythmia is detected. At the same time, the doctor gets the alert, can set a proper diagnosis, and provides valuable recommendations to prevent severe heart damage.

Such a modern AI-based solution to detect and classify dangerous arrhythmia will make our software a more convenient consumer-based medical device with improved accuracy as a diagnostic tool, like a wrist-based blood pressure monitor. This creates new business models for end-user self-diagnosis, which directly addresses patients and real-time monitoring of thousands of outpatients by hospitals.

Customers can subscribe to the monitoring and alerting service (AI Cardiologist) and receive timely and accurate predictions of their heart condition. This gives them peace of mind that their heart is being alerted and preventing severe damage. The user will receive emotionally intelligent messages that minimize stress and offer relaxation tips and gentle advice. Additionally, the doctor can access the monitored data for review and provide personalized medical advice.

The AI Cardiologist solution is a high-potential innovation idea that no one else has (or is available only for hospitalized patients), making it a breakthrough innovation. It addresses a broad spectrum of healthcare, economic, and social problems, offering a vast potential for radically disrupting healthcare for the better. The proposed innovation changes how healthcare is delivered with technology that puts the power back into the hands of patients. Our goal is to improve the cloud-based software solution for real-time monitoring and alerting in the case of an abnormal heart function, bringing it to industrial readiness and maturity for market replication and scalability. The traditional way of healthcare is to use smart devices to measure heart performance and to go to a specialized institution to get a diagnosis from your doctor. Our innovation lies in providing ECG monitoring of the patient's heart function and alerting in case of a detected abnormal heart function that is affordable for everyone, not just hospitalized patients. With the innovation, users will have the luxury of free normal life activities while their hearts are monitored and looked after.

One of the increasing problems is the lack of doctors, and technology-based self-monitoring solutions are evolving. The AI Cardiologist's philosophy is to address patients and give them what they need.

4.2 Business Potential

The global cardiac monitoring market size stood at $8B$ in 2021 and is expected to reach $16B$ by 2030, at a CAGR of 8.2% during 2021–30 [29]. This is ascribed to the rising prevalence of cardiovascular diseases (CVD), which are the leading cause of death (>33%), with a mortality of 17.9 million every year worldwide. Since 80% of premature heart disease and stroke are preventable, there is a vast exploitation potential in a growing market. The number of diagnosed with CVD is estimated to be more than 73.5 M (>1% of the population), and $>1M$ needs long-term (weekly or longer) ECG monitoring.

The continuous cardiac telemetry and ECG monitoring market will increase given that most developed nations have aging societies with an increasing number of those diagnosed with CVD, and mainly attract all those who need long-term ECG monitoring, allowing them to live independently in their homes and feel secure. The added value in our solution is to have real-time access to results, which targets a large segment of buyers and charges premium prices.

Competition is high with the available heart monitor products, including event (on-demand) monitors (smart watches or handheld devices), which can capture smaller ECG measurements (up to 30 s). These uncomfortable Holter devices measure 24 h or 48 h ECG or wearable patches. The trend is steady, typically promoting device quality and performance. Sales are driven by the introduction of new products, and business scale-up is constrained by the regulations for storing medical data in the country of residence, forcing the producers to develop customized solutions and host them in different countries. This project gives us a competitive advantage in creating a remote long-term monitoring solution alerting thousands of patients about dangerous arrhythmia, scaling up the business faster and easier.

Most existing solutions are intended for hospitalized patients, where sensors connect via a WiFi network, or analysis and evaluation are realized after the measurement is finished (postponed). The main difference in the AI Cardilogist's approach to competitive products is that it is the only real-time solution that works for outpatients in their homes and offices.

The business model is based on selling heart monitoring services to patients (B2C) with a subscription fee. The AI Cardilogist's solution matches the vast market opportunity with the competitive advantage for real-time heart monitoring not just locally to the patient but also to the caregiver and doctor for an affordable price. The B2B commercialization strategy addresses medical staff in hospitals, elderly care homes, and insurance companies. It aims to provide better healthcare for dangerous arrhythmias, prevent severe heart damage, and prolong life expectancy.

4.3 Data Processing Limitations

Technology Considerations. As the AI–ECG implementation is still in its infancy [32], a continuously growing clinical investigation agenda will determine the added value of deploying AI tools after thorough validation and verification, promising to transform clinical care. Matching the increasing computational processing power with access to large volumes of data from home monitoring devices, AI increases the day-to-day practice of subspecialised cardiology, including interventional cardiology, electrophysiology, and cardiac imaging [36].

Another disadvantage is the use of statistical learning in developing AI models. Training such models to identify the most likely response to a query, but not necessarily the correct ones, is appropriate for established guideline recommendations and may reinforce existing biases.

One drawback of AI model development with statistical learning is identifying the most probable response to a query rather than the correct one. While this approach aligns with recommended guidelines, it can also perpetuate existing biases.

Training/Testing Data Split. Since the 80/10/10 (train/validate/test) data split or any cross-validation approaches are sensitive to already known signal patterns, we apply patient-wise (P/W) methods, where the training and testing datasets use different records. The development of the train/validate/test datasets is carefully designed and includes rules for the equal proportion of rhythm episode occurrence to avoid potentially unexpected impacts and lousy performance. The P/W method guarantees the solution's robustness.

Data Quality. Performance is highly dependent on data quality. The arrhythmia occurs in a small proportion of the rhythm episodes in ECG datasets, introducing a small positive class ratio versus negatives, which results in overfitting in the training process. Additionally, underfitting causes problems in the training process when developing a precise model. Developing a sophisticated AI model is critical to creating the optimal dataset without over and underfitting.

Biased data used in training is another critical issue since the model will not correspond to the actual medical condition and will not be robust enough to cope with different health conditions and ECG waveforms.

The AI model aims to predict ECG segments where the signal is uninterpretable (corrupted with high noise levels that prevent successful detection and classification performance). This ensures the new model's robustness and security, providing means to filter acceptable noise levels and reject uninterpretable signals. Resilience against faults, defects, or misuse is already referenced in complying with the standards for medical devices.

Model Development. A medical expert needs to analyze the results to avoid biased model development, especially in the case when the model has not been demonstrated in a laboratory or operating environment; we start a procedure to develop a new model by:

- enlarging the training dataset with newly identified case samples,
- developing an improved dataset split,
- selecting different features, and

– selecting different tokenizers.

The procedure ends when the model is successfully performed in the lab and operational environment.

Experimental Results. The results from the experimental approach described in [34] are extracted in Table 1 comparing other authors for approaches using transformers. Sensitivity (SEN), positive predictive value (PPV), and their harmonic value F1-score are suggested as a relevant performance metric due to the imbalanced dataset, where a specific arrhythmia type is present only in, at most, a small portion (<10%) of the data samples. The duration-based method evaluates the detected time frames (durations) versus the durations of the existing rhythm episodes to eliminate the fluctuations (oscillations) close to the limit prediction case. The duration-based method [13] is preferred over the classical approach to evaluating the predicted outcome based on a provided data sample.

Although there are no details about the patient-wise split by other authors in Table 1 [34], we can conclude that the initial results of our approach provide the relevant proof-of-concept guarantee that our system performs consistently throughout its lifecycle.

Table 1. Comparison of AFIB detection performance using transformers [34]

Approach	SEN	F1
Che et al. [6] 7 cls	-	78.60%
Zhou et al. [39]	-	78.00%
Dong et al. [8]	-	82.90%
Our work		
split MITDB/LTAFDB	98.90%	93.33%
split MITDB/AFDB	97.84%	91.97%
split MITDB/TFDB	98.76%	87.37%

5 Conclusion

The AI Cardiologist project outcome improves the existing ViewECG solution with a modern AI-based solution to detect and classify atrial fibrillation as a dangerous arrhythmia that causes severe damage. The possible business domain applications of such a result are the creation of software medical devices for 1) a consumer-based device with improved accuracy as a diagnostic tool and 2) a B2B solution for hospitals, as a very efficient tool for the real-time monitoring of thousands of incoming ECG streams by sophisticated, scalable serverless technology.

The AI Cardiologist project presents an innovative technology for detecting and classifying atrial fibrillation episodes from ECG data. Input data sequences are processed by self-attention to generate contextualized heartbeats and ECG sample representations.

We used Google's BERT model to 1) develop tokens from ECG features, 2) build a heart-based language, and 3) analyze sequences of time-series ECG data with transformers. Training a BERT model with large quantities of tokenized ECG recordings was self-supervised, followed by fine-tuning.

We confirmed that foundation deep learning models for ECG records have the potential to significantly improve the accuracy and efficiency of heart health monitoring and detection of atrial fibrillation, enabling better healthcare and well-being. No similar solution is available on the market outside of hospital environments, giving the project results a high potential for commercialization. This is a non-expensive and non-invasive surveillance method for AFIB detection by patients through commercializing AI in wearable technology.

Acknowledgment. The AI Cardiologist project has received funding from the European Union's Horizon 2020 research and innovation program under grant agreement No 951847.

References

1. Abbas, G.H.: AI-based predictive modeling: applications in cardiology. IJS Glob. Health **7**(2), e0419 (2024)
2. American Heart Association: How to prevent heart disease (2024). https://www.uchicagomedicine.org/forefront/heart-and-vascular-articles/2023/february/heart-month-3-steps-to-a-healthier-heart-right-now. Accessed 08 June 2024
3. Atiea, M.A., Adel, M.: Transformer-based neural network for electrocardiogram classification. Int. J. Adv. Comput. Sci. Appl. **13**(11) (2022)
4. Bhattaru, A., Yanamala, N., Sengupta, P.P.: Revolutionizing cardiology with words: unveiling the impact of large language models in medical science writing. Can. J. Cardiol. (2024)
5. Boonstra, M.J., Weissenbacher, D., Moore, J.H., Gonzalez-Hernandez, G., Asselbergs, F.W.: Artificial intelligence: revolutionizing cardiology with large language models. Eur. Heart J. **45**(5), 332–345 (2024)
6. Che, C., Zhang, P., Zhu, M., Qu, Y., Jin, B.: Constrained transformer network for ECG signal processing and arrhythmia classification. BMC Med. Inform. Decis. Mak. **21**(1), 1–13 (2021)
7. D'Costa, A., Zatale, A.: AI and the cardiologist: when mind, heart and machine unite. Open Heart **8**(2), e001874 (2021)
8. Dong, Y., Zhang, M., Qiu, L., Wang, L., Yu, Y.: An arrhythmia classification model based on vision transformer with deformable attention. Micromachines **14**(6), 1155 (2023)
9. Elul, Y., Rosenberg, A.A., Schuster, A., Bronstein, A.M., Yaniv, Y.: Meeting the unmet needs of clinicians from AI systems showcased for cardiology with deep-learning-based ECG analysis. Proc. Natl. Acad. Sci. **118**(24), e2020620118 (2021)

10. Gala, D., Makaryus, A.N.: The utility of language models in cardiology: a narrative review of the benefits and concerns of ChatGPT-4. Int. J. Environ. Res. Public Health **20**(15), 6438 (2023)
11. Goldberger, A.L., et al.: Physiobank, Physiotoolkit, and Physionet: components of a new research resource for complex physiologic signals. Circulation **101**(23), e215–e220 (2000)
12. Gusev, M.: AI cardiologist at the edge: a use case of a dew computing heart monitoring solution. In: Artificial Intelligence and Machine Learning for EDGE Computing, pp. 469–477. Elsevier (2022)
13. Gusev, M., Boshkovska, M.: Performance evaluation of atrial fibrillation detection. In: 2019 42nd International Convention on Information and Communication Technology, Electronics and Microelectronics (MIPRO), pp. 342–347. IEEE (2019)
14. Hu, R., Chen, J., Zhou, L.: A transformer-based deep neural network for arrhythmia detection using continuous ECG signals. Comput. Biol. Med. **144**, 105325 (2022)
15. Innovation Dooel: AI Cardiologist: Real-time heart monitoring and alerting dangerous arrhythmia by an AI-based detection and classification module-based on RoBERTa LLM (2024). https://aicardiologist.innovation.com.mk/. Accessed 08 June 2024
16. Innovation Dooel: ViewECG: a platform for ECG monitoring and reporting tools (2024). https://viewecg.com/. Accessed 08 June 2024
17. Johnson, K.W., et al.: Artificial intelligence in cardiology. J. Am. Coll. Cardiol. **71**(23), 2668–2679 (2018)
18. Karatzia, L., Aung, N., Aksentijevic, D.: Artificial intelligence in cardiology: hope for the future and power for the present. Front. Cardiovasc. Med. **9**, 945726 (2022)
19. Klang, E., Cohen-Shelly, M., Lopez-Jimenez, F.: Leveraging large language models to enhance digital health in cardiology: a preview of a cutting-edge language generation model. Mayo Clin. Proc. Digit. Health **1**(2), 105–108 (2023)
20. Koulaouzidis, G., Jadczyk, T., Iakovidis, D.K., Koulaouzidis, A., Bisnaire, M., Charisopoulou, D.: Artificial intelligence in cardiology–a narrative review of current status. J. Clin. Med. **11**(13), 3910 (2022)
21. Ledziński, Ł, Grześk, G.: Artificial intelligence technologies in cardiology. J. Cardiovasc. Dev. Dis. **10**(5), 202 (2023)
22. Liu, Y., et al.: Roberta: a robustly optimized BERT pretraining approach. arXiv preprint arXiv:1907.11692 (2019)
23. Lopez-Jimenez, F., et al.: Artificial intelligence in cardiology: present and future. Mayo Clin. Proc. **95**(5), 1015–1039 (2020)
24. Nakamura, T., Sasano, T.: Artificial intelligence and cardiology: current status and perspective. J. Cardiol. **79**(3), 326–333 (2022)
25. Nolin-Lapalme, A., et al.: Maximizing large language model utility in cardiovascular care: a practical guide. Can. J. Cardiol. (2024)
26. Novak, A., et al.: The pulse of artificial intelligence in cardiology: a comprehensive evaluation of state-of-the-art large language models for potential use in clinical cardiology. medRxiv, pp. 2023–08 (2023)
27. OpenAI: ChatGPT (2024). https://chatgpt.com/. Accessed 08 June 2024
28. Pieszko, K., et al.: Clinical applications of artificial intelligence in cardiology on the verge of the decade. Cardiol. J. **28**(3), 460–472 (2021)
29. Prescient and Strategic Intelligence: Cardiac monitoring market report (2024). https://www.psmarketresearch.com/market-analysis/cardiac-monitoring-market. Accessed 08 June 2024

30. Sarraju, A., Ouyang, D., Itchhaporia, D.: The opportunities and challenges of large language models in cardiology. JACC: Adv. **2**(7), 100438 (2023)
31. Shen, Y., et al.: ChatGPT and other large language models are double-edged swords. Radiology **307**(2), e230163 (2023)
32. Siontis, K.C., Noseworthy, P.A., Attia, Z.I., Friedman, P.A.: Artificial intelligence-enhanced electrocardiography in cardiovascular disease management. Nat. Rev. Cardiol. **18**(7), 465–478 (2021)
33. Skalidis, I., et al.: ChatGPT takes on the European exam in core cardiology: an artificial intelligence success story? Eur. Heart J.-Digit. Health **4**(3), 279–281 (2023)
34. Tudjarski, S., Gusev, M., Kanoulas, E.: Heart language model: training and fine-tuning transformer-based foundation models with ECG recordings (2024). https://doi.org/10.21203/rs.3.rs-4575811/v1
35. Wasserlauf, J., et al.: Accuracy of the Apple watch for detection of AF: a multi-center experience. J. Cardiovasc. Electrophysiol. **34**(5), 1103–1107 (2023)
36. Watson, X., D'Souza, J., Cooper, D., Markham, R.: Artificial intelligence in cardiology: fundamentals and applications. Intern. Med. J. **52**(6), 912–920 (2022)
37. World Health Organization: Cardiovascular diseases: Avoiding heart attacks and strokes (2024). https://www.who.int/news-room/questions-and-answers/item/cardiovascular-diseases-avoiding-heart-attacks-and-strokes. Accessed 08 June 2024
38. Zhao, Z.: Transforming ECG diagnosis: an in-depth review of transformer-based deep learning models in cardiovascular disease detection. arXiv preprint arXiv:2306.01249 (2023)
39. Zhou, Y., Diao, X., Huo, Y., Liu, Y., Fan, X., Zhao, W.: Masked transformer for electrocardiogram classification. arXiv preprint arXiv:2309.07136 (2023)

Ambient Assisted Living Sensor-Based Solution for Elderly Self-monitoring

Ivan Kuzmanov(✉) and Nevena Ackovska

Faculty of Computer Science and Engineering, Ss. Cyril and Methodius University,
Ruger Boskovik 16, Skopje 1000, North Macedonia
ivan.kuzmanov@students.finki.ukim.mk, nevena.ackovska@finki.ukim.mk

Abstract. This paper considers the field of ambient assisted living with a focus on elderly care. We develop a proof of concept for a system meant to improve the quality of life for senior citizens. The system consists of both hardware components, a microcontroller with accompanying sensors, in addition to a blood pressure monitoring device and a bluetooth module and software components, an Android application. The hardware collects data, while the software processes it and displays relevant information to the user. The instances of daily life we are focused on are: thermoregulation and temperature anomaly detection, pill differentiation and medication adherence, visible light detection and blood pressure categorization. The goal of the system is to prompt the user to pay attention to their health via easily understandable messages. To test the feasibility and usefulness of the data gathered several experiments have been preformed. The results collected from these experiments help us better understand both the capabilities and limitations of the sensors, as well as the app we have developed. This system can improve the well-being of elderly citizens. However, more research is required before it can be seamlessly integrated into the daily lives of the elderly.

Keywords: ambient assisted living (AAL) · elderly care · sensors · user interface · proof of concept · self-monitoring

1 Introduction

The elderly are considered an at-risk demographic. There are many factors contributing to a decrease of their physical and/or cognitive abilities, which in turn makes them susceptible to chronic diseases. In recent years the elderly population has risen. For example, America's elderly population has increased from 13.1% in 2010 to 16.9% in 2020 and it's projected to climb further [1]. This tendency can be particularly observed in developed countries, as advancements in medical science, better infrastructure and access to healthcare has allowed people to live longer lives. However, this trend puts a strain on the healthcare providers with a fraction of the population utilizing a disproportional percentage of healthcare resources [2].

Ambient assisted living (AAL) is an area of study focused on enabling the independence of people with any kind of impairment within their preferred environment. As such it intersects various fields, such as smart technologies, healthcare, and social sciences. This field of study provides a wide range of solutions that improve users' autonomy, mobility and overall access. Some of these solutions are custom created to meet particular user's needs and meant to directly help them with daily tasks and if necessary provide assistance. Other solutions include a comprehensive alert systems that can notify caregivers or family members in case of emergency. There also exist AAL solutions which primary purpose is to allow healthcare providers to monitor the patient's state remotely, thereby enhancing diagnosis and treatment accuracy [3]. It's important to note that while AAL technologies focusing on healthcare offer promise, various ethical, legal, and security issues may arise. Healthcare data is private, and laws exist to ensure its protection. Legal and regulatory problems can occur when transferring that data, since the local laws about processing healthcare data may differ from one jurisdiction to another [4]. Potential misuse of AAL solutions may cause unequal access to healthcare. The remote capabilities of AAL are not to be used as a replacement for human contact. Any such system should strive to enhance the users' quality of life by providing comfort, safety, and independence, without infringing on their dignity.

In this study we propose a proof of concept for an AAL system meant to be used by the elderly population for self-monitoring. While designing a comprehensive system it's generally considered good practise to consider the perspective of each stakeholder, however since this research is meant to serve as a proof of concept we mainly focus on the developer's aspects. The proposal is structured as follows: The first section introduces the field of study and the existing issue our research is trying to address. The second section considers the AAL domain by exploring existing solutions, both theoretical and applied. The third section explains our vision and provides the methodology we are using to develop our system's architecture. The fourth section explains the nature of the experiments performed and discusses their significance. The fifth section provides a brief overlook of the system's capabilities and limitation, as well as makes suggestions about possible future improvements. The sixth section summarizes the study by emphasizing the main points.

2 Related Work

Internet of things (IoT) refers to equipping non-computing objects with sensor, network connectivity, and computing capability, thus enabling them to interact with their environment. A network formed of things that's in constant communication with other networks via the internet makes up an internet of things. The network's devices can produce, exchange, and consume data [5]. Several successful IoT-based AAL solutions have been developed. They are actively being used to improve the quality of life of its users. An applied IoT AAL solution usually incorporates keep in touch (KIT) technologies, smart objects capable of sensing the environment, and Closed Loop Healthcare Services that process the gathered data and provide communication with the relevant stakeholders [6].

Smart home is a term that refers to a residence in which various devices are connected and allow the user to remotely monitor and/or control the environment. As smart home is mainly focused on devices that are used for automation of tasks within a residence, it may implement IoT for establishing communication networks. To that end a study proposes a "holistic framework which incorporates different components from IoT in order to integrate smart home objects in a cloud-centric IoT based solution" [7]. It defines five levels: smart home, cloud, utility, third party, and user interfaces, with which the key features of IoT solutions for smart homes are generalized. In this study however our focus is more on systems that assist the users in their daily lives, home care system. Another existing research has developed an IoT-based framework meant for home care systems. The study "identifies a hierarchical distributed computing approach that consists of three computing levels: dew computing, fog computing, and cloud computing" [8].

Telemonitoring, remote monitoring of the patient's health is an often-utilized tool for ambient assisted living. A systematic review of telemonitoring focusing on common chronic ailments suggests that this technology provides reliable data, influences the subjects behaviors and has the potential to improve their medical condition [9]. This study finds that effects of telemonitoring on clinical effectiveness outcomes are more consistent for pulmonary and cardiac cases than diabetes and hypertension. Telemonitoring may also be used in places where direct medical care is unavailable, such as remote and inaccessible locations. An experiment conducted on the impact of different alert algorithms on work-load of medical professionals [10]. Here it's important to highlights the importance of the relevancy of data, not all data gathered and flagged is considered meaningful and the utilization of different algorithms may lead to different results. Ultimately the clinical decision depends on the clinicians opinions. The study concludes that advanced alerting algorithms with higher specificity allow for the most effective use of the clinicians time.

Generally one of the most important parameters of assessing the health of the elderly is their mobility [11]. One such, simple, yet effective AAL solution is the sensory alarm system Fearless Comfort System [12]. It consists of multiple sensors that monitor the mobility and automatically, without the need for user intervention, contact the appointed caretakers. It's self-prompting capabilities represent an improvement over more traditional methods. For example, an integrated panic button needs to be pressed to notify caretakers of a problem, however that's not possible if the problem causes the user to lose consciousness.

A single case study conducted on an older person with Alzheimer's disease has had success providing valid and clinically relevant information [13]. By monitoring the subject's daily activities over a long period of time evolving trends were detected and compared with clinical gathered information by a nurse. Most of these were consistent with each other, however some of the monitoring reports highlighted patterns not yet identified by the nurse.

The main objective of a AAL system is to make the user's life easier. Regardless of the designers intentions, a system whose perceived benefits are less than its cost is unlikely to be used in any significant capacity. A study conducted on AAL technologies has outlined 11 acceptance factors: "perceived usefulness, perceived ease-of-use, control & security, financial ability & willingness, privacy versus independence/safety, user involvement, human replacement, awareness, reputation/alignment to current lifestyle, government/politics/legal aspects, and experience" [14].

3 Methodology

3.1 Objectives

The goal of this study is to develop a system for home self-monitoring of at-risk subjects, with the target demographic being elderly people living alone. To that end we propose a system that is capable of sensing both the environment and the users themselves. The system analyzes the data gathered from those sources and extrapolates easily understandable readings that are displayed to the user. This allows the user to be able to make an informed decisions how to modify their behaviour or environment accordingly. Our proposal focuses on the integration of various sensors that monitor both the environment and the patient into a single cohesive system that can function semi-autonomously. Ideally our system needs to be able to answer to various condition that may arise with an appropriate measure, however our focus is on proving the feasibility of such a system by providing a proof of concept. As such we don't go through exhaustive lists of scenarios. Instead we concentrate on a few common use cases.

3.2 System Architecture

The elements of the proposed system can be categorized as hardware and software components. An arduino-based microcontroller serves as a the backbone for the hardware. It's directly connected with the sensors and a bluetooth module that transfers the data gathered by the sensors to the software application. Another hardware component is a cuff-based blood pressure monitoring device which is not connected to the microcontroller. The reading gathered from it are directly inputted into the application. The software components consist of an android application. The app contains a user-interface and methods for processing the type of data gathered in real time. The software and the hardware components communicate via a SSH bluetooth adapter. However, a downside of this approach is that the device running the app needs to be in a close proximity to the arduino bluetooth module to get live readings. We consider this a trivial problem because the system we are designing is meant to be used in the comfort of one's home. Table 1 display the components we are using to develop our system.

Table 1. System Architecture Components

Component	Type	Functions
Microcontroller	Hardware	Central unit controls the flow of information
Bluetooth Module	Hardware	Enables the establishing of a communication's protocol between the arduino and android devices
Humidity and Temperature Sensor	Hardware	Generates output in the form of the current value of temperature and humidity
Light Sensor	Hardware	Generates output in the form of the current value of light intensity
Color Sensor	Hardware	Generates output in the form of the current RGB values in front of the sensor
Blood Pressure and Heart Rate Monitor	Hardware	Calculates values for systolic and diastolic blood pressure, and a value for pulse
Application	Software	Encapsulates several software layers in a coherent whole
Presentation Layer	Software	Defines the user interface
Data Access Layer	Software	Retrieves data from the data storage system
Service Layer	Software	Rules that govern the applications functions

3.3 Hardware Components

The hardware components we use in our study are a microcontroller, a bluetooth module, three sensors and a monitoring devices. Each sensor is focused on sensing a single type of signal, thus making them completely independent of each other. Figure 1 shows an overview diagram of the hardware components and how they are interconnected to one another, with the blood pressure monitoring device being a separate entity not being connected to the rest of the system.

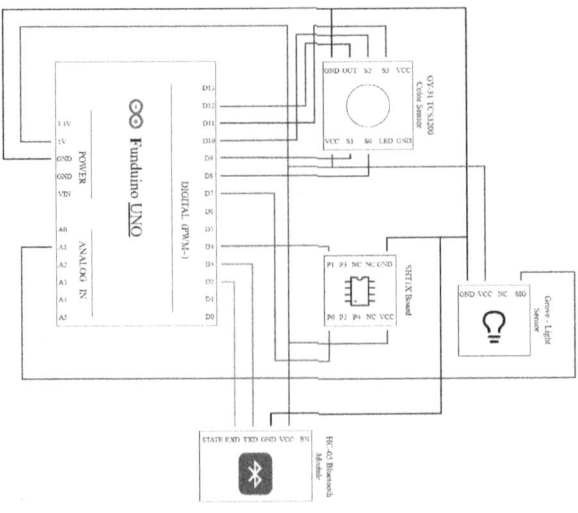

Fig. 1. Overview schema of the hardware components

3.3.1 Microcontroller

The microcontroller used to build our system is a Funduino Uno board. This board is cost-effective alternative to an Arduino Uno. It represents a clone of "Arduino Uno R3 board" that is compatible with Arduino IDE [15]. Our Funduino Uno board is connected to the sensors and the bluetooth module. It's function is to gather the sensor readings and transfer them via the bluetooth. It's important to note that some light preprocessing on the data is done before the information is sent, especially in the case of the color sensor. The microcontroller selects a single color from the list of available colors by analyzing the RGB values the sensor outputs, which is then sent to the app, instead of the RGB values themselves.

3.3.2 Bluetooth Module

In this study the HC-05 bluetooth module is used to enable the microcontroller with bluetooth connectivity [16]. This particular type of module has both a master and a slave mode, meaning that it can both initialize and accept connections from other devices. For our purposes the module is set to a slave mode so that it can approve the communication protocol when the Android device tries to establish contact. In our project once the communication protocol is established the communication is one way. The microcontroller module sends the gathered data, while the Android device receives it.

3.3.3 Temperature and Humidity Sensor

Ambient temperature measurement is an important factor in elderly care since the body's ability to self regulate it's own body temperature wanes with age. Both hypothermia and hyperthermia represent a very real danger to the elderly population. The module we are using is a SHT1x board [17], which contains both a temperature and humidity sensor. The system seeks to notify the user of the sudden onset of a trend of data gathered that indicate that the temperature is drastically rising or falling.

3.3.4 Light Sensor

The light sensor module consists of a sensor that measures the intensity of light. However, it doesn't give precise readings of lumen. The sensor used in this study Grove - light sensor [18] relies on a photo-resistor to measure the intensity of light, it doesn't actually measure the exact lumen of light. The system sends a notification when the light value falls below a preset threshold.

3.3.5 Color Sensor

The sensor used is photoelectric. It functions by emitting light that interacts with an object, which color is being measured. This interaction reflects the light being emitted back to the sensor. The output the sensor provides is the received light intensity of red, green, and blue, which are used to determine an object's color. The module used in this study is the GY-31 TCS3200 Color Sensor Module. In our setup, the color sensor serves a dual purpose. Firstly, it facilitates medication adherence, and secondly, it aids in differentiating pills. Both of these use cases are taken into consideration.

3.3.6 Blood Pressure and Heart Rate Monitor

Blood pressure is an extremely important metric which is often measured along side the vital signs. This measurement is particularly significant if you take into consideration the contemporary lifestyle, that may include an unbalanced diet, excessive amounts of sugar and a sedentary routine with little to no activity, compounded by the fact that the blood vessels stiffen with age [19]. The module includes a cuff monitor for blood pressure. The information obtained may be utilized to detect hypertension, chronic heightened blood pressure, as well as monitor the effectiveness of medications administered to the patient. This module requires active participation of the user, thus there may come times where the user is unable or is unwilling to employ it. There are several standards that define the broad categories of blood pressure monitoring. The most common is set by American College of Cardiology (ACC) and American Heart Association (AHA) [20], which we are using and it's given in Table 2.

Table 2. Blood Pressure Categories

Category	SBP		DBP
Normal	<120	and	<80
Elevated	120–129	and	<80
Grade 1 hypertension	130–139	and	80–89
Grade 2 hypertension	>140	or	>90
Hypertensive Crisis	> 180	and/or	> 120

3.4 Software Components

A software module refers to a complex software program meant to provide a specific functionality. The software module of our system comprises of an Android application and the layers it's made of.

3.4.1 Application

The application developed for our study is written in Kotlin using the Jetpack Compose toolkit. The app is designed using the Model-View-Controller (MVC) architecture. The advantage of this architecture lies in the separation of demands. The app is structured in such way that there is a divide of the roles of the software parts. The model is a representation of data, the view serves as the user

Fig. 2. Data Flow Diagram

interface and the application logic is handled by the controller. In our project the main role of the application is monitoring the parameters and notifying the user when a threshold is exceeded. Figure 2 depicts the way data flows thorough our application.

3.4.2 Presentation Layer

The presentation layer is the user interface. It's the component of the application the user directly interacts with. Since it is the part the user sees and interacts with, it needs to be easy to use and intuitive. UX design needs to be carefully considered, since regardless of the system's capabilities a user won't choose to interact daily with a program they are unable to clearly understand. In this paper our main focus is on developing a prototype proving that the system is feasible instead of the development of a final product. However, we acknowledge the importance of the user interface as a crucial factor for the success of an application. We have worked on several interfaces for specific user groups [21,22]. Figure 3 displays a few screenshots of our application.

3.4.3 Service Layer

The service layer represents the intermediary between the view and the data access layers. This layer handles the user's requests. The defined controllers actually control the flow of both data and logic. They are allowed to manipulate with the data and display it at the user request. In our app this includes the navigation implementation both between views and within the same view.

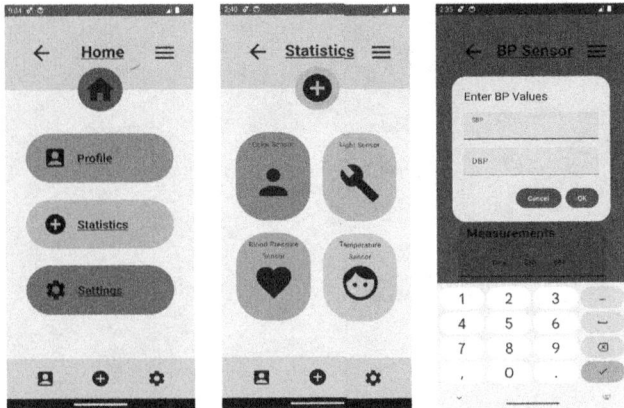

Fig. 3. Screenshots of the user interface

3.4.4 Data Access Layer

In the data access layer, the model encapsulates the interaction with outside sources, for example databases. They abstract the process of communication with data sources and represent an external point of interaction. Other layers may request data from the data access layer. In our case this refers to the data structure definitions and the ways we populate them, as well as methods that make the data available to the other layers. Figure 4 shows plots generated with data drawn from the data access layer. In our app we define view models to load the data. The plot we generate from the sensor data are: line plots for the temperature, humidity, blood pressure and light intensity, and bar plots for the color and light sensor data.

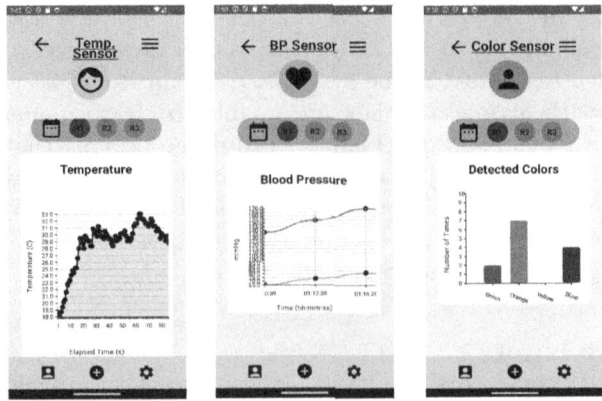

Fig. 4. Screenshots of the data plots

4 Experiments

Several experiments were preformed to test the sensor's fidelity and the system's robustness. An experimental approach allows for iterative improvement of the design and serve to mitigate risk. Each of the performed experiments has lasted exactly 3 h. The first two experiments were made with a 5 s delay between the readings, and the next three are made with a 3 s delay between the recordings for the feature gather automatically. A full table of the experiments are represented in Table 3 where the number of readings made during each trial is displayed. The emphasis in the earlier trials is on fine-tuning the sensor readings. In the later experiments the trials are performed to check whether the chosen use case can be detected, and finally to test the full system interoperability.

Table 3. Performed Experiments Parameters

Name	Count (BP)	Count (Color)	Count (T/H and Light)	Duration
Experiment 1	6	6	1664	3 h
Experiment 2	7	4	1664	3 h
Experiment 3	7	3	2405	3 h
Experiment 4	7	5	2405	3 h
Experiment 5	4	4	2405	3 h

The experiments data for the temperature sensor suggest that both the temperature and humidity are relatively slow changing in an indoor scenario unless prompted by an external event. For example, the event fully opening the window causes the room temperature to drop around 1 C° during experiment 1. Another clear example of this is in experiment 4, in which we are actively trying to cause event such as lighting a candle in the proximity of the sensor, placing a cold object near it, or heating it with body temperature by holding it. Even so the temperature returns to the baseline relatively quickly after the event. It should be noted that temperature is highly correlated with humidity. As such a change in one is likely to cause a change in other. Comparison between temperature and humidity measurements of different experiments can be seen at Fig. 5.

Experimentally we have also concluded that the light sensor we are working with has show in some cases to have a low intermediary sensitivity. To that end a simple filter may be applied to get more salient results. It's worth mentioning that the trials with a lit candle had more effect on the light sensor instead of the temperature one. Figure 6 presents violin plots of the light intensity cross multiple experiments.

The color sensor needs to be manually initialized when the user attempts to differentiate color. Here the system returns one of five color: blue, green, orange, yellow, and white (baseline). In some of the earlier experiments the values for green and blue, and orange and yellow were mixed up, and there was a case where

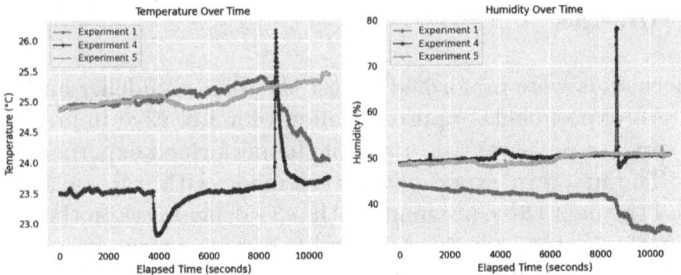

Fig. 5. Plots of the temperature and humidity for experiments 1, 4, 5

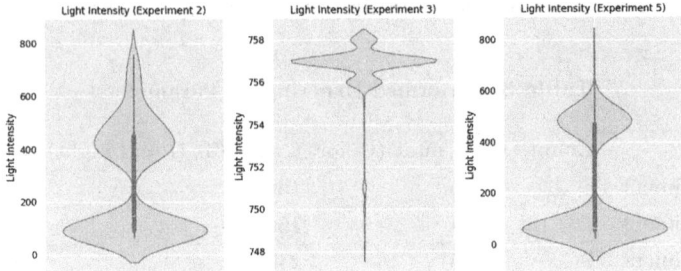

Fig. 6. Plots of light intensity for experiments 2, 3, 5

an strong external light makes the colors indistinguishable. After fine tuning the color sensor with each iteration of the previous ones, we have reached our near optimal parameters that allowed the sensors to function without a single mistake in the latter experiments.

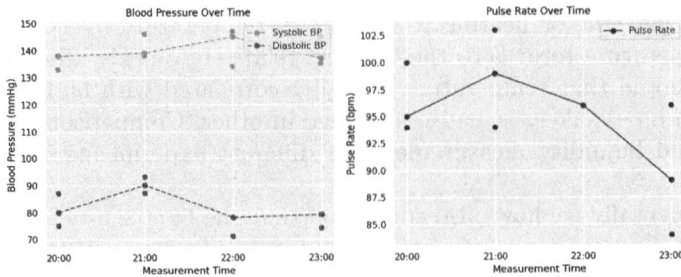

Fig. 7. Plots of the blood pressure values for experiment 5 (Color figure online)

Blood pressure is measured in our case with a cuff-based at-home monitoring device. Our experiments have brought into question the fidelity of the it. There is no a doubt of the necessity to include blood pressure. However, if the device fails to reliably provide information with accuracy and precision, it doesn't merit

Table 4. Statistical Analysis of the Sensor Data

Sensor	Metric	Mean	SD	Range
Temperature	C°	25.04	0.99	(22.81, 42)
Humidity	%	43	2.82	(15.54, 78.06)
Light	lux	490.69	124.35	(24, 758)
Systolic BP	mmHg	138.39	10.39	(118, 163)
Diastolic BP	mmHg	80.32	7.61	(67, 97)
Pulse	bpm	91.70	13.06	(47, 111)

it's use. To that end during experiment 5 we measured blood pressure three consecutive times per measurement. As can be seen at Fig. 7 while some discrepancy exists between consecutive measurement it's within 5 mmHg, which is usually considered a guideline for the accuracy of blood measurement devices.

Table 4 shows the statistical features for the data gathered across all experiments. There exist an uneven number of samples per experiment due to different sampling rates. This is taken into account when generating Table 4 by first calculating the chosen metrics for each trial separately, and then averaging them out. After the experiments were finished, a selection of several traceable events were chosen for our app to monitor and notify the user over. Those are:

1. Temperature is not in the range (18, 24) C°;
2. A sudden drastic change of temperature
3. Drastic change of the light intensity;
4. An object is placed in front of the color sensor;
5. A color sensor not being initialized at the allocated time;
6. Entering blood pressure values
7. A blood pressure not being inputted at the allocated time.

The system is able to detect all of these events and alert the user of the changes. Hence the system is meant for the senior citizens to monitor their health parameters and the environment parameters with the help of our integrated sensors.

The goal of this study is the development of a prototype. Thorough the performed experiments we demonstrate the capabilities of our system. Overall, our priority is providing a proof of concept, as such for the purpose of this paper we don't dwell on the ethically dilemmas about the trust and reliability. However, we recognize that these are very important components especially in health applications. For example, a detection error may potentially endanger human life by misclassifying a pill.

5 Discussion

The blood pressure cuff itself can be considered somewhat intrusive. However, the system considers that many elderly people prefer not to wear sensor devices, such as smart watches and other patches. Still, since the information about blood pressure is very relevant to the healthcare of the users, we consider that it is important for the users to monitor it. And indeed the elderly tend to measure blood pressure almost regularly, and our system benefits form those measurements. However, we acknowledge that there is room for improvement. One possible way to lessen the strain of use is by applying optical character recognition. This would allow the user the option to take a picture of the values of the monitor, as an alternative to manual imputation.

Another avenue of improvement are estimation algorithms for blood pressure measurements from ECG and PPG. Cuff-based tool like the one used in this study are considered invasive and uncomfortable. Both machine learning [23] and deep learning [24] models have been developed capable of interpolating blood pressure from ECG and/or PPG with a certain degree of accuracy. The inclusions of such an algorithm, alongside the accompanying senors, may allow for a more seamless integration without the need for active user interaction, as well as continuous coverage.

The current system works in such a way that readings from the sensors are processed in real time and run through the rule based algorithm to provide immediate feedback. The conditions are preset and the system waits for a signal to trigger an event, resolves it and goes back to waiting. However, the real benefit of the generated data can only be obtained by evaluation over a longer periods of time. Each person has their own unique rhythm and characteristics. For example, circadian rhythm analysis and sleep pattern detection would be ideal for our purposes, however their use depends on complex algorithms and 24/7 coverage. Different values are considered the norm for different target groups. The information obtained by analysing the underlying patterns of data, may provide insight in the patient specific parameters and with enough monitoring a baseline can be made which can be used to annotate deviations from the norm. This personalization is the basis of future work and improvement of the system.

6 Conclusion

The role of AAL is to allow it's users independence and control within their preferred living space. Various solutions have been proposed utilizing the latest technologies to answer to the needs of their users to the best of their ability. The needs considered in each system vary based on its users, for example people with mobility issues would benefit the most out of a fall detection system, that can automatically notify caregivers if necessary, while people with cognitive disabilities would benefit more of cognitive assistance and behavioral monitoring. For the purpose of our study our interests are on developing tools for self-monitoring and self-regulation. Our target demographic are the elderly. The metrics our system calculate provide assurance that there is nothing wrong with the environment of the user. In addition the system actively monitors the user's health and well being.

The work done so far creates a prototype of an ambient assisted living sensor-based solution for elderly self-monitoring system. Our proposal offers a simple ambient assisted living system consisting of arduino-based hardware and android-based software. The available sensors, temperature and humidity, light, color and blood pressure, generate data which is then processed by the software part. The main functions of our app include monitoring parameters, triggering events, and communicating with the user. We experimentally tested the capabilities of the developed system. We also interpolated relevant information from the experiments. The system is technically feasible and it's operational costs is not high, since the sensors used are low costs. Expanding its versatility by increasing the number and type of sensors is the next step of the evolution of the system. Our proposed solution holds the potential to improve the quality of people's lives especially since it could learn from the patterns of each individual using it.

References

1. Statista: U.S. - seniors as a percentage of the population 1950–2050 (2023). https://www.statista.com/statistics/457822/share-of-old-age-population-in-the-total-us-population/. Accessed 28 Aug 2023
2. NZ Government: New Zealand Ministry of Health (2016) Our changing population (2016). https://www.tewhatuora.govt.nz/our-health-system/data-and-statistics/nz-health-statistics/health-statistics-and-data-sets/older-peoples-health-data-and-stats/our-changing-population. Accessed 29 Aug 2023
3. Tun, S., Madanian, S., Mirza, F.: Internet of things (IoT) applications for elderly care: a reflective review. Aging Clin. Exp. Res. **33**, 855–867 (2021)
4. Rose, K., Eldridge, S., Chapin, L.: The Internet of Things: an overview. Internet Soc. (ISOC) **80**, 1–50 (2015)
5. Arif, I., Ackovska, N.: IoT aided smart home architecture for anomaly detection. In: Data Science and Internet of Things: Research and Applications at the Intersection of DS and IoT, pp. 1–19 (2021)
6. Dohr, A., Modre-Opsrian, R., Drobics, M., Hayn, D., Schreier, G.: The Internet of Things for ambient assisted living. In: 2010 Seventh International Conference on Information Technology: New Generations, pp. 804–809 (2010)
7. Stojkoska, B., Trivodaliev, K.V.: A review of internet of things for smart home: challenges and solutions. J. Clean. Prod. **140**, 1454–1464 (2017)
8. Risteska Stojkoska, B., Trivodaliev, K., Davcev, D.: Internet of Things framework for home care systems. Wirel. Commun. Mob. Comput. **2017**(1), 8323646 (2017)
9. Paré, G., Jaana, M., Sicotte, C.: Systematic review of home telemonitoring for chronic diseases: the evidence base. J. Am. Med. Inform. Assoc. **14**(3), 269–277 (2007)
10. Cuba Gyllensten, I., Crundall-Goode, A., Aarts, R.M., Goode, K.M.: Simulated case management of home telemonitoring to assess the impact of different alert algorithms on work-load and clinical decisions. BMC Med. Inform. Decis. Mak. **17**(1) (2017)
11. Scanaill, N., Carew, C., Barralon, S., Noury, P., Lyons, N., Lyons, D.: A review of approaches to mobility telemonitoring of the elderly in their living environment. Ann. Biomed. Eng. **34**(4), 547–563 (2006)

12. Programme, F.-A.: Fearless Comfort System (2012). https://www.aal-europe.eu/projects/fearless/. Accessed 05 Feb 2024
13. Lussier, M., et al.: Using ambient assisted living to monitor older adults with Alzheimer disease: Single-case study to validate the monitoring report. JMIR Med. Inform. **8**(11), 20215 (2020)
14. Weegh, H., Kampel, M.: Acceptance criteria of ambient assistant living technologies. In: Assistive Technology, pp. 857–864 (2015)
15. Funduino Uno Documentation. https://www.funduino.de/Arduino-tutorials-08092014.pdf. Accessed 29 Apr 2024
16. Bulebots: HC05 Bluetooth Module Documentation. htttps://bulebots.readthedocs.io/en/latest/hc05_bluetooth.html. Accessed 30 Apr 2024
17. Arduino, P.: SHT1x Library for Arduino. https://github.com/practicalarduino/SHT1x. Accessed 05 Feb 2024
18. Seeed Studio: Grove Light Sensor - Seeed Studio. https://wiki.seeedstudio.com/Grove-Light_Sensor/. Accessed 05 Feb 2024
19. Kuzmanov, I., Ackovska, N., Lehocki, F., Bogdanova, A.M.: Implementation of the time series and the convolutional vision transformers for biological signal processing-blood pressure estimation from photoplethysmogram. In: International Conference on ICT Innovations, pp. 46–58. Springer, Cham (2023)
20. Guirguis-Blake, J.M., Evans, C.V., Webber, E.M., et al.: Screening for hypertension in adults: an updated systematic evidence review for the U.S. preventive services task force. In: Evidence Synthesis, vol. 197. Agency for Healthcare Research and Quality (US), Rockville, MD (2021)
21. Ilijoski, B., Ackovska, N., Zorcec, T., Popeska, Z.: Extending robot therapy for children with autism using mobile and web application. MDPI (2022)
22. Ilijoski, B., Ackovska, N.: Developing applications for children with special needs into a project based learning approach at human-computer interaction course. In: 2022 IEEE Global Engineering Education Conference (EDUCON), pp. 1322–1329. IEEE (2022)
23. Kuzmanov, I., Zdravevski, E., Lamenski, P., Stojkoska, B., Bogdanova, A.M.: A study on appropriate segment length for generalized cuff-less blood pressure estimation from ECG features. In: 2024 47th MIPRO ICT and Electronics Convention (MIPRO), pp. 1181–1186. IEEE (2024)
24. Kuzmanov, I., Bogdanova, A.M., Kostoska, M., Ackovska, N.: Fast cuffless blood pressure classification with ECG and PPG signals using CNN-LSTM models in emergency medicine. In: 2022 45th Jubilee International Convention on Information, Communication and Electronic Technology (MIPRO), pp. 362–367. IEEE (2022)

Classification of Autism and Typical Development Children Based on EEG Signals

Aleksandar Tenev[1](✉), Silvana Markovska-Simoska[2], and Igor Mishkovski[1]

[1] Faculty of Computer Science and Engineering, Sts Cyril and Methodius University,
Skopje 1000, Republic of North Macedonia
aleksandar.tenev@finki.ukim.mk
[2] Macedonian Academy of Sciences and Arts,
Skopje 1000, Republic of North Macedonia

Abstract. Autism spectrum disorder (ASD) is a neurodevelopmental condition that affects the brain's function. Electroencephalography (EEG) is a non-invasive technique that measures the electrical activity of the brain and can reveal its dynamics and information processing. We obtained the data during an eyes-opened resting state EEG recordings and it consisted of 52 children with autism (ASD) and 39 typically development (TD) children. In this study, we trained and compared 5 different classifiers for ASD and TD based on the EEG signals: the Dummy classifier, a Random Forest classifier, a K Nearest Neighbours classifier, a Support Vector Machine and an XGBoost classifier.

Keywords: ASD · Classification models · Machine Learning · XGBoost · SVM · KNN

1 Introduction

Autism Spectrum Disorder (ASD) is characterized by a diverse array of behavioral symptoms that develop in early childhood. Ongoing research on ASD is progressing quickly, revealing a broadening understanding of its causes and developmental paths. This has led to the recognition of ASD as having varied cognitive, behavioral, and neural patterns and subtypes [1].

Studies on high-risk infant siblings have shown that the key behavioral characteristics of ASD begin to appear in the latter part of the first year and into the second year of life [2]. Since ASD is defined by behavioral rather than biological criteria, making a formal diagnosis before the age of three is challenging [3,4]. As a result, children often do not receive a diagnosis until preschool or later [5]. Detecting milder forms of ASD early is particularly difficult because many neurodevelopmental symptoms overlap with those of other conditions [6]. Despite diagnostic uncertainties, reliable biomarkers that can identify early ASD symptoms could be valuable for creating effective early interventions [3].

Since atypical brain development that results in ASD symptoms likely occurs months or even years before atypical behaviors manifest, relying solely on behavioral features for diagnosis or screening could cause a crucial developmental window for early intervention to be missed. This concern has driven the search for early neural markers or biological indicators that can detect ASD in its prodromal stage.

EEG, being a non-invasive technique, allows for the recording of electrical activity produced by the brain's neuronal networks. In recent years, an increasing amount of research has focused on identifying distinctive EEG patterns associated with ASD [7-9], aiming to deepen our understanding of the neurological mechanisms underlying its varied clinical presentations. This exploration of EEG data in relation to ASD has revealed a range of abnormalities in neural processing. Researchers have examined connectivity patterns, changes in frequency bands, and event-related potentials (ERPs) to understand the intricate interactions of neural circuits involved in ASD pathology [10-15].

Machine learning can significantly aid in the diagnosis of ASD by analyzing complex and high-dimensional data, such as EEG signals, to identify patterns that are not easily discernible through traditional methods. Machine learning algorithms can be trained on large datasets to recognize specific neurophysiological markers and atypical brain activity associated with ASD. These models can then classify individuals based on these patterns with high accuracy. By leveraging advanced feature extraction and selection techniques, machine learning can enhance the sensitivity and specificity of ASD diagnosis, potentially enabling earlier detection and intervention, even before behavioral symptoms become apparent.

The application of machine learning algorithms in the classification of neurodevelopmental disorders using EEG data has seen a significant rise and expansion in the last decade. These algorithms excel at processing and interpreting the vast amounts of complex data generated by EEG recordings, identifying subtle patterns and biomarkers that may indicate the presence of conditions such as Autism Spectrum Disorder (ASD), ADHD, and epilepsy [16-24]. By employing advanced techniques like deep learning, support vector machines, and random forests, researchers can enhance the precision and reliability of diagnoses, often surpassing traditional diagnostic methods. This extensive use of machine learning not only improves the accuracy of detecting these disorders but also facilitates earlier intervention, personalized treatment plans, and a deeper understanding of the underlying neural mechanisms, paving the way for better clinical outcomes.

2 Methods

2.1 Data

We have obtained EEG data from 39 children with typical development and 52 children that are diagnosed with autism by a professional psychiatrists and clinical psychologists. The mean age of the ASD group was 6.18 years with standard deviation of 1.98 years, and the TD group had a mean age of 5.35 years and standard deviation of 2.31 years. The gender was excluded because of the lack of significant gender ratio between the groups. Because of the nature of the disorder, the EEG signals were obtained in the resting state condition with eyes opened, using the 10–20 EEG montage system with 19 electrode channels. The EEG recordings were done using the Mitsar WinEEG Software. The sampling frequency of the signal was 250 Hz. We obtained a time series of 3 min for the TD and variable length in the range of 7 to 30 min for the ASD group. It is important to note that the length of the signal in ASD group was different because of the challenging environment before and during the recording. However, we managed to obtain a valuable signals with the help of the parents.

2.2 Preprocessing

The presence of noise and artefacts is a standard occurrence in this type of signal. Noise in EEG refers to all the unnecessary modifications of the signal recorded by EEG, but not generated by the brain, such as heart beating signals, nearby electromagnetic field noise eye movements, such as blinks, and other face muscles etc. To deal with this type of modification, we initially applied a band-pass filter that helped in isolating specific frequency ranges, relevant for further analysis. Generally used band-pass filters on EEG signals are from 1 Hz to 40 Hz. So, after this filtering, we no longer have frequencies below 1 Hz and above 40 Hz.

On the other hand, for removing the artefacts that are in the frequency range we used the ICA algorithm. ICA is a computational method for separating a multivariate signal into its underlying components. Using ICA, we can extract the desired component from the amalgamation of multiple signals. We went through all of the files and manually extracted the components we believed were external noise. ICA effectively removed the EOG artefacts in our dataset. From here on, the data is filtered and with enhanced quality, making it ready for the further part of the prepossessing which involves the utilization of the windowing-based method.

The windowing-based method involves segmenting the continuous EEG signal into shorter, overlapping time windows. The use of overlapping windows helps to maintain continuity and ensures that important information is not lost at the boundaries of individual windows. We decided to use this method on our data, because the duration varied among the instances, with some lasting for 3 min, others for 9 min, and a few lasting as long as 17 min. By trying a few different lengths, we landed on 30 s being our window size. With this technique, we got N instances out of one recording, where N is calculated as the duration of the recording in seconds divided by the 30 s window size.

In order to monitor which of the N instances are in which set (train or test) we introduced an additional column to the data named 'fileId'. This way we ensured that instances that belong to the same recording (fileId) were strictly in the test or train sets. After implementing this, we encountered the problem of imbalanced data. The ASD class was significantly larger than the normal class, approximately comprising 80% ASD instances and 20% TD instances as shown in Fig. 1. To resolve this issue, we opted to use the SMOTE technique. SMOTE is an oversampling technique that generates synthetic samples from the minority class (in our case the normal class). It is used to obtain a synthetically class-balanced or nearly class-balanced training set, which is then used to train the classifier.

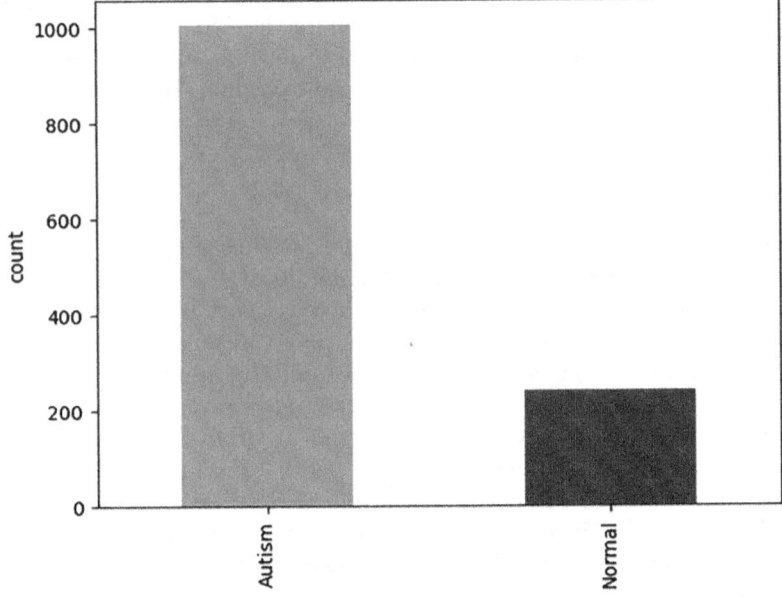

Fig. 1. Dataset before applying SMOTE

After implementing these techniques the data was perfectly balanced (Fig. 2 and the signals were cleared from the noise, so we moved to the feature extraction phase.

We decided on extracting four features for each of the 19 channels, so we ended up with a total of 76 features. The four features we extracted are min, max, mean and std. All of these features were later provided to the model for the training.

3 Results

We trained 5 different classifiers for this model, the Dummy classifier, a Random Forest classifier, a K Nearest Neighbours classifier, a Support Vector Machine and an XGBoost classifier. The models were validated using 10-fold cross-validation technique.

For our baseline results, we used a Dummy classifier with the 'most frequent' strategy which means that the method always returns the most frequent class

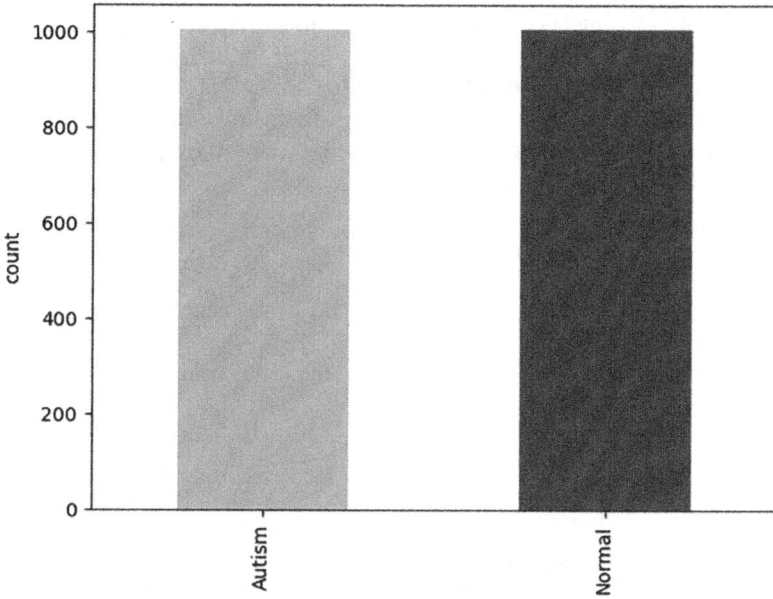

Fig. 2. Dataset after applying SMOTE

in the observed y argument from the test set passed to fit. It resulted in 43% accuracy which is to be expected.

A random forest classifier is a popular machine learning algorithm that belongs to the ensemble learning family of algorithms, which combines multiple individual models to make more accurate predictions. The random forest classifier consists of a collection of decision trees where each one is built using a subset of the training data and a random selection of features. To make a prediction, the random forest classifier combines the predictions of all the individual trees and outputs the class that receives the most votes. Using this classifier we got a result of 89% accuracy.

In the case of the KNN classifier, the algorithm classifies a new data point by considering the class labels of its k nearest neighbours. In this model, we assigned k to have a value of 3, and we managed to get an accuracy of 83%.

The main idea behind SVM is to find an optimal hyperplane that separates the different classes in the dataset and maximises the margin, which is the distance between the hyperplane and the closest data points of each class. The kernel that we used was the default, radial basis function kernel. This method gave a result of 87% accuracy.

An XGBoost classifier is a machine-learning algorithm that belongs to the family of gradient-boosting methods. XGBoost stands for Extreme Gradient Boosting and is an optimised implementation of the gradient boosting algorithm. The idea behind gradient boosting is to iteratively train new models that correct the mistakes made by the previous models, with a focus on minimising the overall prediction error. This method proved to be the most useful, as it gave an accuracy score of 92%.

Another metric we used to determine the success of our model was the F1 score of the model. The F1 score combines precision and recall into a single value, providing a balanced measure of a model's accuracy. The formula used to calculate the F1 score is the following:

$$F_1 = \frac{2 \cdot \text{Precision} \cdot \text{Recall}}{\text{Precision} + \text{Recall}} \tag{1}$$

The XGBoost algorithm showed to be the best according to this metric as well, giving an F1 score of 0.91.

The evaluation of the performance metrics for our models are depicted in Table 1.

Table 1. Performance Metrics for Predictive Models

Model	F1 Score	Accuracy	Precision	Recall	ROC AUC
Dummy	0.610478	0.439344	0.439344	1.00000	0.50000
Random Forest	0.880143	0.890164	0.845361	0.917910	0.893166
KNN	0.786957	0.839344	0.942708	0.675373	0.821605
SVM	0.840164	0.872131	0.931818	0.764925	0.860533
XGBoost	0.911111	0.921311	0.904412	0.917910	0.920944

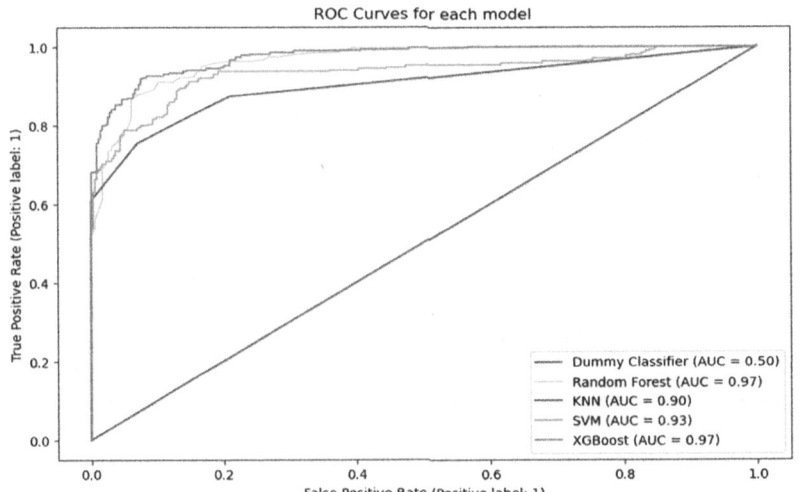

Fig. 3. ROC curves for each model

4 Conclusion

In this study, we build models for classification of ASD and TD children on the basis of raw EEG signals which could improve the time complexity of similar applications since most of the classification models are build using features from the spectral domain of the EEG signal. We emphasised several methods of preprocessing EEG signals, that included applying a band-pass filter, ICA algorithm, windowing-based method and finally the SMOTE technique to balance the dataset. By using these techniques collectively, we were able to extract useful features from the EEG signals and successfully developed a model that gave acceptable predictive performance in determining the target value.

The results showed that the best performing model is the XGBoost model, achieving the highest F1 Score (0.911) and highest ROC AUC (0.921), which is a significant improvement over the other models as seen in Fig. 3. Since these results were developed on a relatively small dataset, it remains an open question for future research whether these models will fit large datasets with many different types of ASD, which would result in different patterns on the EEG.

References

1. Amaral, D.G., et al.: In pursuit of neurophenotypes: the consequences of having autism and a big brain. Autism Res. **10**(5), 711–722 (2017)
2. Ozonoff, S., et al.: A prospective study of the emergence of early behavioral signs of autism. J. Am. Acad. Child Adolesc. Psychiatry **49**(3), 256–266 (2010)
3. Reiersen, A.M.: Early identification of autism spectrum disorder: is diagnosis by age 3 a reasonable goal? J. Am. Acad. Child Adolesc. Psychiatry **56**(4), 284–285 (2017)
4. Sheldrick, R.C., Garfinkel, D.: Is a positive developmental-behavioral screening score sufficient to justify referral? A review of evidence and theory. Acad. Pediatr. **17**(5), 464–470 (2017)
5. Steiner, A.M., Goldsmith, T.R., Snow, A.V., Chawarska, K.: Practitioner's guide to assessment of autism spectrum disorders in infants and toddlers. J. Autism Dev. Disord. **42**, 1183–1196 (2012)
6. Pettersson, E., Anckarsäter, H., Gillberg, C., Lichtenstein, P.: Different neurodevelopmental symptoms have a common genetic etiology. J. Child Psychol. Psychiatry **54**(12), 1356–1365 (2013)
7. Boutros, N.N., Lajiness-O'Neill, R., Zillgitt, A., Richard, A.E., Bowyer, S.M.: EEG changes associated with autistic spectrum disorders. Neuropsychiatric Electrophysiol. **1**(1), 1–20 (2015)
8. Milovanovic, M., Grujicic, R.: Electroencephalography in assessment of autism spectrum disorders: a review. Front. Psych. **12**, 686021 (2021)
9. Hashemian, M., Pourghassem, H.: Diagnosing autism spectrum disorders based on EEG analysis: a survey. Neurophysiology **46**(2), 183–195 (2014)
10. Chan, A.S., Leung, W.W.: Differentiating autistic children with quantitative encephalography: a 3-month longitudinal study. J. Child Neurol. **21**(5), 391–399 (2006)
11. Cornew, L., Roberts, T.P., Blaskey, L., Edgar, J.C.: Resting-state oscillatory activity in autism spectrum disorders. J. Autism Dev. Disord. **42**, 1884–1894 (2012)

12. Sutton, S.K., et al.: Resting cortical brain activity and social behavior in higher functioning children with autism. J. Child Psychol. Psychiatry **46**(2), 211–222 (2005)
13. Kutas, M., Hillyard, S.A.: Reading senseless sentences: brain potentials reflect semantic incongruity. Science **207**(4427), 203–205 (1980)
14. Kutas, M., Hillyard, S.A.: Brain potentials during reading reflect word expectancy and semantic association. Nature **307**(5947), 161–163 (1984)
15. Donchin, E., Coles, M.G.: Is the p300 component a manifestation of context updating? Behav. Brain Sci. **11**(3), 357–374 (1988)
16. Bosl, W.J., Loddenkemper, T., Nelson, C.A.: Nonlinear EEG biomarker profiles for autism and absence epilepsy. Neuropsychiatric Electrophysiol. **3**, 1–22 (2017)
17. Bosl, W.J., Tager-Flusberg, H., Nelson, C.A.: EEG analytics for early detection of autism spectrum disorder: a data-driven approach. Sci. Rep. **8**(1), 6828 (2018)
18. Tenev, A., Markovska-Simoska, S., Kocarev, L., Pop-Jordanov, J., Müller, A., Candrian, G.: Machine learning approach for classification of ADHD adults. Int. J. Psychophysiol. **93**(1), 162–166 (2014)
19. Hurt, E., Arnold, L.E., Lofthouse, N.: Quantitative EEG neurofeedback for the treatment of pediatric attention-deficit/hyperactivity disorder, autism spectrum disorders, learning disorders, and epilepsy. Child Adolesc. Psychiatric Clin. **23**(3), 465–486 (2014)
20. Baygin, M., et al.: Automated ASD detection using hybrid deep lightweight features extracted from EEG signals. Comput. Biol. Med. **134**, 104548 (2021)
21. Chen, H., Chen, W., Song, Y., Sun, L., Li, X.: EEG characteristics of children with attention-deficit/hyperactivity disorder. Neuroscience **406**, 444–456 (2019)
22. Rahman, M.M., Usman, O.L., Muniyandi, R.C., Sahran, S., Mohamed, S., Razak, R.A.: A review of machine learning methods of feature selection and classification for autism spectrum disorder. Brain Sci. **10**(12), 949 (2020)
23. Hyde, K.K., et al.: Applications of supervised machine learning in autism spectrum disorder research: a review. Rev. J. Autism Dev. Disord. **6**, 128–146 (2019)
24. Duda, M., Ma, R., Haber, N., Wall, D.: Use of machine learning for behavioral distinction of autism and ADHD. Transl. Psychiatry **6**(2), 732 (2016)

NATO Workshop

Detecting the Unseen: Exploiting Radar-Sonar Sensor Fusion for Visual Detection of Low-Profile Naval Drones

Giacomo Longo[1]($^{\boxtimes}$), Alessandro Cantelli-Forti[2], and Enrico Russo[1]

[1] Department of Informatics, Bioengineering, Robotics and Systems Engineering, University of Genova, Genoa, Italy
`giacomo.longo@dibris.unige.it, enrico.russo@unige.it`
[2] Radar and Surveillance Systems National Laboratory, National Inter-University Consortium for Telecommunications (CNIT), Parma, Italy
`alessandro.cantelli.forti@cnit.it`

Abstract. Recent advancements in naval technology have introduced agile and stealthy anti-ship drones. Characterized by rapid movements and compact above-water hulls, these drones often exhibit small radar cross sections (RCS), making them difficult to detect using traditional methods. To counter this emerging challenge, this research explores the integration of radar and sonar sensors, harnessing their complementary strengths to counter such threats via enhanced visualization capabilities. We illustrate how combining measurements from both radars and sonars, despite their different operational ranges and angular coverages, can significantly increase the surveillance effectiveness beyond their individual contributions. The proposed approach is expected to enhance the accuracy and reliability of visual detection systems, enabling better detection and tracking capabilities for these elusive naval drones.

Keywords: Drones · Radar detection · Sonar · Sensor fusion

1 Introduction

In the annals of naval history, the sinking of the Japanese heavy cruiser Haguro [1] in May 1945 by the British Royal Navy's 26th Destroyer Flotilla in Operation Dukedom exemplifies the efficacy of smaller, agile vessels to sink larger adversaries. Today, this age-old tactic has endured and evolved by introducing a novel class of naval drones that boast impressive speeds and maneuverability. They stay barely above the water to evade visual detection and generate minimal noise. Equipped with ultra-thin structures and radar-absorbing materials, they maintain a low Radar Cross-Section (RCS), helping them avoid radar detection.

By leveraging their superior capabilities, they can stealthily approach enemy vessels without being noticed until they are dangerously close. This tactic gives attacking forces a significant advantage but also introduces complex challenges for defensive strategies to counter similar raids from adversaries.

This latter aspect has forced navies to also focus on developing counter-drone capabilities. The effectiveness of these solutions lies in their capacity to detect and locate such malicious small crafts early on, enabling maneuvers to minimize the threat or to neutralize them with point defense weapons.

Our paper focuses on a novel counter-drone proposal that aims to enhance these timely detection capabilities. It involves deploying a constellation of smaller, autonomous, friendly naval drones that operate alongside the main ship, extending its radar and sensory range beyond its systems. This approach aims to increase the observability of the surrounding space by combining the mother ship's radar capabilities with localized sonar data from these auxiliary drones.

We summarize our contributions as follows.

- We propose a sensor drone constellation targeted at enhancing anti-drone surveillance around a mother ship.
- We describe how to shape such a constellation in accordance with the expected threat profile.
- We formalize how to integrate the measurements from the constellation with the mother ship's radar sensor in a generic and numerically stable manner.
- We detail a technique to allow efficient rendering of the additional information introduced by the constellation.
- We perform extensive experiments to assess the effectiveness and measure the performances.

Structure of the Paper. The rest of the paper is organized as follows. In Sect. 3, we describe our methodology. In Sect. 4, we evaluate our solution using a simulated environment. Finally, we conclude the paper in Sect. 5.

2 Related Work

Prior research has extensively explored the utilization of drones for enhanced surveillance capabilities, demonstrating their effectiveness in augmenting situational awareness. Studies have investigated leveraging aerial [2-7] and marine [8-10] drone platforms, individually or in conjunction [11,12], to extend surveillance coverage. Furthermore, systems incorporating sensor data from these drone networks are increasingly recognized as important components of modern surveillance infrastructure [13,14]. However, existing solutions often necessitate significant modifications to underlying surveillance systems and require substantial development efforts for sensor fusion algorithms [15-17]. This work proposes a novel, cost-effective approach that leverages existing visual/blob detection infrastructure for target identification. Our system minimizes the need for intrusive modifications or complex algorithm development. This characteristic enables seamless integration with previous drone-based surveillance proposals while simultaneously providing a readily deployable solution for immediate operational needs, bridging the gap until more sophisticated systems are implemented.

3 Methodology

The primary objective of this article is to describe a system aimed at enhancing the radar coverage of the mother ship by using a constellation of drones equipped with sonar sensors. This sonar network, being potentially closer to incoming drones and capable of detecting objects beneath the sea surface, can significantly augment surveillance capabilities.

3.1 Drone Network

Identifying the optimal arrangement of these drones is crucial for the success of this strategy. This optimal arrangement depends on drone specifications such as the sonar viable horizontal aperture (A_H) and range (A_R), as well as on contextual factors like prior intelligence indicating a potential attack from a specific direction [18–20]. The choice of A_H and A_R values is influenced by the desired characteristics of the constellation. For instance, a simplistic model might use the entire range and horizontal aperture, assuming the sensors have a uniform probability of detection across the entire coverage. Alternatively, more complex schemes, such as selecting those values based on a chosen probability of detection, are also viable options.

We propose two schemas based on the assumption that each of the N drones in the constellation possesses identical characteristics.

The first schema, illustrated in Fig. 1, aims at maximizing surveillance coverage against unknown threats. In this configuration, the drones are positioned along a circumference of radius d w.r.t. the mother ship, with their sonars oriented outward relative to the mother ship. Each drone is evenly spaced at an angular distance of $\frac{2\pi}{N}$.

To estimate the surveillance coverage of this formation, we calculate the ratio between the perimeter of the hull enclosing the entire formation with the perimeter covered by any sonar.

$$\alpha_i = O + \frac{i \cdot 2\pi}{N} \quad \alpha_{B_i}, \alpha_{E_i} = \alpha_i \mp \frac{A_H}{2}$$
$$P_{O_i} = \begin{bmatrix} d\sin(\alpha_i) & d\cos(\alpha_i) \end{bmatrix}$$
$$P_{B_i} = \begin{bmatrix} A_R \sin(\alpha_{B_i}) & A_R \cos(\alpha_{B_i}) \end{bmatrix} + P_{O_i} \quad (1)$$
$$P_{E_i} = \begin{bmatrix} A_R \sin(\alpha_{E_i}) & A_R \cos(\alpha_{E_i}) \end{bmatrix} + P_{O_i}$$

This estimation involves constructing a perimeter that includes the approximate endpoints P_B and P_E of each drone's sonar coverage, calculated as shown in Eq. 1, and the lines that connect each drone start of coverage point P_{B_i} to its predecessor end of coverage point $P_{E_{i-1}}$. Within Eq. 1, α_i denotes each drone angle, O is a fixed offset allowing to shift the drone grid w.r.t. the axis orientation, and P_{O_i} is the sonar origin for the i-th drone.

The second disposition, shown in Fig. 2, is applied when there is specific information about the most likely direction of an attack **a**. In this case, the drones are arranged in a line perpendicular to the anticipated direction.

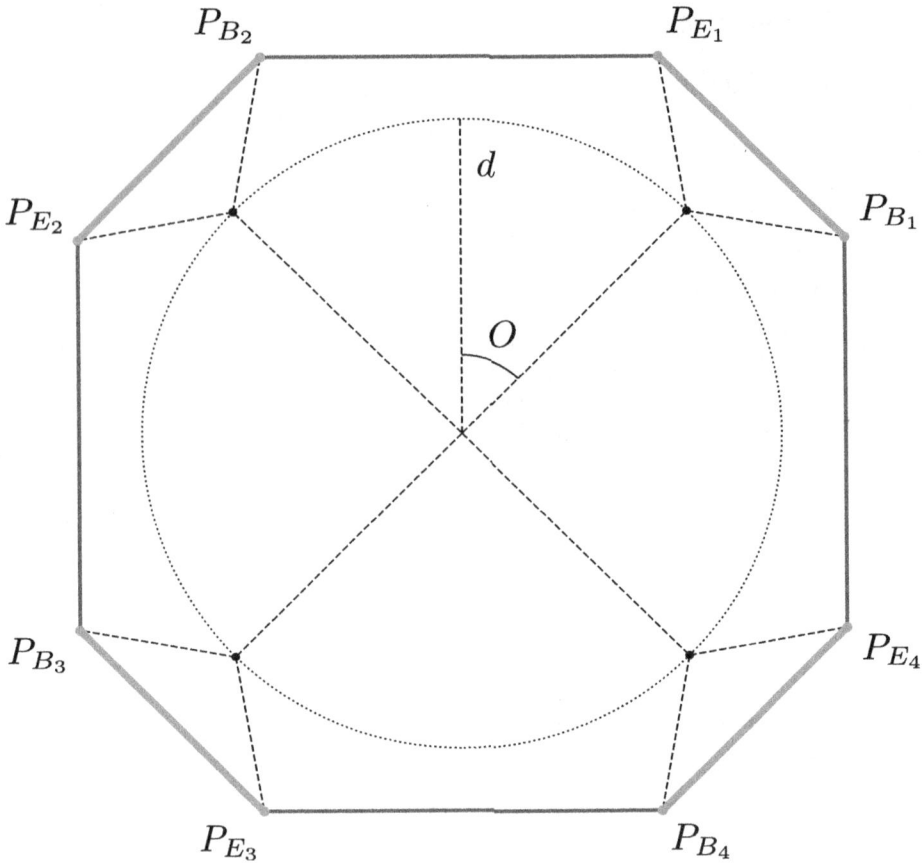

Fig. 1. Constellation for unknown threats ($N = 4$).

$$D_o = [d\cos O \; d\sin O \; 0] \qquad (2)$$

The first drone of the formation is placed at position D_o shown in Eq. 2, with O being the angle between the mother ship bow and **a**.

$$\langle D_{l_k}, D_{r_k} \rangle = D_o \pm 2k \left(\frac{D_o}{||D_o||} \times [0\;0\;1] \right) A_R \sin \frac{A_H}{2} \qquad \text{where } k \geq 1 \qquad (3)$$

The other drones are instead spaced so that their viable sonar coverage area just touches that of its neighbor. Equation 3 details the calculation for determining such drones anchor points. Those anchor points are calculated two drones at a time, with k indicating the distance from D_o, and D_l, D_r denoting the origins obtained by decomposing the $\pm$ operator in two separated equations.

$$\langle PB_i, PE_i \rangle = D_i \pm A_R \sin \frac{A_H}{2} + \frac{D_o}{||D_o||} \qquad (4)$$

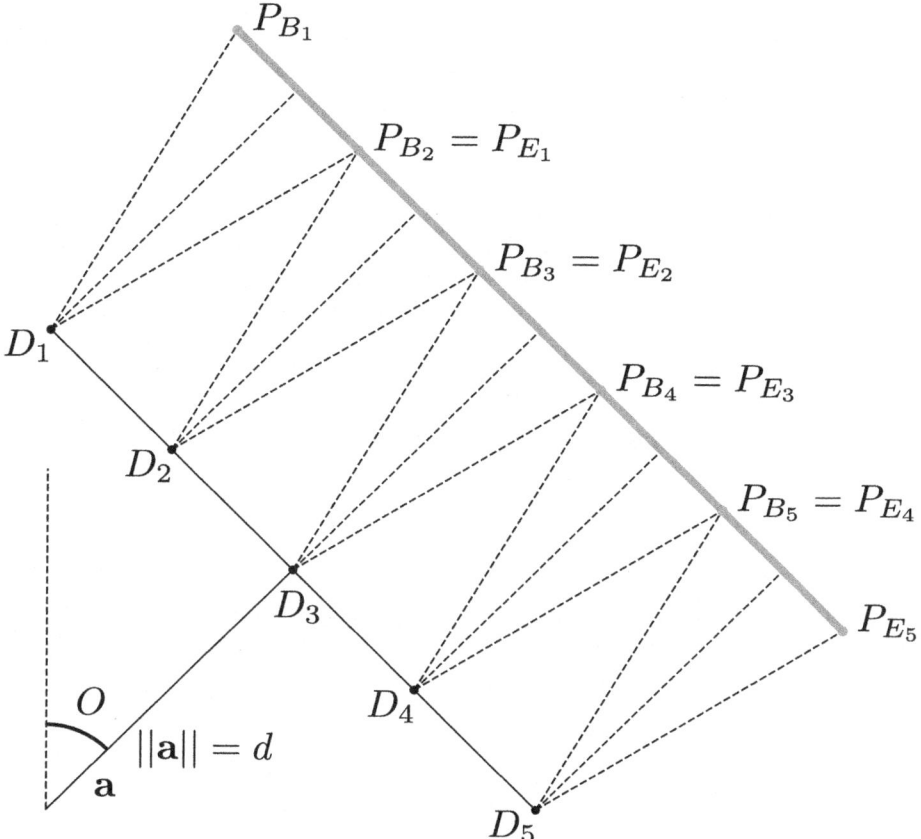

Fig. 2. Constellation for threats from direction **a** ($N = 5$).

The start and end points of each drone sonar coverage is calculated as in Eq. 4, with D_i indicating the origin of a generic drone i. The covered perimeter in this formation is quantified by the aggregation of norms of the vector $P_E - P_B$ for each drone. Unlike the first placement strategy, this method does not allow for an assessment of the extent of the area not under surveillance.

3.2 Image Fusion

To enhance the radar image displayed on the plan position indicator (PPI), the drone-borne sonar scans need to be integrated with the original image.

This process involves two key steps:

1. Converting each recorded echo coordinate into the mother ship's PPI space.
2. Rasterizing the transferred echoes efficiently alongside existing data.

The first step, coordinate conversion, requires projecting all echoes from the drone-mounted sonars back onto a common reference orientation used by the mother ship PPI. This reference uses a point known as the Consistent Common Reference Point (CCRP) as its origin and an orientation coplanar with the sea surface [21].

Echoes Coordinate Conversion. Each echo received by the drones' constellation sonars is reported in polar coordinates, expressed in terms of bearing Θ and range ρ relative to the center of the sonar antenna array on each drone. Given this, each origin needs to undergo two coordinate shifts: from the original sonar space into a local coordinate system specific to each drone, and then into the master ship's local coordinate system used by the PPI. To achieve such transformations, we leverage a quaternion-based approach for rotations, followed by a translation. Considering the various combinations of attacker positions, drone positions, and their relative orientations with respect to the mothership, quaternions are necessary to prevent gimbal lock that could occur with the use of Euler angles [22].

$$\mathbf{t} = \begin{bmatrix} X & Y & Z \end{bmatrix} = \begin{bmatrix} \rho & 0 & 0 \end{bmatrix} \tag{5}$$

The procedure starts by defining **t**, a direction vector pointing outwards from a right-handed coordinate system centered in the sonar origin. As seen in Eq. 5, its magnitude corresponds to the echo range ρ.

$$\hat{Q_D} = \begin{bmatrix} w \\ x \\ y \\ z \end{bmatrix} = \begin{bmatrix} \cos(\frac{\Theta}{2}) \\ 0 \\ 0 \\ \sin(\frac{\Theta}{2}) \end{bmatrix} \qquad Q_D = \frac{\hat{Q_D}}{||\hat{Q_D}||} \tag{6}$$

The first quaternion Q_D, covering the rotation from the sonar-local space to the drone-local space is constructed using the *Body 3-2-1* sequence (Eq. 6), with the echo bearing Θ as its rotation w.r.t. the yaw axis.

As all quaternions in this article denote pure rotations, we refer to them after normalization, i.e. the quaternion is always divided by its norm after each operation.

$$\hat{Q_S} = \begin{bmatrix} \cos(\frac{\phi}{2})\cos(\frac{\theta}{2})\cos(\frac{\psi}{2}) + \sin(\frac{\phi}{2})\sin(\frac{\theta}{2})\sin(\frac{\psi}{2}) \\ \sin(\frac{\phi}{2})\cos(\frac{\theta}{2})\cos(\frac{\psi}{2}) - \cos(\frac{\phi}{2})\sin(\frac{\theta}{2})\sin(\frac{\psi}{2}) \\ \cos(\frac{\phi}{2})\sin(\frac{\theta}{2})\cos(\frac{\psi}{2}) + \sin(\frac{\phi}{2})\cos(\frac{\theta}{2})\sin(\frac{\psi}{2}) \\ \cos(\frac{\phi}{2})\cos(\frac{\theta}{2})\sin(\frac{\psi}{2}) - \sin(\frac{\phi}{2})\sin(\frac{\theta}{2})\cos(\frac{\psi}{2}) \end{bmatrix} \tag{7}$$

$$Q_S = \frac{\hat{Q_S}}{||\hat{Q_S}||}$$

A second quaternion Q_S covers instead the rotation from the mother ship local space to the one of the drone. Similar to Q_D, this quaternion is assembled out of the same sequence (Eq. 7) with the drone reported roll ϕ, reported pitch θ, reported yaw relative to mother ship $\psi = \psi_D - \psi_S$ as its inputs[1]. Here, ψ_x denotes the angle between vessel x bow and true north, with D indicating the drone, and S the mother ship.

$$Q = Q_S \cdot Q_D \tag{8}$$

The two rotations are then consolidated into a single quaternion Q by multiplication of Q_D with Q_S (Eq. 8).

$$\mathbf{x} = \mathbf{R}\mathbf{t} + \begin{bmatrix} \Delta X & \Delta Y & \Delta Z \end{bmatrix} \tag{9}$$

Finally, the echo position in the mother ship PPI is calculated as in Eq. 9. Where $\mathbf{R}$ is the rotation matrix corresponding to the quaternion rotation Q, and ΔX, ΔY, ΔZ is the distance between the mother ship CCRP position and the sonar origin.

Efficient Rasterization of Echoes. Given the high volume of elements needing redraw in a PPI, often tens of thousands of cells per second, it is crucial to efficiently integrate the sonar echoes from the sonar constellation without significantly increasing the computation required for redraws. To address this problem, we have devised a strategy aimed at minimizing the number of additional draw calls.

Our procedure is structured as follows: each echo received from the drone network is broken down into a series of approximated cells. Given the fixed angular and range resolutions, each possible echo from a drone can be captured in a rectangular matrix of known size.

For example, echoes from a drone equipped with a sonar having 100 azimuthal cells and 300 longitudinal cells can be stored within a 100×300 matrix. By combining this matrix with information about the currently set maximum range, the boundaries of each cell with respect to azimuth and distance, delineating the four sides of each cell, are precisely known.

Based on this data, a spatial indexing structure, such as an R*-tree [24], is constructed to map which indices in the matrix correspond to specific azimuths and ranges in sonar-local space.

More formally, the tree maps axis-aligned bounding boxes (AABBs) in sonar-local space to three indices: the drone identifier and a pointer to the cell within the matrix.

$$\mathbf{x}_S = \mathbf{R}^T \mathbf{x} - \begin{bmatrix} \Delta X & \Delta Y & \Delta Z \end{bmatrix} \tag{10}$$

When it is time to draw a particular cell on the PPI, the rasterizer first transforms the cell centroid to each sonar-local space using the inverse transformation shown in Eq. 10. There, $\mathbf{x}_S$ is the position converted to sonar-local space and $\mathbf{R}^T$ is the inverse[2] of $\mathbf{R}$. It then queries the sonar-local cell index to check if that

---

[1] Here the same definition for Euler angles as [23] is assumed.
[2] By construction, the inverse of a rotation matrix is its transpose.

cell is also covered by a sonar from the drone network. If the query is positive, the outputs from both radar and sonar are combined.

This approach is advantageous because it requires the spatial indexes associated with each drone to be recalculated only with each change in maximum range – a relatively infrequent operation. These spatial indexes thus help to amortize the rasterization costs associated with determining if a cell being drawn is shared with any sonar in the constellation, significantly reducing computational load w.r.t. more naive approaches, i.e., constantly redrawing every updated cell.

3.3 Processing of Fused Information

Following the procedure given in the previous section, the system generates an augmented image incorporating the echoes recorded by the drone constellation.

$$I = \max\left(0.0, r + \sum_{i=1}^{N} s_i\right) \quad (11)$$

Each cell in this new PPI image has an intensity I obtained by merging the original radar echo intensity r with the sum of all sonar returns attributed to that cell s_i. This combination, shown in Eq. 11, is performed in decibels relative to full scale ($dBFS$) units. As a result, this new intensity value loses any physical significance w.r.t. its originating sources.

Track Generation. The system is designed to aid in visual detection, aiming to change the appearance of what is shown on the screens. As such, automatic detection and tracking of targets can be performed as usual, utilizing techniques such as blob detection applied to images, CFAR techniques on cell values [25], and similar methods. Still, track generation and maintenance, along with their associated particle filters, such as Extended Kalman Filters [26], must account for the differences when tracking a target belonging solely to the radar, the sonar constellation, or a combination of both. These differences can be addressed by adjusting the detection covariance based on the composition of each echo.

$$\sigma = \underbrace{\frac{r}{r + \sum_{i=1}^{N} s_i} \sigma_r}_{\text{Contribution by radar}} + \underbrace{\sum_{i=1}^{N}\left(\frac{s_i}{r + \sum_{j=1}^{N} s_j} \sigma_{s_i}\right)}_{\text{Contribution by the constellation}} \quad (12)$$

Equation 12 illustrates this calculation, where the detection covariance for a particular cell σ is obtained by summing the individual sensor covariances (σ_r and σ_{s_i}), weighted by their respective intensities (r and s_i). Thanks to this provision, the filtering algorithms can account for the varying levels of accuracy associated with each sensor.

4 Evaluation

We evaluate the proposed method in a modified version of MaCySTe [27] containing proprietary additions tackling physics-based simulation of sonar and primary radar systems. Our PPI is implemented using the Godot game engine [28]. We equip the mother ship with a C-band radar system with a refresh rate of $1Hz$, with an angular resolution of $0.08°$ and range resolution of 5 m. Each connected drone is instead equipped with a sonar system covering a range of 50 m, horizontal aperture of $130°$, and an update rate of 5 Hz. Those characteristics have been inspired by the specifications of actual radar and sonar systems used by navies around the world. We model the attacking drone as a 6m long, 2 m wide, 0.7 m height above water, vessel capable of reaching speeds in excess of 60 knots. Its convex hull (46 triangles) is depicted in Fig. 3.

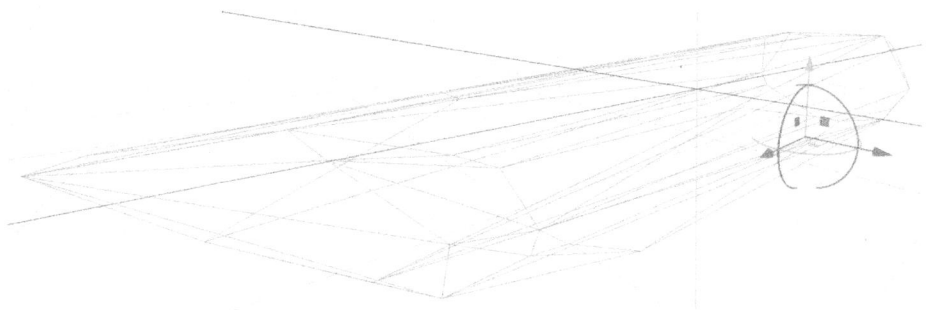

Fig. 3. Convex hull of the attacking drone.

4.1 Increase in Detection Performance

To evaluate the improved detection performance relative to the baseline scenario, we conducted 100 Monte-Carlo trials where an attacking drone was randomly positioned 6 nautical miles away from the mother ship. The attacking drone was programmed to continuously change its heading as it approached the mother ship. In defense, the mother ship was protected by a formation of four drones, arranged as described in Sect. 3.1, without prior knowledge of the attacking drone's direction. Tracking-wise, we employ a multi-stage image processing pipeline designed to mimic real-world implementations. Starting from intensities calculated as per Sect. 3.3, initial clutter rejection is achieved through a low-pass filter. Subsequently, CFAR techniques are utilized to identify illuminated cells against the background noise floor. Blob detection algorithms then unify adjacent potential target signatures spanning across multiple cells. Finally, an alpha-beta particle filter maintains and updates individual tracks over time.

For each generated PPI frame, we tracked how often the drone echo was recorded solely by the drone constellation, indicating the number of times the constellation detected an otherwise undetectable target. Additionally, in frames where the drone was detected by both the mother ship's radar and the drones' sonars, e.g., as in Fig. 4, we measured the increase in echo intensity relative to full scale, as per Eq. 11. During our experimental campaign, the surveillance tracking performance achieved an overall rate of 98% in terms of the time the target remained visible to any sensor. In 7% of these instances, the sonar constellation was the sole sensor with target visibility. Furthermore, the sonar sensor network detected the target sooner than the radar in 5% of the trials. Additionally, when radar and sonar echoes were present simultaneously, the sonar contributed to around a 3 dB increase in the reported echo strength. This signal-to-noise ratio increase allowed a target to be distinguished from sea clutter by in around 2% of such cases.

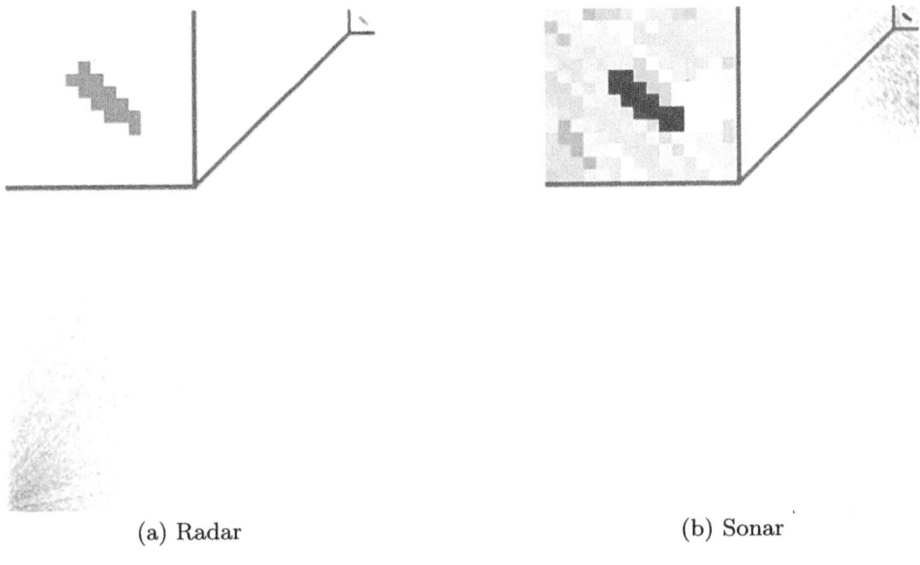

(a) Radar (b) Sonar

Fig. 4. Simultaneous detection by both radar and sonar (in original PPI space).

Overall, these results confirm that this sensor fusion technique can improve maritime surveillance capabilities.

4.2 Rendering Performance

We compared the performance of the procedure described in Sect. 3.2 with a naive approach in which every new echo received from the drone network triggers a redraw of the PPI. To ensure deterministic execution, we conducted this trial in "Movie Maker" mode, which maintains all time deltas used by the rendering engine at a fixed value and by using a pre-recorded set of echoes. We recorded the performance metrics from the game engine's built-in monitoring facilities once

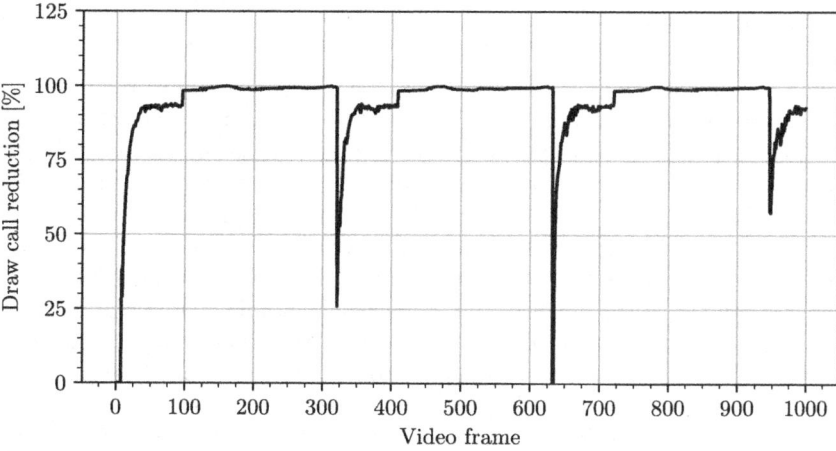

Fig. 5. Effect on draw calls of the optimizations.

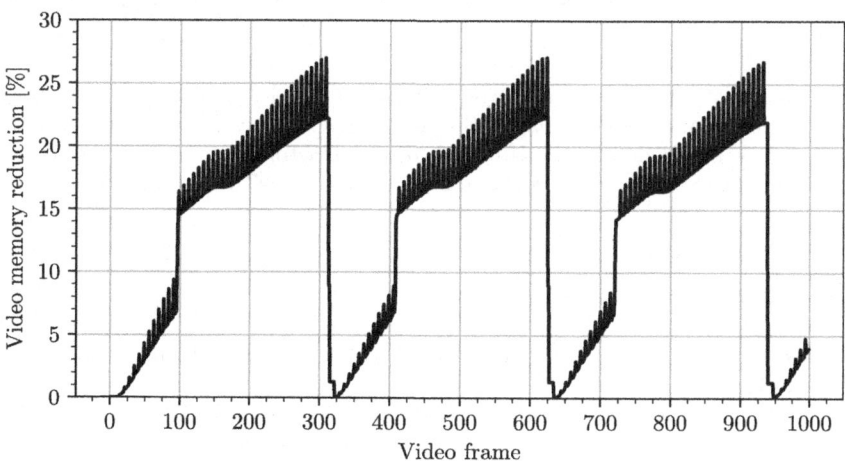

Fig. 6. Effect on video memory of the optimizations.

per video frame. Figure 5 illustrates a comparison between the draw calls made by our method and those made by the naive implementation, demonstrating that our method reduces the number of required draw calls by half. Figure 6 displays the amount of memory allocated to the renderer. Here, our method achieves memory savings of approximately 15% compared to the naive approach. Those figures suggest that the proposed approach succeeds in reducing the performance overhead of joining together the echoes from the drone constellation with the ones from the mother ship.

5 Conclusions

Throughout this article, we have explored the characteristics of next-generation naval drones and the challenges they pose to traditional surveillance methods. To address these challenges, we proposed the use of a sensor drone constellation to enhance the visibility zone around the protected mother ship. We also explained how to configure this constellation based on the expected threat profile and how to calculate the coverage area associated with it. We also detailed a method for integrating the data from the constellation into the plan position indicator of the mother ship in a generic and numerically stable manner and discussed how to optimize this integration for rendering. Our experiments confirmed that this approach can aid in detecting such drones and that our rendering method succeeds in reducing the computational overhead compared to more straightforward solutions. The results are promising, yet they remain preliminary. Future research will need to address the practical issues related to controlling, orchestrating, and communicating with these drone swarms. Similarly, the applicability of this kind of solution to semi-submersible crafts should be subject to further research.

Acknowledgements. This work was partially funded by the NextGenerationEU project "Security and Rights in CyberSpace" (SERICS). It was carried out while Giacomo Longo was enrolled in the Italian National Doctorate on Artificial Intelligence run by the Sapienza University of Rome in collaboration with the University of Genoa.

References

1. Morse, J.: Naval War Coll. Rev. **34**(6), 113–114 (1981). http://www.jstor.org/stable/44636075
2. Bauk, S., Kapidani, N., Lukšić, V., Rodrigues, F., Sousa, L.: Review of unmanned aerial systems for the use as maritime surveillance assets. In: 2020 24th International Conference on Information Technology (IT), pp. 1–5 (2020). https://ieeexplore.ieee.org/abstract/document/9070718
3. Huang, J.-L., Cai, W.-Y.: UAV low altitude marine monitoring system. In: 2014 International Conference on Wireless Communication and Sensor Network, pp. 61–64 (2014). https://ieeexplore.ieee.org/abstract/document/7061695
4. Jeon, I., et al.: A real-time drone mapping platform for marine surveillance. Int. Arch. Photogram. Remote Sens. Spat. Inf. Sci. **XLII-2-W13**, 385–391 (2019)
5. Ribeiro, C.G., Raptopoulos, L., Dutra, M.S.: An autonomous airship swarm for maritime patrol. In: Rocha, A., Pereira, R.P. (eds.) Developments and Advances in Defense and Security, pp. 307–320. Springer, Singapore (2020)
6. Sendner, F.-M.: An energy-autonomous UAV swarm concept to support sea-rescue and maritime patrol missions in the mediterranean sea. Aircr. Eng. Aerosp. Technol. **94**(1), 112–123 (2021)
7. Yang, S., Lin, D., He, S., Hussain, I., Seneviratne, L.: Aerial swarm search for GNSS-denied maritime surveillance. IEEE Trans. Aerosp. Electron. Syst. **60**(3), 3442–3453 (2024)

8. Ferri, G., et al.: Cooperative robotic networks for underwater surveillance: an overview. IET Radar Sonar Navig. **11**(12), 1740–1761 (2017)
9. Terracciano, D., Bazzarello, L., Caiti, A., Costanzi, R., Manzari, V.: Marine robots for underwater surveillance. Curr. Robot. Rep. **1**(4), 159–167 (2020)
10. Zou, J., Gundry, S., Kusyk, J., Sahin, C.S., Uyar, M.U.: Bio-inspired topology control mechanism for autonomous underwater vehicles used in maritime surveillance. In: 2013 IEEE International Conference on Technologies for Homeland Security (HST), pp. 201–206 (2013). https://ieeexplore.ieee.org/abstract/document/6699000
11. Orfanidis, G., et al.: Autonomous swarm of heterogeneous robots for surveillance operations. In: Tzovaras, D., Giakoumis, D., Vincze, M., Argyros, A. (eds.) Computer Vision Systems, pp. 787–796. Springer, Cham (2019)
12. Stolfi, D.H., Brust, M.R., Danoy, G., Bouvry, P.: UAV-UGV-UMV multi-swarms for cooperative surveillance. Front. Robot. AI **8** (2021). https://www.frontiersin.org/journals/robotics-and-ai/articles/10.3389/frobt.2021.616950/full
13. Bürkle, A., Essendorfer, B.: Maritime surveillance with integrated systems. In: 2010 International WaterSide Security Conference, pp. 1–8 (2010). https://ieeexplore.ieee.org/abstract/document/5730231
14. Li, X., Wang, W., Gu, H., Jin, K.: Exploring an early warning system for maritime security risks: an approach based on compressed sensing. Expert Syst. Appl. **249**, 123670 (2024)
15. Jing, X., Xiukun, W., Jiasheng, H., Jing, Z.: Multi-platform bearings-only tracking fusion of maritime targets. In: 2001 CIE International Conference on Radar Proceedings (Cat No.01TH8559), pp. 1112–1114 (2001). https://ieeexplore.ieee.org/abstract/document/984905
16. Mallick, M., Chang, K.-C., Arulampalam, S., Yan, Y., La Scala, B.: Heterogeneous track-to-track fusion in 2D using sonar and radar sensors. In: 2019 22th International Conference on Information Fusion (FUSION), pp. 1–8 (2019). https://ieeexplore.ieee.org/document/9011399
17. Rao, S.K., Murthy, K., Rajeswari, K.R.: Data fusion for underwater target tracking. IET Radar Sonar Navig. **4**(4), 576–585 (2010)
18. Maresca, S., Braca, P., Horstmann, J.: Detection, tracking and fusion of multiple HFSW radars for ship traffic surveillance: experimental performance assessment. In: 2013 IEEE International Geoscience and Remote Sensing Symposium - IGARSS, pp. 2420–2423 (2013). https://ieeexplore.ieee.org/abstract/document/6723308
19. Maresca, S., Braca, P., Horstmann, J., Grasso, R.: Maritime surveillance using multiple high-frequency surface-wave radars. IEEE Trans. Geosci. Remote Sens. **52**(8), 5056–5071 (2014)
20. Karagiannidis, L., et al.: Ranger: radars and early warning technologies for long distance maritime surveillance (2019). http://www.theseus.fi/handle/10024/266999. Accepted: 2019-12-18T18:56:16Z
21. International Maritime Organization. Resolution msc.192(79) - adoption of the revised performance standards for radar equipment (2004). https://wwwcdn.imo.org/localresources/en/KnowledgeCentre/IndexofIMOResolutions/MSCResolutions/MSC.192(79).pdf. Accessed 3 May 2024
22. Henderson, D.M.: Shuttle Program. Euler angles, quaternions, and transformation matrices working relationships, no. NASA-TM-74839, nTRS Author Affiliations: McDonnell Douglas Tech. Services Co., Inc. NTRS Document ID: 19770024290 NTRS Research Center: Legacy CDMS (CDMS) (1977). https://ntrs.nasa.gov/citations/19770024290

23. Martelli, M., Viviani, M., Altosole, M., Figari, M., Vignolo, S.: Numerical modelling of propulsion, control and ship motions in 6 degrees of freedom. Proc. Instit. Mech. Eng. Part M: J. Eng. Maritime Environ. **228**(4), 373–397 (2014)
24. Beckmann, N., Kriegel, H.-P., Schneider, R., Seeger, B.: The R*-tree: an efficient and robust access method for points and rectangles. In: Proceedings of the 1990 ACM SIGMOD international conference on Management of data. In: SIGMOD 1990, pp. 322–331. Association for Computing Machinery, New York (1990). https://dl.acm.org/doi/10.1145/93597.98741
25. Scharf, L.L., Demeure, C.: Statistical Signal Processing: Detection, Estimation, and Time Series Analysis. Addison-Wesley Publishing Company (1991)
26. Kalman, R.E.: A new approach to linear filtering and prediction problems. J. Basic Eng. **82**(1), 35–45 (1960). https://asmedigitalcollection.asme.org/fluidsengineering/article/82/1/35/397706/A-New-Approach-to-Linear-Filtering-and-Prediction
27. Longo, G., Orlich, A., Musante, S., Merlo, A., Russo, E.: Macyste: a virtual testbed for maritime cybersecurity. SoftwareX **23**, 101426 (2023)
28. Godot engine - free and open source 2D and 3D game engine. https://godotengine.org/. Accessed 3 May 2024

Evaluating Killer Drone Defense: NATO SPS Project "Anti-Drones" Field Trials

Alberto Lupidi[1(✉)], Francesco Mancuso[1,2], Giulio Meucci[1,2], Edmond Jajaga[3], Veton Rushiti[3], and Alessandro Cantelli-Forti[1]

[1] Radar and Surveillance Systems National Laboratory, CNIT, 56124 Pisa, Italy
{alberto.lupidi,francesco.mancuso,giulio.meucci,
alessandro.cantelli.forti}@cnit.it
[2] Department of Information Engineering, University of Pisa, 56122 Pisa, Italy
[3] Mother Teresa University, 1000 Skopje, North Macedonia
{edmond.jajaga,veton.rushiti}@unt.edu.mk

Abstract. The growing number of Unmanned Aerial Vehicles (UAVs) in both military and civilian sectors necessitates advanced tracking and detection systems. This paper presents the development and field trial results of a high-resolution radar and optical system for UAV detection conducted within the NATO SPS project "Anti-Drones". By using a multichannel linear frequency modulated continuous wave (FMCW) radar architecture, the system demonstrates robust capabilities in identifying and tracking various UAV types, including "mini" and "micro" drones. Field trials evaluated detection accuracy, tracking reliability, and payload recognition. Advanced signal processing and machine learning (ML) algorithms enhanced real-time recognition and reduced false alarms. Integrating radar data with optical camera tracking further improved performance in multi-target scenarios. This study highlights the system's potential for military and civilian applications in addressing UAV threats. Future research will focus on refining data fusion algorithms and extending operational range based on trial insights.

Keywords: UAV · Drones · Radar · Camera · Detection · Recognition · Tracking

1 Introduction

Unmanned Aerial Vehicles (UAVs), also called drones, have become a significant component of modern warfare and surveillance systems. Their ability to operate in hostile environments without risking human lives has made them indispensable in military operations. However, the widespread availability of UAV technology has also raised security concerns, as these devices can be used for malicious purposes by non-state actors and terrorists. This has created an urgent need for reliable and effective detection and tracking systems, especially for "mini" and

---

A. Lupidi, F. Mancuso, G. Meucci, E. Jajaga, V. Rushiti and A. Cantelli-Forti—These authors contributed equally to this work.

"micro" UAVs, which pose unique challenges in detection and elimination [1,2]. Recent developments in drone warfare [3–5] have highlighted the transformative role of UAVs on the battlefield. Various forces utilize drones for reconnaissance, precision strikes, and long-range attacks, prompting strategic adjustments such as repositioning heavy machinery and fortifying defenses against aerial threats. These fleets usually include a variety of commercially available drones modified for military purposes. The conflict has driven rapid advancements in drone technology, making them a cost-effective alternative to traditional artillery. For instance, First-person view (FPV) drones can cost less than a single artillery shell while providing higher accuracy. Both sides have developed drones capable of long-range attacks, targeting infrastructure deep within enemy territory, and expanding the battlefield far beyond the front lines. Drones are also mentioned by the NATO S&T Technology Trends 2023–2043 reports [6,7] also highlight drones as key components for intelligence collection and intelligent surveillance and reconnaissance (ISR), as well as potential future hypersonic threats. Air surveillance for UAV detection involves four phases: detection (identifying objects), recognition (categorizing objects), identification (determining affiliation and characteristics), and localization (tracking coordinates). These phases occur rapidly and are facilitated by systems like Identification Friend or Foe (IFF). Integrated within the C2 architecture, this process operates in real-time through C4ISR (Command, Control, Communications, Computers, Intelligence, Surveillance, and Reconnaissance) systems. It is possible to find relevant literature regarding the use of radar [8–11], optics and acoustic sensor [12–15], but often not work in synergy. While radar is the primary sensor, integrating optical, acoustic, and laser sensors is crucial for improving detection accuracy and reducing false alarms. Sensor fusion, which combines data from multiple sensors, will be vital for enhancing overall system performance and reliability in UAV detection and tracking. This paper presents the results of field trials conducted to evaluate a high-resolution radar and optical system designed specifically for UAV detection. The radar sensor utilizes a multichannel linear frequency modulated continuous wave (FMCW) architecture. Advanced signal processing techniques, coupled with artificial intelligence (AI) algorithms, enhance the system's ability to distinguish UAVs from other objects, thereby reducing false alarms and improving overall detection accuracy. The integration of radar data with optical camera tracking further enhances performance in multi-target scenarios.

2 The Radar Sensor

The radar architecture, specifically designed for detecting and recognizing moving targets, features one transmitting channel and three receiving channels, with an optional fourth receiving channel available for future research. The external placement of antennas from the main cabinet allows for diverse acquisition geometries, facilitating the application of various radar processing techniques such as interferometry and monopulse. This configuration also supports radar fusion techniques to enhance detection and recognition capabilities. The radar system consists of several key hardware components: a radar transceiver, a power amplifier, and a digital board. The power amplifier (Mini-circuits ZVE-3W-183+ is equipped with a heatsink. The digital unit includes a Trenz Electronic TEBF0808 carrier board with a ZU3EG-1E Zynq UltraScale+, 2 GB DDR4 RAM, an Abaco FMC 168 Mezzanine Card 250 MS/s, and a Samsung 970EVO Plus NVMe with 1 TB capacity. The system boots from an SD card containing a Linux kernel image and firmware, operating on a customized version of Petalinux for Xilinx FPGA. The PL design incorporates an FMC interface, a trigger generator, and a CIC filter for down-sampling high-speed data rates. The core of the system, a .bit file, is generated by the Vivado software. The radar components are arranged on two metal planes within the cabinet: the top plane houses the digital radar unit, power supply, power amplifier, and RF board, while the bottom plane contains the IF board. The radar's transmit waveform was tested for phase noise and spurious levels at maximum power in continuous wave mode at 9.6 GHz. The phase noise spectrum has a floor of -103 dBc at 4 MHz offset and the highest spur at -59 dBc. When configured to transmit a 500 MHz linear chirp centered at 9.6 GHz, the effective frequency chirp spanned from 9.362 GHz to 9.850 GHz with transmit power around 33 dBm and in-band fluctuations of 1.7 dB. The radar supports various sweep bandwidths and PRFs. This comprehensive overview of the radar sensor details its design, components, interfaces, data management, and functional tests, emphasizing its capabilities and versatility in target detection and recognition applications.

2.1 High-Resolution Radars for Drone Detection and Tracking

High-resolution radars are optimized for detecting and tracking drones by processing reflected electromagnetic signals and comparing them to a database for drone characterization. Advanced processing techniques, including Machine Learning (ML) and Artificial Intelligence (AI), reduce false positives by filtering out non-drone objects like birds, enhancing detection accuracy and reliability. The system provides real-time tracking by calculating the range and angle of the target relative to the radar. Detection range varies with the Radar Cross Section (RCS) of the drone; larger RCS drones are detectable at greater distances. Typically, drones like the Phantom 4 can be detected up to one mile away. Weather conditions, such as rain and fog, may slightly affect the detection range, especially at higher frequencies. Radars operating from X-Band and above are suitable for detecting UAVs due to their dimensions and equivalent RCS, offering satisfactory accuracy for coordinate measurement with compact antenna sizes.

2.2 Detection Capabilities

Radars can detect various types of drones without relying on their communication signals, unlike systems that intercept WiFi or cellular signals. While very small toy drones might evade radar, they pose minimal threat due to limited payloads. Optimal detection requires a radar network with 360-degree coverage, influenced by factors like Radar Cross Section (RCS) and Minimum Detectable Velocity (MDV). Detecting slow or hovering drones in cluttered environments is challenging, but micro-Doppler modulation improves accuracy. Advanced techniques like Space-Time Adaptive Processing (STAP) and Multiple-Input Multiple-Output (MIMO) beamforming enhance performance by canceling clutter. Adjusting radar settings, such as Pulse Repetition Frequency (PRF) and bandwidth, allows detection of drones at varying speeds and distances, with higher antenna gains and longer integration times improving range.

3 Tracking Algorithms Description

Our approach combines radar and camera data for faster and more accurate localization and tracking. Radar data positions the camera, which then uses YOLOv4 and CAMSHIFT for continuous tracking. The Center-Based algorithm for motor movement ensures minimal occlusion by keeping the object centered. The integrated tracking system leverages radar for initial detection and a camera for continuous tracking. Radar data provides the camera with the object's position and speed, enhancing tracking accuracy. The combined use of clustering, Kalman filters, and object recognition algorithms ensures robust multi-target tracking, which is crucial for monitoring small, fast-moving drones.

3.1 Radar-Based Multi-target Tracking

Radar-based tracking systems rely on sensors to measure target range and speed, using range and Doppler data. However, azimuth measurements are often imprecise, complicating multi-target detection. To improve tracking accuracy, a probabilistic algorithm, enhanced with clustering and a modified expectation-maximization method, is used for data association. Multi-sensor data is essential for precise 2D localization, though it is computationally demanding. The system employs K-Means clustering, Kalman filtering, and Global Nearest Neighbor (GNN) algorithms to manage noise, clutter, and track initiation. Gating helps filter unlikely track associations, ensuring robust tracking in dense environments with multiple targets.

3.2 Camera-Based Object Tracking

Camera-based object tracking monitors the movement of objects in video frames, which is crucial for surveillance applications. This system requires a camera capable of rotating 360° horizontally and 180° vertically. The system uses the OpenCV library and the YOLOv4 platform for object recognition.

- **Dominant Color:** This algorithm tracks objects based on their dominant color. It identifies and follows the most prominent color within the object's region in each frame, making it suitable for objects with distinct and consistent colors.

- **CAMSHIFT (Continuously Adaptive Mean Shift):** An adaptation of the Mean Shift algorithm, CAMSHIFT dynamically adjusts the search window size and position based on the object's color distribution. It is particularly effective in tracking objects that change in size, shape, or orientation over time.
- **CMT (Consensus-based Matching and Tracking of Keypoints):** This algorithm tracks objects by detecting and matching key points within the object's region. It is robust against partial occlusion and scale changes and can handle large datasets efficiently. CMT combines keypoint tracking with a consensus mechanism to maintain accurate tracking even with dynamic inputs.
- **Boundary-Based:** This algorithm controls the stepper motor to move the camera when the tracked object moves beyond predefined boundaries within the frame. It ensures that the object remains within the camera's field of view by repositioning the camera as needed.
- **Center-Based:** This algorithm aims to keep the tracked object centered in the frame. It minimizes occlusion and maintains optimal visibility by adjusting the camera's position to keep the object at the center of the view. This approach is particularly useful in dynamic environments where objects move unpredictably.

4 Recognition Algorithms Description

4.1 Radar-Based Recognition

Radar-based recognition for Unmanned Aerial Vehicles (UAVs) primarily focuses on differentiating UAVs from non-UAV targets and distinguishing between different UAV models. Due to the low Radar Cross Section (RCS) of UAVs, detecting and tracking them can be challenging. This difficulty is compounded by the fact that micro-UAVs can hover or move slowly, resulting in a low Doppler frequency return that is often indistinguishable from clutter by standard surveillance radars Practically, if an object is slow or not moving, it does not produce a Doppler return, so the radar can confuse it with static returns like the return from the ground. An effective way to discriminate these targets from natural objects like birds is by analyzing the micro-Doppler signature of UAV blades, which move at high speeds. Achieving fine Doppler resolution requires a long coherent integration time. This need has led to the development of staring radars, which use arrays to detect and recognize UAVs. To optimize feature representation in target classification, several micro-Doppler analysis approaches are evaluated:

- **Spectrogram:** This popular technique reveals the time-frequency variation of spectral content. It involves segmenting raw data into overlapping time frames and performing FFT on each frame.
- **Cepstrogram:** Obtained by taking the logarithm of the STFT magnitude and performing inverse FFT along the Doppler dimension, this method contains periodicity information.

- **Cadence Velocity Diagram (CVD):** Achieved by taking the logarithm of the STFT magnitude and performing another FFT along the time dimension.

These processed images are then used for subsequent feature extraction and classification. The data processing chain involves several steps to ensure accuracy and robustness:

1. **Stacking and Normalization:** The 2-D training data matrix is converted into a 1-D vector, normalized to zero-mean and unit-variance. Principal Component Analysis (PCA) is then applied to reduce dimensionality and remove unreliable signal sub-space.
2. **Clustering:** To better characterize and discriminate between the classes "drone present" and "drone absent," clustering within each class is performed.
3. **Further Subspace Analysis:** PCA is conducted again using within-cluster and between-class matrices to improve discriminability.
4. **Feature Extraction:** Mahalanobis distances to the cluster centers are calculated, normalized, and used as features for classification using a Support Vector Machine (SVM).

The Doppler signature of UAVs varies by frequency band, which aids in target characterization. Lower offset frequencies (up to 1 kHz) correspond to blade root returns, mid offset frequencies (1 kHz to 2 kHz) are usually unused, and high offset frequencies (2 kHz to 5 kHz) are associated with blade tip returns. While high-resolution Doppler is essential for identifying UAVs, ultra-high resolution offers little benefit due to the incoherent nature of micro-UAV Doppler signals. The key features extracted for classification include velocity or body radial velocity, total bandwidth of the Doppler signal, offset of total Doppler, bandwidth without micro-Doppler, cadence/cycle frequency, and target RCS. Using SVM classification, a confusion matrix can be generated to assess performance, with fixed-wing drones being the hardest to classify due to their lack of moving parts.

4.2 Camera-Based Recognition

Various deep learning frameworks for object recognition were considered, including YOLO [16], TensorFlow, and PyTorch. The typical workflow involves several key steps: examining and understanding the data, building an input pipeline, constructing and training the model, and iteratively testing and improving the model. In image classification, these models assign labels to detected objects and provide localization information, identifying and bounding the objects in images or video streams. Each framework offers unique models and precision characteristics tailored to different needs. TensorFlow utilizes an object detection API with models trained on the COCO 2017 dataset. This framework supports flexible architecture and deployment across various platforms. It also offers easy model building with Keras, support for multiple languages, and efficient cloud deployment. PyTorch, based on the Torch library, is suitable for computer vision and natural language processing. It is known for its transparency and straightforward modeling. PyTorch supports transparent modeling with direct execution without session interfaces, offering both eager and graph modes, making it suitable for cloud support. YOLO is renowned for its speed and accuracy, capable of processing up to 67 FPS and detecting small objects with low hardware requirements. It is favored for its high accuracy, real-time processing capability, open-source nature, and efficiency in detecting small objects. Specifically, YOLOv4 [17] offers significant improvements in average precision and FPS.

Dataset Challenges and Methodology
Building a comprehensive dataset for UAV detection presents challenges due to the lack of existing data and the quality of images. Key considerations for dataset creation include:

- Enriching the dataset with classes like drones, birds, fixed-wing drones, and copters.
- Ensuring diverse backgrounds and target sizes.
- Utilizing various open-source datasets and manually annotating additional images.

The dataset preparation involves the following steps:

1. **Image Extraction and Selection:** Extracting frames from videos and manually selecting relevant images.
2. **Annotation Adaptation and Normalization:** Converting annotations to the YOLO format and normalizing them.
3. **Image Reduction:** Removing redundant frames.
4. **Train-Test Distribution:** Organizing images into training and testing sets.
5. **Fixing Annotations:** Correcting class labels and ensuring annotation consistency.

YOLO-Based Approaches
Several state-of-the-art YOLO-based approaches were analyzed, focusing on network architecture, dataset and annotations, and classes and accuracy. These approaches are based on Darknet and fine-tuned for specific applications, combining real and synthetic data with manual annotations. Typically, they focus on single classes like drones, achieving high precision and recall. However, it is important to note that most approaches do not detect drones with payloads or handle multiple classes effectively. In our previous efforts, described in [18], a number of challenges were encountered when building up a rather qualitative dataset. In general, the challenges can be considered as i) technical ones and ii) the ones related to the model overfitting and underfitting. The technical challenges were related to image extraction and selection, annotation adaptation and normalization, as well as image reduction. On the other side, enriching the dataset with fixed-wing drones and drones appearing in complex backgrounds and far distances mark the cases when model overfitting and underfitting got experienced.

5 Experimental Results

The campaign conducted at Stenkovec Training Field near Skopje (MK) had several objectives. The primary goal was to assess the detectability of small and medium-sized drones. The secondary objective was to evaluate the feasibility of tracking these UAVs in a noisy environment. Lastly, the campaign aimed to determine the potential for recognizing different types of UAVs.

Fig. 1. (a) Targets and setup for the experiment, (b) Drone Helicopter with payload (circle) and birds passing by (rectangle)

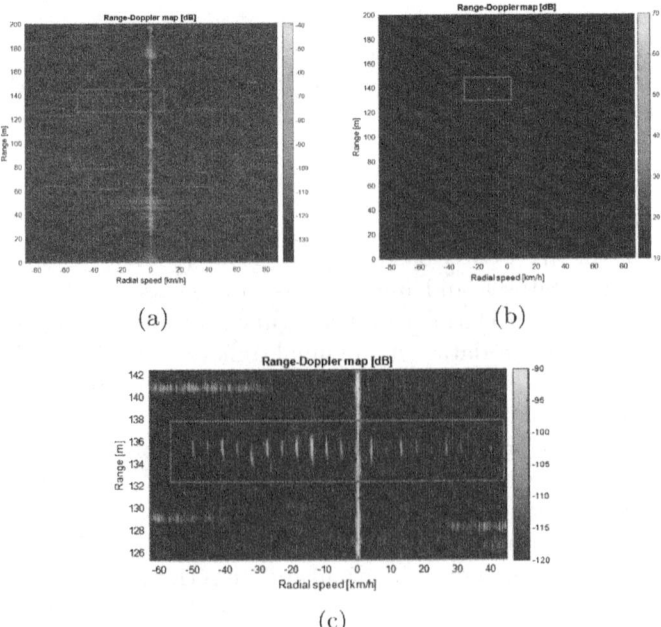

Fig. 2. Quadcopter: (a) range-Doppler map, (b) detection after cleaning, (c) micro-doppler features

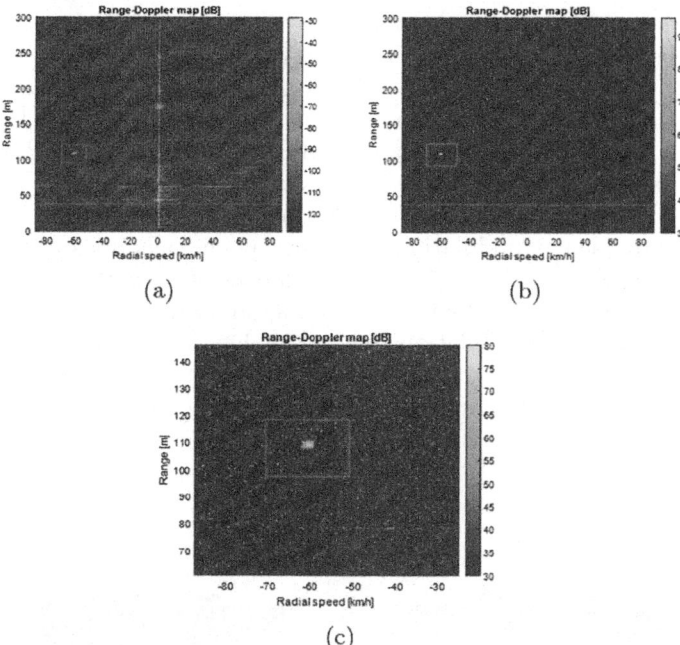

Fig. 3. Fixed-wing: (a) range-Doppler map, (b) detection after cleaning, (c) micro-doppler effect is absent here

5.1 Radar Trials

The observation campaign took place in an open area with minimal buildings and trees. The clutter mainly came from short grass, primarily affecting the radar's sidelobes, which were slightly oriented toward the sky. The test targets included a hexacopter, a quadcopter, and a fixed-wing drone, as illustrated in Fig. 1a. Additionally, during the data acquisition of the payload-mounted drones, several birds were recorded (Fig. 5), providing further information for discrimination analysis. It is interesting to see how the fixed-wing drone (shown in Fig. 3), as expected, has a completely different behavior compared to what we observe in Fig. 2. While moving much faster, it is still possible to detect it, and no micro-Doppler modulations are present. The most interesting acquisition was performed to check if the radar is able to detect the presence of payload with the drones, as shown in Fig. 4. The figure set illustrates that the helicopter was hovering close to the clutter line and deployed its payload in the second frame. Despite the low RCS, the radar successfully a) detected the micro-Doppler signature of the helicopter and b) observed the payload detaching, highlighted in the circle. It's important to note that the payload fell perpendicular to the radar's line of sight, resulting in only a small component being projected in the radial direction. In the last figure, it passes the clutter line as the radial component changes sign.

An ulterior analysis can be performed with radar data to extract more features, namely, the spectrogram of the target. By analyzing a time series of appropriate length (maximum 1 s to include several blade rotations) it is possible to detect the signature of the blade tips which is quite peculiar for the UAS (also, it is possible to infer the number of blades and rotation period), and differentiation between various behaviors. Figure 6a shows the spectrogram of the quadcopter, where sinusoidal signature due to the harmonic motion of the blades is produced by the frequency modulation, while this phenomenon is quite reduced for the birds, as shown in Fig. 6b. The line is sloped because of the motion of the bird's body. Finally, in 6c there is the spectrogram of the fixed wing UAS, which has no Doppler, micro-modulation, because it has no rotation parts, but only the Doppler frequency due to the speed of the main body is present. The line is almost flat because the acquisition time is very small (0.25 s) so there is no migration both in Doppler and range.

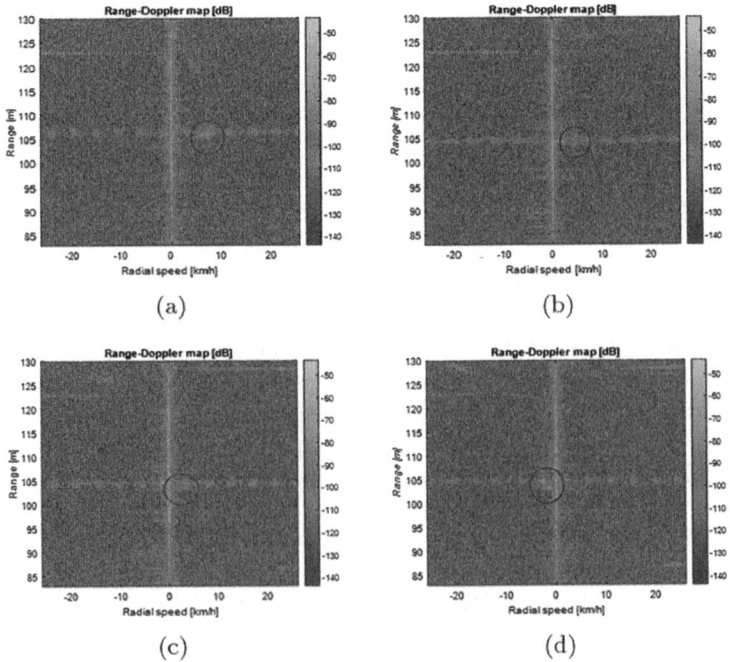

Fig. 4. Payload detachment. Time spacing between frames equal to 0.5 s

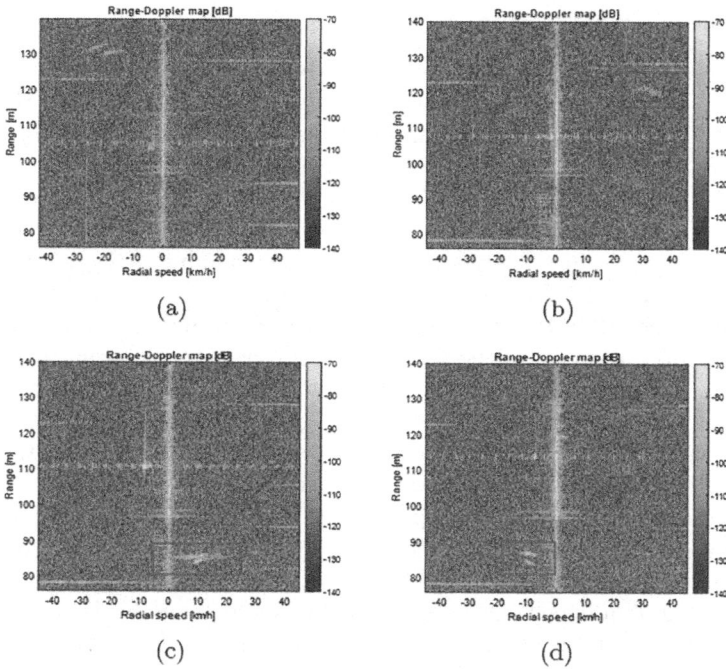

Fig. 5. Flyby of birds couple. Time spacing between frames equal to 0.5 s

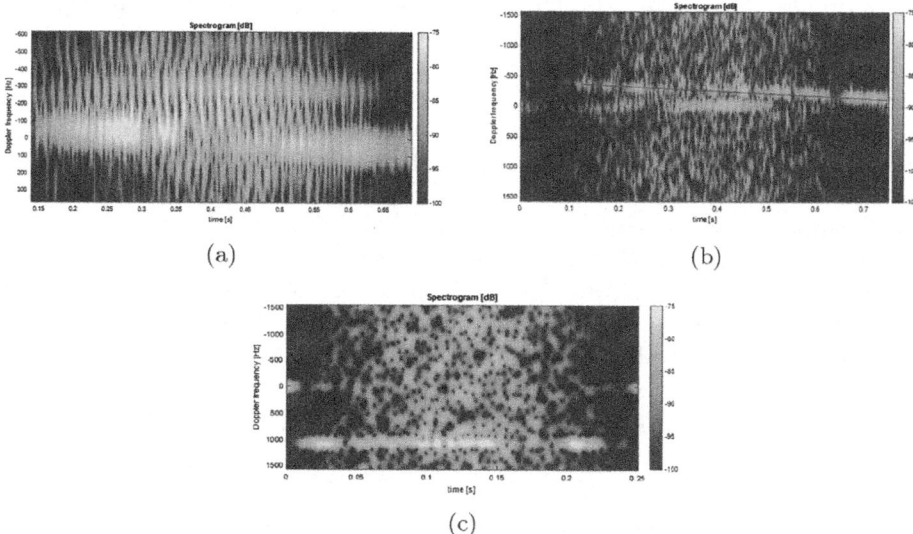

Fig. 6. Spectrograms: (a) Quadcopter, 0.7 s acquisition, (b) Birds, 0.75 s acquisition, (c) Fixed-wing, 0.25 s acquisition

5.2 Camera Trials

The proposed optical model supports also real-time detection and recognition. Its first test in real environment was performed in Stenkovec (North Macedonia) together with the radar solution in front of the North Macedonia Army end user. The solution was tested in a PC-laptop machine with an Intel Core i7-10750H CPU, 8 GB RAM, 512 GBSSD, NVIDIAGeForce GTX1650, and Windows 10 Pro x64. A typical web camera with standard high definition (HD Ready or 720p) resolution was used per real-time performance. Two flying drones, DJI Phantom 4Pro v2 and DJI Mini 3 Pro, were investigated by both radar and camera sensor. The model detected and recognized the drones. Eventually, the model was able to recognize both of them at the same time. However, the recognition performed only in near distances i.e., less than 100 m, with 30 Based on the analysis of the results it can be observed that there is a significant need for improvements of the optical model. The main challenges encountered while dealing with neural networks are related on the network model and dataset fine-tuning. Regarding the former one, the results vary on the methodology of training the model. It is recommended to first train the model with a different and more general dataset for a similar problem e.g., ImageNet [19]. This provides better initial points than random ones for the parameters of the network and especially useful when the training data is scarce e.g., drone with payload images. Additionally, the visual detection method offers strong resistance to jamming; however, due to the intricate characteristics of small drones, its reliability is reduced. [19]. As authors of [20] recommend, one should consider increasing the resolution of images to greater than in input size on the testing configuration. In terms of dataset generation, we have observed that our approach does not support fixed-wings drones. This is an expected result, since it was trained with multi rotor drones. Another issue appeared in mixed birds and drone images, in which it was difficult to accurately recognize drones and birds. On the other side, it was obvious that clear backgrounds provided better results. As a recommendation for improving our dataset we plan to train the model with higher resolution images and more images of multi rotors drones and drones with payload. Additionally, the poor dataset of its first version implied too many false alarms, e.g., the form of the lights confusing the model. The model is mainly trained with drone images appearing in open sky backgrounds. Thus, the light similar to drone residing in a sky background caused this ambiguity. We have managed to overcome this issue in the final version. However, there is a need to address the kinds of these issues by applying different techniques. Namely, combining different backgrounds with drones and birds with synthetically generating the dataset, like the experience provided in [19], or coupling CNNs with additional processing steps used in moving object detection tasks, like the winners of the Drone-vs-Bird Challenge.

6 Data Fusion Algorithms Description

The next challenge in the project's evolution is establishing an efficient data fusion algorithm, crucial at both high and low operational levels. At the high level, the system must integrate radar and camera data, while at lower levels, data fusion occurs within individual data sources. The Fusion Center Unit merges data from separate transmitting and receiving stations. For high-level data fusion, we adopt a strategy similar to Liu et al. [20], combining SVM radar data with YOLO camera data instead of camera and acoustic data. The system merges target attributes from radar and optical camera sources. Radar data includes range, speed, and radar cross-section, while optical data involves photo and video images. Typically, the radar scans within a 3 km radius, detecting any objects entering the secured area and continuously measuring their attributes. This generates a coefficient, C1, indicating the likelihood of the object being a drone. If C1 exceeds 90%, the system immediately acts to prevent the drone. If not, the camera activates, aligns based on radar data, and captures images or videos of the object. These are processed to generate a second coefficient, C2. If C2 also exceeds 90%, the drone is prevented. If not, the mean value of C1 and C2 is calculated. If this average is above 80%, the object is deemed a potentially harmful drone; otherwise, it is classified as non-drone, and the system resets.

7 Conclusion

The radar and optical system developed within this project demonstrated significant potential for UAV detection and tracking. Field trials confirmed the system's robust detection capabilities, reliable tracking performance, and effective recognition of UAVs carrying payloads. The integration of advanced signal processing techniques and machine learning (ML) algorithms has played a crucial role in enhancing the system's overall performance, particularly in complex and dynamic environments. The synergy created by combining radar and optical data resulted in improved accuracy and reliability, especially in multi-target scenarios where distinguishing between multiple UAVs and other objects is critical.

However, while the system shows promise, there are specific areas where further refinement could lead to even greater performance gains. One area for improvement is the data fusion algorithms that integrate radar and optical inputs. By refining these algorithms, the system could achieve more seamless integration of data, leading to faster and more accurate decision-making. Moreover, extending the system's operational range will be essential to increase its applicability in diverse environments, including those with challenging topographical features or varying weather conditions that may affect sensor performance.

Another potential advancement lies in the continuous development of machine learning models tailored to the evolving characteristics of UAV threats. As UAV technology advances, so too must the algorithms used to detect and track them. This includes not only adapting to new UAV designs and materials that may affect radar and optical signatures but also enhancing the system's ability to recognize and predict UAV behavior patterns. Incorporating adaptive learning techniques could allow the system to self-optimize over time, improving its effectiveness against a broader range of threats.

The insights gained from these field trials will be invaluable in guiding future enhancements. For instance, trial data could be used to identify specific scenarios where the system's performance was suboptimal, such as under conditions of extreme visual clutter or in detecting small, low-altitude UAVs. Addressing these challenges may involve integrating additional sensors, such as infrared or acoustic sensors, to provide complementary data that can further improve detection and classification accuracy.

Moreover, the integration of emerging technologies, such as quantum radar or advanced photonic systems, holds the potential to revolutionize the capabilities of UAV detection systems. Quantum radar, for example, could significantly improve the system's ability to detect stealth UAVs or those operating in challenging environments where traditional radar might struggle.

This architecture will therefore demonstrate its full potential in applications that demand high sensitivity, reliability, and adaptability. In particular, it will be invaluable in scenarios where both threat detection and identification are critical, such as urban protection, densely populated areas, or anti-terrorism efforts. By continuously refining the system based on trial insights and leveraging emerging technologies, this research can make a significant and lasting contribution to the development of comprehensive surveillance systems capable of addressing the growing UAV threat.

Acknowledgements. We would like to thank the NATO Emerging Security Challenges Division Science for Peace and Security Programme for funding the activities named "Anti-Drones - Innovative Concept to Detect, Recognize and Track Killer-Drones" SPS.MYP.G5633 and "Ai4CUAV - Innovative AI-framework to enable the Detection, Classification and Tracking of Killer-Drones" SPS.MYP.G6246. We sincerely thank Prof. Walter Matta for his invaluable expertise and scientific guidance, which were instrumental in the successful completion of this research. We also extend our gratitude to ECHOES s.r.l. for providing the radar sensor.

References

1. Lupidi, A., Jajaga, E., Matta, W.: An artificial intelligence application for a network of LPI-FMCW mini-radar to recognize killer-drones (2022)
2. Semenyuk, V., Kurmashev, I., Lupidi, A., Cantelli-Forti, A.: Developing the GoogleNet neural network for the detection and recognition of unmanned aerial vehicles in the data Fusion System. Eastern-Eur. J. Enterp. Technol. **122**(9) (2023)
3. Council on Foreign Relations: How the Drone War in Ukraine Is Transforming Conflict. https://www.cfr.org/article/how-drone-war-ukraine-transforming-conflict. Accessed 23 July 2023
4. Reuters: How Drone Combat in Ukraine Is Changing Warfare. https://www.reuters.com/graphics/UKRAINE-CRISIS/DRONES/dwpkeyjwkpm/. Accessed 23 July 2023
5. Policy, F.: The army is building a new way to fight. https://foreignpolicy.com/2021/03/30/army-pentagon-nagorno-karabakh-drones/. Accessed 23 July 2023
6. CESMAR: Science & Technology Trends 2023–2043, Volume 1

7. NATO: Science & Technology Trends 2023-2043, Volume 2. https://www.nato.int/nato_static_fl2014/assets/pdf/2023/3/pdf/stt23-vol2.pdf. Accessed 23 July 2023
8. Park, S., Kim, H.T., Lee, S., Joo, H., Kim, H.: Survey on anti-drone systems: components, designs, and challenges. IEEE Access **9**, 42635–42659 (2021)
9. Drozdowicz, J., et al.: 35 GHz FMCW drone detection system. In: 2016 17th International Radar Symposium (IRS), pp. 1–4 (2016)
10. Jahangir, M., Baker, C.: Robust detection of micro-UAS drones with l-band 3-D holographic radar. In: Signal Processing for Defence (SSPD), pp. 1–5 (2016)
11. Nemer, I., Sheltami, T., Ahmad, I., Yasar, A.U.-H., Abdeen, M.A.: RF-based UAV detection and identification using hierarchical learning approach. Sensors **21**(6), 1947 (2021)
12. Sapkota, K.R., et al.: Vision-based unmanned aerial vehicle detection and tracking for sense and avoid systems. In: 2016 IEEE/RSJ International Conference on Intelligent Robots and Systems (IROS), pp. 1556–1561 (2016)
13. Zhao, Z.-Q., Zheng, P., Xu, S.-T., Wu, X.: Object detection with deep learning: a review. IEEE Trans. Neural Netw. Learn. Syst. **30**(11), 3212–3232 (2019)
14. Pawełczyk, M.L., Wojtyra, M.: Real world object detection dataset for quadcopter unmanned aerial vehicle detection. IEEE Access **8**, 174394–174409 (2020)
15. Meng, W., Tia, M.: Unmanned aerial vehicle classification and detection based on deep transfer learning. In: 2020 International Conference on Intelligent Computing and Human-Computer Interaction (ICHCI), pp. 280–285 (2020)
16. Redmon, J., Divvala, S., Girshick, R., Farhadi, A.: You only look once: unified, real-time object detection (2016). https://arxiv.org/abs/1506.02640
17. Bochkovskiy, A., Wang, C.-Y., Liao, H.-Y.M.: YOLOv4: optimal speed and accuracy of object detection (2020). https://arxiv.org/abs/2004.10934
18. Jajaga, E., et al.: an image-based classification module for data fusion anti-drone system. In: International Conference on Image Analysis and Processing, pp. 422–433 (2022)
19. Aker, C., Kalkan, S.: Using deep networks for drone detection (2017). https://arxiv.org/abs/1706.05726
20. Liu, H., Wei, Z., Chen, Y., Pan, J., Lin, L., Ren, Y.: Drone detection based on an audio-assisted camera array. In: 2017 IEEE Third International Conference on Multimedia Big Data (BigMM), pp. 402–406 (2017)

STEM Workshop

Academic Career and Gender Balance Perceptions Among Bachelor's Students in Computer Science: A Case Study

Ozge Buyukdagli[✉][iD], Amal Mersni[iD], and Merjem Talic

Faculty of Engineering and Natural Sciences, International University of Sarajevo,
Sarajevo, Bosnia and Herzegovina
{obuyukdagli,amersni}@ius.edu.ba

Abstract. The field of Information Technology (IT) is constantly evolving, and both industry and academia are experiencing a shortage of human resources in this sector. As interest in IT-related studies increases worldwide, there is a growing demand for academics in this field. Understanding student perceptions of academia as a career path is crucial for addressing this workforce gap. This study focuses on the gendered aspect of this discussion, exploring the perceptions of young Computer Science (CS) students, particularly women, regarding the pursuit of an academic career in CS. It also examines their personal views of gender disparity in the IT industry and academia. The aim of this study is to enrich global understanding of gender diversity, academic progression, and career choices in the IT sector, using Bosnia and Herzegovina (BiH) as a case study. The results indicate the need for greater diversity, fairness, and inclusivity, which are directly related to the concept of responsible computing.

Keywords: academic career · gender · survey · computer science

1 Introduction

The expansion of the technology sector presents many opportunities, but it also highlights the critical need for a diverse and skilled workforce to drive innovation and sustain the industry's advancement. The underrepresentation of women in IT fields, specifically in CS, exacerbates this challenge [2,8]. Numerous studies over the years [5,12,15] have confirmed the field's reputation as traditionally male-dominated. In academia, particularly in Europe, women have shown a declining interest in pursuing academic careers in CS [12]. Globally, efforts have been made to increase inclusivity in the field, with various countries conducting analyses to define the scope of female underrepresentation in their domestic contexts.

One example is the CS Department at the University of Crete in Greece, which conducted a data analysis of the students accepted into their program.

The analysis revealed a persistent gender disparity, consistent with global trends of women's underrepresentation in CS, and underscored the need for further efforts to understand and address this phenomenon [15].

Similarly, a study conducted at Tecnologico de Monterrey, a private university in Mexico, analyzed historical data from a STEM program and conducted a three-month survey to determine the causes of student retention and attrition. The historical data showed that only 17% of candidates admitted to the program were female. The survey, which aimed to investigate factors behind decisions to drop out or remain in the program, identified motivating factors such as inspirational faculty staff, while demotivators included the challenging nature of the environment and perceived lack of support from teaching staff [13].

A study highlighting the underrepresentation of women in CS in the United States found that various factors impact recruitment and retention strategies in this field. After a comprehensive literature review that distinguished between recruitment and retention challenges, their study focused on the latter. The researchers proposed a retention model that includes individual, institutional, and societal factors, emphasizing the necessity of sustainable retention strategies [14].

In Romania, research on women in academia within the CS discipline sought to identify factors influencing their decisions to stay or leave the field. The study included female graduate students as well as faculty members. The researchers developed a replication package that can be used in any country to assess the situation in a broader context. The authors recommend applying the package across several countries and analyzing the results at the European level, which would facilitate the formulation of public policies that will increase women's participation in the IT world [12].

The situation in the IT sector in BiH aligns with global trends, with an increasing demand for skilled computer engineers proficient in advanced computer systems and programming. Particularly in the capital city, Sarajevo, the IT sector is experiencing a considerable boom, as outlined in the Sarajevo Canton Development Strategy for 2021–2027 [10]. However, this fast-growing sector faces a significant challenge, as the supply of skilled labor is insufficient to meet demand [9]. The gender gap further exacerbates the situation. For instance, in 2022, only 35.2% of graduates in Information and Communication Technologies (ICT) were women, compared to 64.8% men [4]. Men also dominate research and academic IT positions, at 58.3%, with women representing only 41.7% of researchers in Engineering and Technology, including ICT [4].

To address the shortage of IT specialists as well as bridge the gender gap in this field, higher education institutions (HEIs) must bolster their IT-related programs by enhancing both the quality of instruction and teaching capacity. This places significant pressure on HEIs to hire more qualified academic staff, a resource that is already in very limited supply.

This challenge furnishes one of the key motivations for the present study. By shedding light on how CS bachelor's students in BiH perceive academic careers, this study aims to furnish insights that can be used to increase interest in this professional path. A secondary goal of the study is to explore students' views on gender diversity in both academia and the IT industry. We have also sought to determine motivational and demotivational factors influencing students' decisions to pursue academic careers. With these objectives in mind, the following research questions have been formulated:

RQ1: What are the perceptions of Bosnian CS bachelor's students regarding academic careers in their field?

RQ2: What are the perceptions of Bosnian CS bachelor's students on gender diversity in both industry and academia?

To address these research questions, a short, structured survey was conducted, investigating the perceptions of CS bachelor's students at a Bosnian university on academic careers and gender diversity in the IT field.

The remainder of this paper consists of three sections. The Methodology and Survey section describes the survey design, data collection, and the data cleaning process, as well as the quantitative and qualitative analysis. In the Results section, the collected data is presented and analyzed. Finally, the Conclusion discusses the implications of the study findings, their limitations, and prospective directions for further research.

2 Methodology and Survey

This study addresses the important issue of student perceptions of academic careers in CS, with a particular focus on gender diversity to promote a more inclusive environment in the IT sector. A mixed methods research methodology has been used, incorporating a structured survey in English to collect data on student perceptions and provide a solid basis for analysis.

2.1 Survey Design

The survey was designed to elucidate the views of predominantly bachelor-level CS students regarding the pursuit of an academic career in this field. The questions also sought to capture their opinions on gender balance in both the IT industry and academia. Additionally, the study examines gender-based differences in these perceptions, which will be discussed in Sect. 3.

The short survey consisted of thirteen questions (see Table 1) in total. We used one open and twelve closed questions. There are eight Simple Multiple-Choice (SMC) questions, and four Multi-Select Multiple-Choice (MMC), allowing respondents to choose all that apply to them among the provided alternatives. Finally, we included one Open question to gather additional participant comments or insights.

Questions 1 to 4 (**Q1–Q4**) collected basic demographic data such as age, gender, nationality, and current academic level. Questions 5 and 6 (**Q5–Q6**) aimed to understand the participants' motivations for choosing the CS field and their career aspirations.

Questions 7 to 9 (**Q7–Q9**) focused on students' perspectives on academic careers, exploring the factors that encourage or discourage them from pursuing such a career. **Q7** pertained to their conceptual associations with an academic path, while **Q8** and **Q9** explored motivational and demotivational factors.

Before answering questions **Q10–Q13**, the content of the questions was explained to the participants, and the participants were directed to provide their answers based on their lived experiences and perceptions of the IT sector in BiH. Questions **Q10** through **Q12** seek to interpret the students' attitudes and levels of concern about gender disparity and female representation in the IT field. Question **Q13**, on the other hand, targets the students' insights into gender inclusivity in academic environments compared to the industry.

Table 1. Survey questions

ID	Question	Type
Q1	Please select your age:	SMC
Q2	Please select your gender:	SMC
Q3	What is your nationality?	SMC+other
Q4	What is your current academic year or level?	SMC
Q5	What influenced your choice of field of study in Computer Science?	MMC+other
Q6	What are you planning to do after graduation?	MMC+other
Q7	Please describe "academic career" in your own words. What do you know about academic career? What comes to your mind?	Open
Q8	What do you think are the motivational factors to pursue for an academic career in Computer Science?	MMC+other
Q9	What do you think are the demotivational factors to pursue academic career in Computer Science?	MMC+other
Q10	Do you think there is a gender gap in the IT sector in BiH?	SMC
Q11	What are your thoughts on the representation of women in your field:	SMC
Q12	Are you personally concerned about the presence of gender bias in your potential future work environment?	SMC
Q13	Which of the following statements best reflects your perception of gender inclusivity in academic environments compared to the industry?	SMC

2.2 Data Collection and Cleaning

The target group for this research was bachelor's students majoring in IT-related programs in BiH. The survey was distributed among the students at various levels of undergraduate IT studies at the International University of Sarajevo (IUS), which was selected for the expected responsiveness of its student body to the research team. No financial incentives or other benefits were offered to participants, relying solely on the students' willingness to support their professors and peers.

Both open and closed invitations were used to reach the participants. In November and December 2023, the Computer Science and Engineering department mass-emailed students from the IT field through their official university accounts. Professors were also encouraged to share the survey with their students during class, and the survey link was disseminated through additional channels like dedicated Microsoft Teams groups to ensure it reached as broad a base as possible.

Data was collected between the Fall of 2023 and the Spring of 2024, with the survey closing in June 2024 after receiving 187 responses. Following a preliminary analysis, the 30 non-Bosnian respondents were excluded to focus on a more homogeneous group, i.e., citizens of BiH. This decision was made to avoid skewing the results, as foreign students might have different motivations and perspectives, considering they are living abroad. Since all questions were compulsory, the dataset had no missing values.

The final dataset included 157 Bosnian students, consisting of 69 females and 88 males, from various academic years. Figure 1 shows the distribution of participants by academic year. Most respondents were freshmen (41), sophomores (47), and juniors (41), with a smaller number of seniors (23) and graduate students (5).

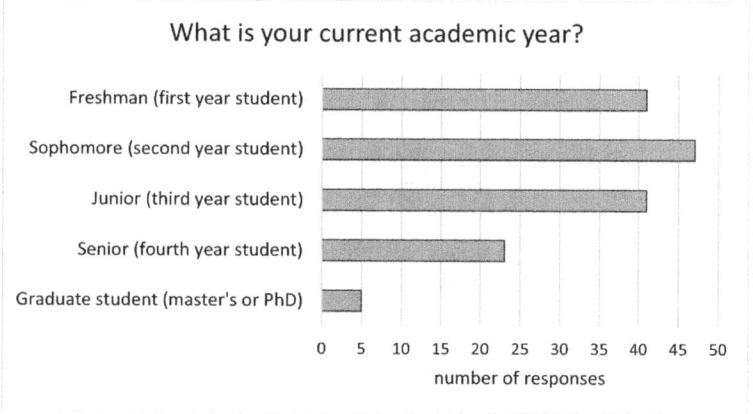

Fig. 1. Current academic study years of respondents

Threats to Validity. The survey was distributed among students from a single university, which presents a potential bias toward the perceptions and experiences of this specific group. A more diverse sample could provide a more balanced view. In future research on the matter, the study would benefit from a larger sample size, including other universities in the country. Additionally, the survey should be disseminated over multiple media channels and translated into the local language to ensure its accessibility to a broader range of students.

2.3 Quantitative and Qualitative Analysis

The survey included three types of questions-SMC, MMC, and open-form-each of which required distinct analytical approaches. Descriptive statistics were reported, and hypothesis tests were applied for some questions to determine if there were significant differences between female and male responses. For the SMC questions, the chi-square test of independence was used with a significance level of $p < 0.05$. For one MMC question (Q6), the response options were considered separate binary variables to overcome potential dependencies, and a chi-square test was conducted for each hypothesis. Since participants could select multiple options, introducing potential dependencies, a Bonferroni correction [3] was applied to minimize the risk of Type I errors (false positives) as a result of conducting multiple tests. Cramer's V value [7] was also calculated to assess the effect size and better understand the strength of associations.

For the only open question in the survey, we conducted open coding [11], a method used in grounded theory [6] for qualitative analysis. The first researcher read all responses and identified the main keywords, grouping similar responses into codes. These codes were then categorized into broader labels, considering repeated keywords, similarities, and differences. The second researcher then reviewed all answers and suggested adjustments to the labels. After both researchers reached an agreement, the labels were validated and finalized.

For questions with an 'other' option, some participants provided additional feedback in their own words. Noteworthy responses are shared in Sect. 3. The notation of R#ID was used, where R stands for 'respondent' and #ID is the ID number of the respondent in the dataset.

3 Results

Using the survey data, both quantitative and qualitative analyses were conducted. In this section, these analyses are presented in detail, starting with descriptive statistics obtained from multiple-choice questions. Where applicable, the hypothesis tests are also introduced to assess the significance of differences, and their results are reported. Further discussion of the findings and concluding insights is presented in Sect. 4.

Influencing Factors to Study CS. In question Q5, participants were asked what influenced their decision to study in the CS field. Students could select multiple options, with an 'other' option for free expression. Of the 157 responses,

115 respondents, 49 were females and 66 males, cited *personal interest* as their main motivation, while 105 (46 females and 59 males) listed *job opportunities*. A minority of 23 respondents (9 females and 14 males) identified *family influence* as the primary factor. Some participants also mentioned the influence of friends and teachers, such as R62 - *"my best friend"* and R104 - *"Recommendation by friends and teachers"*.

Plans After Graduation. In question Q6, participants were asked about their plans after graduation. A total of 111 respondents expressed intentions to seek employment in the IT industry, while 65 indicated interest in further education at the graduate level, and 59 outlined plans to start their own businesses. A minority of 9 respondents were considering a career as a university professor, and 12 remained undecided. It should be noted that this question allowed students to select multiple responses, as students might plan to both pursue a master's degree and work in the industry simultaneously.

For this question, we wanted to see if these plans differ meaningfully between female and male respondents. However, it should be noted that this question is an MMC type, allowing participants to select multiple future plans, which creates dependencies between the choices. To handle this, we treated each response option as a binary variable (chosen or not chosen) and then performed separate hypothesis tests for each option. Bonferroni correction was also applied to control for false positives, which may occur when multiple statistical tests are conducted.

Hypotheses were formulated for each career choice n ($n \in \{1, 2, ..., 5\}$) where 1. *Find a job in the industry*, 2. *Start my own business*, 3. *Become a university professor*, 4. *Continue my graduate studies (master's)*, or 5. *I don't know yet*. The aim of this test was to determine whether preferences between females and males for each career choice differed significantly. The hypotheses are as follows:

$H_{0\_Q6_n}$: *There is no significant difference between the number of females and males choosing the career option n*.

$H_{1\_Q6_n}$: *There is a significant difference between the number of females and males choosing the career option n*

Among all career choices, only "Continue my graduate studies (master's)" ($p = 0.00066$) was found to be significantly different between female and male preferences, even after applying the Bonferroni correction, which provides a stricter adjusted significance level.

Academic Career Perceptions. Question Q7, the only open question in the survey, was included to better understand participants' opinions and perceptions of academic careers. Respondents were asked to describe an 'academic career" in their own words and were further asked *what do you know about academic career?*. This was the first question related directly to academic careers, and it was placed intentionally to avoid biasing participants' responses with the themes and concepts solicited by subsequent questions (Q8 and Q9) about pursuing an academic career. The aim was to allow participants to independently formulate their responses based on their own perspectives and experiences.

This question was compulsory, ensuring that all participants provided their comments. Out of 157 responses, 14 were excluded due to irrelevant content such as punctuation or symbols. For the remaining 143 responses, qualitative data analysis [16] was implemented utilizing the open coding technique [11]. The resulting labels and the keywords, as well as the total number of comments assigned to each label, are given in Table 2.

Table 2. Labels and keywords for academic career perceptions (Q7)

Label	Keywords	# of comments
Professional Growth	expert, expertise, improvement, mentors, development, growth	14
Educational Commitment	school, education, studying, continuous study, lifelong learning	19
Research Involvement	research, discovery	13
Teaching and Mentoring	teach, teaching, tutor, mentoring, helping students	21
Perceptions and Attitudes	positive, good, interest, feel, nothing, opinion, passion, satisfaction	11
Career Opportunities	jobs, career, financial benefits, employment, opportunities	27
Challenges and Rewards	difficult, challenging, success, rewards, recognition	12

In addition to the number of comments for each label, Fig. 2 provides the gender distributions, illustrating the frequency of certain themes mentioned by female and male participants. The most frequently observed label for both genders was *career opportunities*. For male participants, the next most common label was *perceptions and attitudes*, which included comments expressing personal feelings, such as whether academia was perceived as *good* or *boring*. For females, the most common labels after *career opportunities* were *teaching and mentoring* and *educational commitment*.

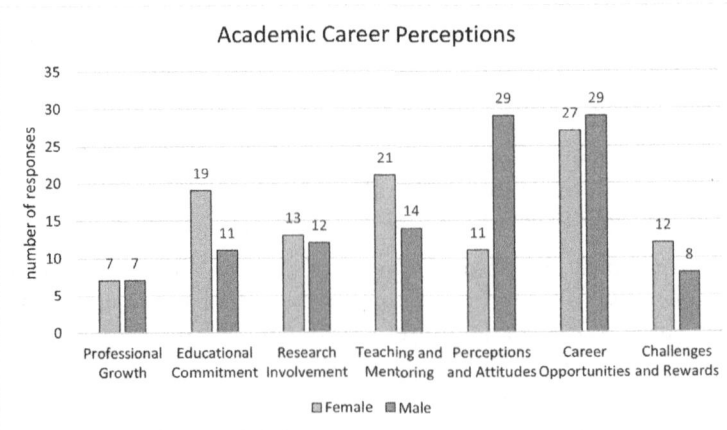

Fig. 2. Labels for academic career perceptions

Motivational Factors to Pursue Academic Career in CS. The participants were given some factors and asked to evaluate if they believe they are motivational factors to pursue an academic career in the CS field or not (question Q8). These factors were mostly selected among the motivational factors that are given in [1], but additional factors such as *flexible work environment* or *social status* were also included. These additional factors were not presented as main motivational factors in [1] but were mentioned in qualitative analyses as part of participants' comments. Many participants selected multiple factors since it is an MMC type of question.

Figure 3 illustrates the number of selections for each factor. Interestingly, the most frequently chosen factor was the *flexible work environment*. The second and third most selected factors were *interest and confidence in research* and *passion for teaching*, respectively.

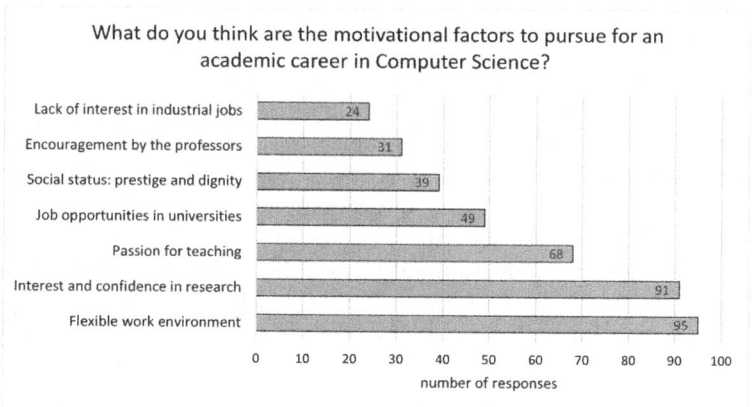

Fig. 3. Frequency of selected options for Q8

In addition to the given factors, there was also an *other* option in this question. Not many participants responded, but we believe some of the comments are worth sharing. R101 wrote *"Passion for learning more"*, highlighting the nature of working in academia that includes continuous learning. R104 mentioned the advantage of being independent in choosing the research topic to work on in academia as follows: *"Ability to work on projects that one finds personally significant with less restraint than in industry. In academia, one is not bound by the profit motive and market viability of some research avenue"*.

Demotivational Factors to Pursue Academic Career in CS. Similar to the previous section, where motivational factors were discussed based on the results of question Q8, a similar approach was followed for demotivational factors. Some of the demotivational factors mentioned in [1] were included in the question, along with an additional 'other' option (question Q9).

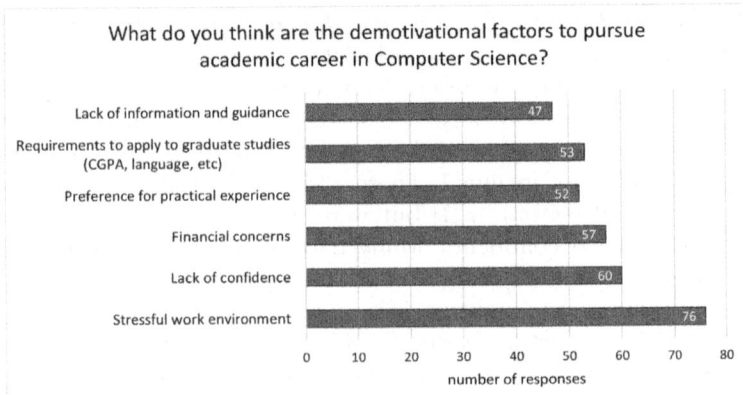

Fig. 4. Frequency of selected options for Q9

For this question, participants selected the provided factors in a more uniform distribution, resulting in small variations in the frequency of chosen options. (see Fig. 4). Most participants stated that *stress* would be a demotivational factor in pursuing an academic career in CS. Some of the 'other' responses mentioned the personality factor, stating that an academic career is not for everyone, like R40: *"I don't think everyone has the personality for being a good professor, like myself"*. Some other respondents highlighted the fact that working on too much theory may be demotivational for them such as the following comments: R101 - *"To much theory and unneeded stuff during the studying process"* and R186 - *"Advance math concepts in computer science"*.

Perception of Gender Representation in the IT Field. To better understand the participants' perceptions of gender dynamics in the IT sector, we asked two related questions (questions Q10 and Q11). Question Q10 focuses on respondents' perceptions of the existence of a gender gap within the IT sector. The definition of *gender gap* was also provided as a note in the question to introduce the term to the ones who do not know and to increase awareness. In question Q11, participants were further asked about their opinions on women's representation in the IT field. Three options were selected: *Women are underrepresented, Women are well-represented* and *Not sure*. This question aims to narrow the focus of respondents to women's representation by making them think of their direct experience or observation about women in the field.

66 of the participants believe that there is a gender gap in the IT sector in BiH, whereas 53 of them believe the opposite. 38 participants stated that they were not sure about this statement. Most survey participants (68) stated that they think women are well-represented in the field, whereas 52 believe they are underrepresented. 37 participants say they are not sure. We then further investigated these results by examining the distributions of answers from female and male participants. We formed our hypotheses for Q11 as follows:

H_{0_Q11}: *There is no statistically significant difference between gender and perception of women's representation in the field.*
H_{1_Q11}: *There is a statistically significant difference between gender and perception of women's representation in the field.*

The chi-square test of independence is applied to examine the relationship between answers of female and male survey participants for question Q11, and the resulting statistics are presented in Table 3. As can be observed from the table, the relation between answers for question Q11 and the gender of the participant is statistically significant ($p < 0.05$). Namely, female participants had a higher level of agreement with the statement *Women are underrepresented* than males. Also, the effect size indicates that gender has a large, statistically significant effect on the perception of women's representation in the field.

Table 3. Chi-Square test for the answers of Q11: *What are your thoughts on the representation of women in your field?*

Q11 answers	# of female	# of male	χ^2	p value	Effect size
Women are underrepresented	42	10	44.849	**0.000**	0.534
Women are well-represented	14	54			
Not sure	13	24			

Concerns About Gender Bias in Future Work Environment. Participants were asked if they were personally worried about the presence of a gender bias in their potential future work environment (question Q12). This question requires a *yes* or *no* answer with an additional *not sure* option. The summary of the answers is illustrated in Fig. 5.

Over half of the survey participants (80 respondents) stated they were not concerned about gender bias. It is important to note that 70% of these respondents were male students. To statistically test the significance of gender on these answers, we again conducted a chi-square test with the following hypotheses:

H_{0_Q12}: *There is no statistically significant difference between gender and being concerned about the potential gender bias in the future work place.*
H_{1_Q12}: *There is a statistically significant difference between gender and being concerned about the potential gender bias in the future workplace.*

The resulting statistics are presented in Table 4. The statistical analysis shows a significant relationship between gender and responses to question Q12 with a medium effect size, suggesting that concerns regarding gender bias in future work environments vary between male and female CS students.

Perception of Gender Inclusivity in Academia vs. Industry. The last question of the survey (question Q13) aims to assess students' perceptions of gender inclusivity in academic environments versus the industry. Specifically, it

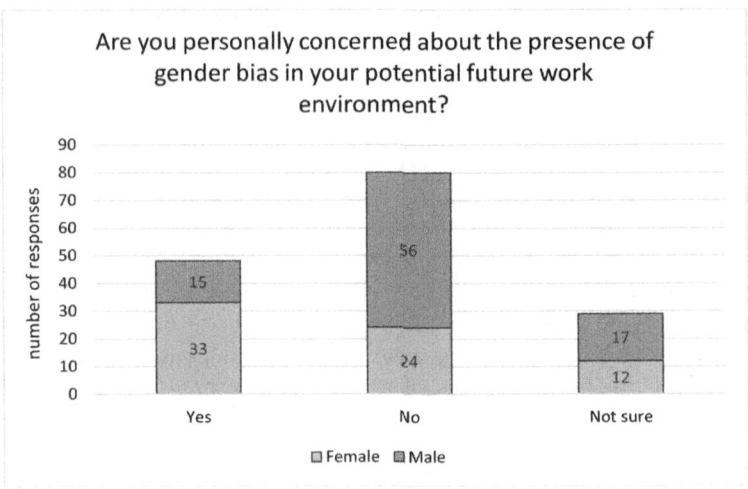

Fig. 5. Answer distribution for Q12

Table 4. Chi-Square test for the answers of Q12: *Are you personally concerned about the presence of gender bias in your potential future work environment?*

Q12 answers	# of female	# of male	χ^2	p value	Effect size
Yes	33	15	18.38	**0.000**	0.342
No	24	56			
Not sure	12	17			

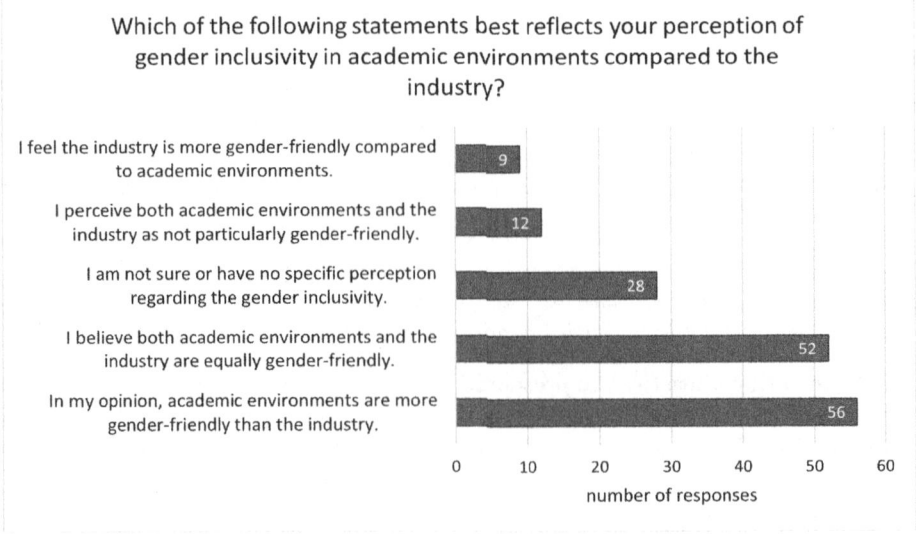

Fig. 6. Answer distribution for Q13

aims to understand how students view the level of gender inclusivity in academic environments compared to the industry.

Survey participants were given five statements and asked to select only one. These statements were designed to be mutually exclusive and exhaustive to represent all possible opinions on this question. 56 respondents (36%) believe that academic environments are more gender-friendly than the industry, making it the most popular answer. The second most agreed-upon statement was that both academic environments and the industry are equally gender-friendly, with 52 respondents (33%) agreeing. Only 9 respondents (6%) stated that they believe the industry provides a more gender-friendly environment for their employees. Distribution of the all responses is illustrated in Fig. 6.

4 Discussion and Conclusions

Statistics show that men continue to outnumber women in academic careers in IT-related fields in BiH. Understanding the main reasons behind this trend is essential for developing corrective measures to close the gender gap. For this reason, we conducted a short, structured survey in English at our university to investigate the perceptions of CS bachelor's students on academic careers and gender diversity in the IT field. The ultimate goal is to extend this study to other universities in BiH. We hope to better understand the factors contributing to this trend and develop recommendations for addressing the gender gap in BiH's academic IT sector.

However, when asked about their career plans after graduation, we see an obvious preference for work in the industry and a pronounced disinterest in academia. The difference in future career plans by gender is statistically significant. While there is interest, especially among women, in seeking a master's degree after graduation, this does not translate into a career in academia. They wish to have a master's degree.

To assess students' understanding of what it means to work in academia, we asked them to describe academia in their own words. The most frequently mentioned theme by both genders was **career opportunities**. Among male students, another popular theme was **perceptions and attitudes** toward academia, whereas females more often mentioned **teaching and mentoring** and **educational commitment**.

Regarding the motivating factors for pursuing an academic career, respondents seem to view academia as a **flexible work environment**, followed by **research interest** and a **love of teaching**. Although not widely applicable, one student's mention of being able to choose personally significant research projects is notable. On the other hand, their view of demotivating factors is somewhat contradictory. They view academia as a **stressful work environment**, in direct opposition to saying it is flexible previously. While conjectural, this may be because they see the working hours and atmosphere as flexible but the work itself as stressful. This may relate to the workload, such as publishing, teaching, and other duties. The next most significant concerns are **lack of confidence** and **financial concerns**.

When asked about the gender imbalance in IT, the majority of respondents either did not believe there was a gender gap or were uncertain. However, a clear majority of female students felt that women are underrepresented in the field, while most male students disagreed. This shows a significant difference in perceptions, with men generally being unaware of the gender gap and viewing women as well-represented in IT. Regarding gender bias in the workplace, female students expressed greater levels of concern than their male counterparts. However, when comparing academia and industry, the majority of students felt that academia offered a more gender-inclusive environment, while the second largest group believed them to be equally gender-friendly.

Our study provides a snapshot of the current perceptions of academia and the industry. Based on these insights, we propose the following remedial strategies:

Outreach Events and Role Models. For respondents, primarily women, who perceive gender underrepresentation and are concerned about gender bias, encouraging initiatives can be undertaken to highlight potential role models. Increasing the visibility of successful female academicians and industry leaders could contribute to addressing these concerns and inspiring students to pursue this field.

Workshops and Training. Many students, particularly women, expressed interest in pursuing a master's degree as part of their future career aspirations, likely due to the potential for higher salaries or better promotion prospects. However, workshops and training sessions that highlight academic career opportunities could encourage more students to consider this path.

Early-Stage Research. Many students are not fully aware of what an academic career truly entails. Engaging these students in early-stage research and academic work with professors could spark their interest in academia and, potentially, a future academic career.

Acknowledgements. This work was supported by COST Action CA19122 - EUGAIN (European Network for Gender Balance in Informatics).

References

1. Abraham, E., et al.: Why do women pursue a PhD in computer science? J. Syst. Softw. (2024, Submitted)
2. André, B.F.: Gender balance in computer science: how do women view the transition into a PhD? Master's thesis, Universidade Nova de Lisboa, Lisbon, Portugal (2022). https://run.unl.pt/bitstream/10362/155510/1/Andre_2022.pdf
3. Bonferroni, C.: Teoria statistica delle classi e calcolo delle probabilita. Pubblicazioni del R istituto superiore di scienze economiche e commericiali di firenze **8**, 3–62 (1936)
4. for Statistics of Bosnia, A., Herzegovina: Women and Men in Bosnia and Herzegovina. Agency for Statistics of Bosnia and Herzegovina, Sarajevo, Bosnia and Herzegovina (2023). https://bhas.gov.ba/data/Publikacije/Bilteni/2024/FAM_00_2023_TB_1_EN.pdf. Accessed 12 July 2024

5. Commission, E., for Research, D.G., Innovation: She figures 2021 - Gender in research and innovation - Statistics and indicators. Publications Office (2021). https://doi.org/10.2777/06090
6. Corbin, J., Strauss, A.: Basics of Qualitative Research: Techniques and Procedures for Developing Grounded Theory. Sage publications (2014)
7. Cramér, H.: Mathematical Methods of Statistics, vol. 26. Princeton University Press (1999)
8. Europe, I.: Students in informatics bachelor's programs (all semesters) - ratio per million people (2007-2023). https://www.informatics-europe.org/dataportal/page=statistics/bachelor_all_semesters_ratio.html
9. Europe, I.: Students in informatics bachelor's programs (all semesters) - ratio per million people (2019). https://www.undp.org/bosnia-herzegovina/publications/software-industry-skills-needs-assesment-bosnia-and-herzegovina
10. Government of Sarajevo Canton: Sarajevo canton development strategy 2021–2027 (2021). https://zpr.ks.gov.ba/sites/zpr.ks.gov.ba/files/sarajevo_canton_development_strategy_2021-2027.06.07.pdf. Accessed 11 July 2024
11. Khandkar, S.H.: Open coding. Univ. Calgary **23** (2009)
12. Motogna, S., Alboaie, L., Todericiu, I.A., Zaharia, C.: Retaining women in computer science: the good, the bad and the ugly sides. In: Proceedings of the Third Workshop on Gender Equality, Diversity, and Inclusion in Software Engineering, pp. 35–42 (2022)
13. Ortiz-Martínez, G., Vázquez-Villegas, P., Ruiz-Cantisani, M.I., Delgado-Fabián, M., Conejo-Márquez, D.A., Membrillo-Hernández, J.: Analysis of the retention of women in higher education stem programs. Humanit. Soc. Sci. Commun. **10**(1), 1–14 (2023)
14. Pantic, K., Clarke-Midura, J.: Factors that influence retention of women in the computer science major: a systematic literature review. J. Women Minorit. Sci. Eng. **25**(2) (2019)
15. Papadakis, S., Tousia, C., Polychronaki, K.: Women in computer science. The case study of the computer science department of the University of Crete, Greece. Int. J. Teach. Case Stud. **9**(2), 142–151 (2018)
16. Seidel, J.V.: Qualitative data analysis (1998)

Author Index

A
Ackovska, Nevena 206, 238
Arsov, David 34
Atanasovski, Simon 51
Avramovski, Kiril 63

B
Baran, Melis 159
Bogdanova, Ana Madevska 195, 206
Buyukdagli, Ozge 295

C
Cantelli-Forti, Alessandro 263, 277
Cetin, Ilhan Mert 159

D
Dag, Hasan 51
Dimitrova, Vesna 51
Dimitrovski, Ivica 106, 121
Dobreva, Jovana 3
Donev, Filip 206

E
Eftimov, Tome 19
El-Gazzar, Rania 174
Ercan, Gulce Berfin 159

F
Filipovska, Elena 3

G
Gusev, Marjan 223

H
Huseini, Admir 63

I
Ilievska, Natasha 81

J
Jajaga, Edmond 277
Jovanov, Mile 34

K
Kitanovski, Dimitar 3
Kitanovski, Ivan 106, 121
Kitanovski, Teo 34
Konca, Ecem 159
Korkmaz, Ilker 159
Koteska, Bojana 195
Kuzmanov, Ivan 238

L
Lameski, Petre 3
Longo, Giacomo 263
Lupidi, Alberto 277

M
Mancuso, Francesco 277
Markovska-Simoska, Silvana 253
Mersni, Amal 295
Meucci, Giulio 277
Mirceva, Georgina 95
Mishkovska, Bojana 195
Mishkovski, Igor 253
Mitrev, Atanasko Boris 95, 136
Mitrov, Goran 3
Mladenovska, Ana 3
Mollakuqe, Elissa 51

N
Nweke, Livinus Obiora 174

P
Pandilova, Ema   121
Petrov, Marko   121
Petrovikj, Nenad   195

R
Ristovska, Vesna Dimitrievska   147
Rushiti, Veton   277
Russo, Enrico   263

S
Saiti, Art   63
Sasanski, Darko   19
Sekuloski, Petar   147

Spasev, Vlatko   106, 121
Stojanov, Riste   19
Stojkoska, Biljana Risteska   136

T
Talic, Merjem   295
Tenev, Aleksandar   253
Todorovski, Andrej   19
Trajanov, Dimitar   19
Trpeski, Bojan   19

Z
Zdravevski, Eftim   3

Made in the USA
Monee, IL
03 May 2026

49438454R00181